Fisheries and Aquaculture Economics

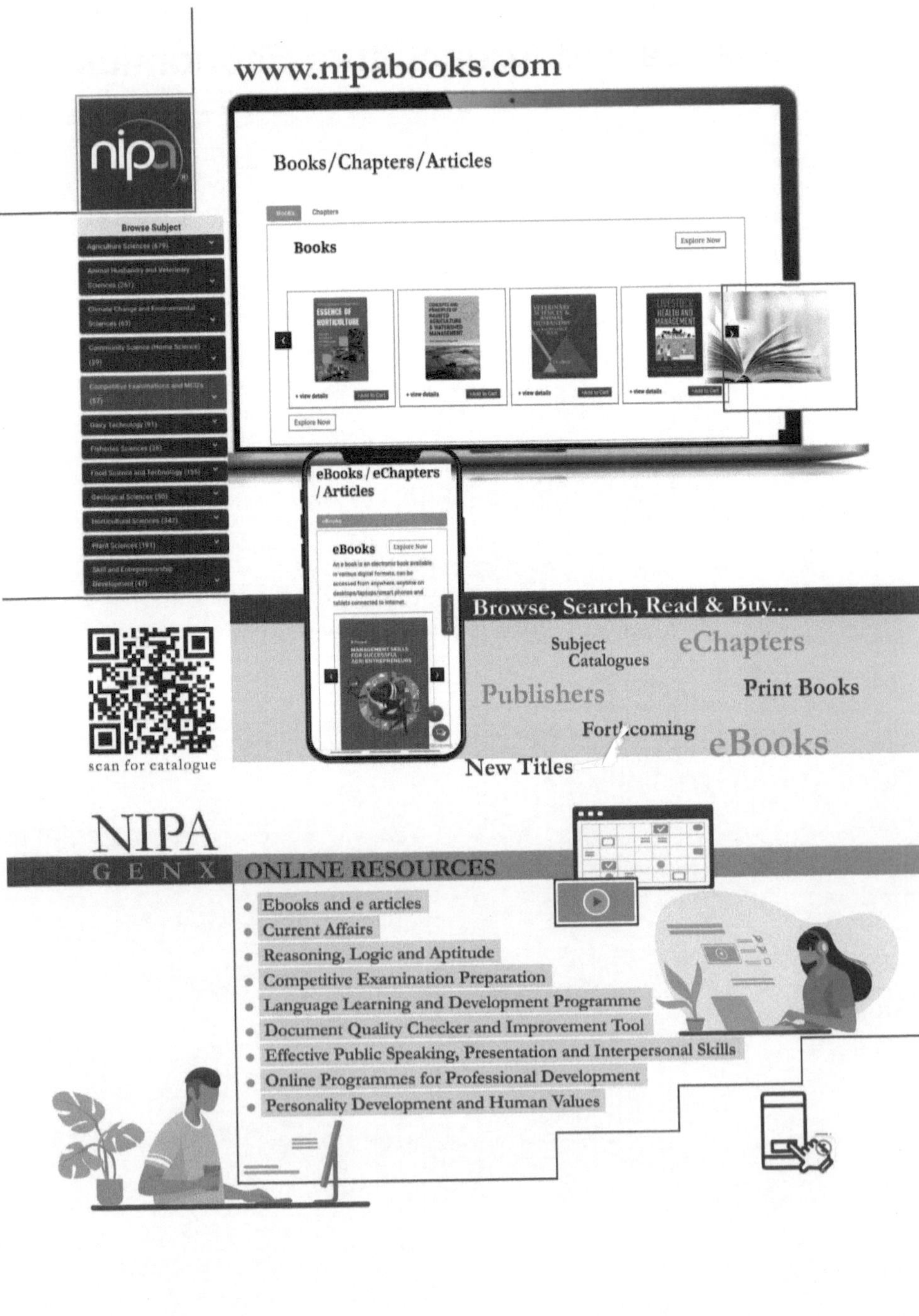

www.nipabooks.com
nipa
Books/Chapters/Articles
Books
Explore Now
eBooks / eChapters / Articles
eBooks
Browse, Search, Read & Buy...
Subject Catalogues
eChapters
Publishers
Print Books
Forthcoming
eBooks
New Titles
scan for catalogue
NIPA
GENX
ONLINE RESOURCES
Ebooks and e articles
Current Affairs
Reasoning, Logic and Aptitude
Competitive Examination Preparation
Language Learning and Development Programme
Document Quality Checker and Improvement Tool
Effective Public Speaking, Presentation and Interpersonal Skills
Online Programmes for Professional Development
Personality Development and Human Values

Fisheries and Aquaculture Economics

Editors

A.D. Upadhyay

Associate Professor (Fisheries Economics)
Department of Extension and Social Science
College of Fisheries, Central Agricultural University
Lembucherra-799210, Tripura (W)

A.K. Roy

Ex. National Consultant (Impact Assessment)
NAIP (World Bank Funded), ICAR
New Delhi - 110012

Pramod Kumar Pandey

Dean, College of Fisheries, CAU (I)
Lembucherra, Tripura-799210

NEW INDIA PUBLISHING AGENCY

New Delhi – 110 034

NEW INDIA PUBLISHING AGENCY
101, Vikas Surya Plaza, CU Block, LSC Market
Pitam Pura, New Delhi 110 034, India
Phone: + 91 (11) 27 34 17 17 Fax: + 91 (11) 27 34 16 16
Email: info@nipabooks.com
Web: www.nipabooks.com

Feedback at feedbacks@nipabooks.com

ISBN: 978-93-87973-92-3

Composed, Designed and Printed in India

Dedication

Dedicated to the parents of the editors who has shown the path of education

Preface

The book is an outcome of decade long teaching & research experiences of the editors and authors in the field of Fisheries and Aquaculture Economics. While teaching conventional economic theories applied in fisheries and aquaculture to the students, it was felt necessary to equip the students with advances that have taken place recently for sustainable fisheries. The fisheries and aquaculture is key sub sector of Agriculture in India which registered exceptionally higher growth rate since last three decades. Since this sector has pivotal role in food and nutritional security, gainful employment, gender equity, entrepreneurship development, agro based industrial developments and foreign earnings etc. therefore, sustainable development, of the sector is very important. Along with technological development the proper planning and policy support is also equally important for the sustainable development of the sector. In this book the different aspects related to fisheries and aquaculture economics have been presented which are based on research studies and expertise of researchers particularly in big data analytics and Artificial Intelligence deserve special mention here. The State of World Fisheries and Aquaculture projects that by 2030 combined production from capture fisheries and aquaculture will grow to 201 million tons from the fish production of 171 million tons in 2016.To achieve the production level, conservation management of the commercial marine fisheries has to be attended to maximize long-term benefits from the exploitation of the fish resources keeping in mind the maximum sustainable average catch and the maximum net economic yield as these are mutually exclusive. In the capture fisheries sector, over the years much has been done to improve fisheries management in different countries. The reported current situation in capture fisheries is challenging because of illegal, unreported, and unregulated (IUU) fishing is posing as a threat to the sustainability of fisheries. In coastal region countries IUU occurs in both small-scale and industrial fisheries, in marine and inland water fisheries, as well as in fishing zones falling under national jurisdictions and on the high seas. Carrying on with business as usual will lead to low fisher incomes, lost economic potential, and undue environmental damage given the delicate balance of marine ecosystems. Aquaculture and other seafood businesses continue to grow rapidly around the world. According to the latest State of World Fisheries and Aquaculture report, nearly 60 million people worldwide, out of which 14 per cent women are directly employed in fisheries and the aquaculture sector. Besides, the fisheries sector is crucial in meeting hunger and malnutrition, and its contribution to economic growth and the fight against poverty is growing, Further, fish account for about 17 per cent of animal protein consumed around the world, providing around 3.2 billion people on earth with nearly 20 per cent of their animal protein needs as fish represent a highly nutritious food that is especially helpful in counteracting important deficiencies in dietary intake. In this context, Fisheries & Aquaculture Economics and Financing plays an important role in deciding as to how to how to monitor and evaluate economic performance, manage capital, labour, and business risk including financing, marketing, and developing a business. Presently the fastest growing sector is facing a lot of challenges in efficiency of resource utilization influenced by wider social and economic factors including globalization, urbanization, factor prices and consumer demand.

Against this backdrop, it is attempted to bring together a few chapters in the form of a Book broadly on various issues of fisheries & aquaculture economics. The book contains nineteen chapters covering topics like economics of production, economic significance of resource use and species diversification, analysis the value chain of processed fish products, elasticity of income and expenditure behavior, analysis of food security, livelihood and income generation, constraints of fish marketing practices, socio-economic and impact assessment tools for evaluating fisheries technologies etc. The main focus and highlight on the following recent technological advancement like modeling neural network forecasts in the areas of Feed Conversion Ratio (FCR), Specific Feeding Ratio (SFR), Key Performance Indicators (KPI): Application of Big Data analytics in aquaculture for in depth understanding; Use of ANN to Forecast Water Quality and Temperature; Artificial Intelligence Systems for increased process efficiency; reduced energy and water losses; reduced labor costs; reduced stress and disease and better understanding of the process. Partial Budget Analysis for efficient use of the resources for aquaculture and women's participation in ornamental fish accessories manufacturing and trade are also highlighted. Finally, tips for designing of surveys, data collection, decide on the right type of statistical test to be undertaken and statistical analysis performed is covered exclusively in seperate chapter.

This unique book explores a wide range of analytical issues centered on the aquaculture process management. What distinguishes this book most from previous works is that it gives a holistic, global-scale focus on challenges of IUU and technology initiatives to face the challenge. The book is targeted for both professional economists and students of economics who are engaged with fisheries research, development and management is expected to be of immense help in updating in the areas of recent technological advances and use for sustainable management of fisheries and aquaculture particularly the resource managers and researchers working with big data to advance more sustainable fisheries practices.

A.D. Upadhyay
A.K. Roy
Pramod Kumar Pandey

Contents

List of Contributors

Dr. Aswathy N
Senior Scientist, Fisheries Economics and Extension Division, Central Marine Fisheries Research Institute, Kochi-682 018

Dr. R Narayana Kumar
Principal Scientist & Head, Fisheries Economics and Extension Division, Central Marine Fisheries Research Institute, Kochi-682 018

Dr. Ajit Kumar Roy
Ex. National Consultant (Impact Assessment), NAIP (World Bank Funded), ICAR, New Delhi; Ex. Consultant (Statistics), College of Fisheries, Central Agricultural University, Agartala, India; Ex. Principal Scientist & Head Social Science Section, Central Institute of Freshwater Aquaculture (CIFA), Kausalyaganga, Bhubaneswar-751002, India; Ex. Computer Specialist & Head Computer Division SAARC Agricultural Information Centre (SAIC), Dhaka-1215, Bangladesh

Dr. Neha W. Qureshi
Scientist, Fisheries Economics, Extension and Statistics Division, ICAR-Central Institute of Fisheries Education, Panch Marg, Off Yari Road, Versova, Mumbai

Dr. Sivaramane N.
ICAR-Central Institute of Fisheries Education, Panch Marg, Off Yari Road, Versova, Mumbai

Dr. M. Krishnan
Principal Scientist and Head, Education Systems Management Division ICAR-National Academy of Agricultural Research and Management, Hyderabad, Telangana, India

Dr. A.D. Upadhyay
Associate Professor, Department of Extension and Social Sciences, College of Fisheries, Central Agricultural University (I), Lembucherra, Tripura-799210, India

Dr. A.B Patel
Professor, Department of Aquaculture, College of Fisheries, Central Agricultural University (I), Lembucherra, Tripura-799210, India

Dr. D.K. Pandey
Associate Professor, Extension Education, College of Horticulture and Forestry, Central Agricultural University(I), Pasighat-Arunachal Pradesh

Mrs. Bahni Dhar, Assistant Professor (Fish Processing Technology), Department of Fish Processing Technology and Engineering, College of Fisheries, Central Agricultural University (I), Lembucherra, Tripura-799210, India

Dr. B. Nightingale Devi
Assistant Professor, Fisheries Extension Department. College of Fisheries, Kawardha, Raipur Chhattisgarh

Mr. Nilesh Pawar
Assistant Chief Technical Officer, ICAR-Central Marine Fisheries Research Institute, Cochin

Smt. Ambalika Ghosh
Assistant Director of Fisheries, Department of Fishery, Govt. of West Bengal, India

Dr. B.K Mohapatra
Principal Scientist, ICAR-CIFE Kolkata Centre, GN-Bl, Salt Lake, India

Late Dr. M.C Nandisha
Former Dean, College of Fisheries, Fisheries College and Research Institute, Tamil Nadu Dr. J. Jayalalithaa Fisheries University, Thoothukudi, Tamil Nadu

Dr. S.K. Mishra
Principal Scientist, Extension, Communication and Training Division, ICAR-Central Rice Research Institute, Cuttack

Dr. B.K. Singh
Assistant Professor (SS), Process and Food Engineering, College of Agricultural Engineering and Post Harvest Technology (CAU), Central Agricultural University (Imphal), Ranipool, Gangtok Sikkim

Dr. T. Umamaheswari
Assistant Professor, Department of Fisheries Extension, Economics and Statistics, Fisheries College and Research Institute, Tamil Nadu Dr.J. Jayalalithaa Fisheries University Thoothukudi – 628 008

Dr. G. Sugumar
Dean, Fisheries College and Research Institute, Tamil Nadu, Dr. J. Jayalalithaa Fisheries University, Thoothukudi, Tamil Nadu

Dr. Pradip C. Bhuyan
Associate Professor, Department of Fisheries Extension, Economics and Statistics, College of Fisheries, Assam Agricultural University, Raha – 782103, Assam, India

Mr. Narendra Kumar Vrema, Assistant Professor, College of Fisheries, Kishanganj, Bihar Animal Sciences University, Patna, Arrabari (Kishanganj) - 855107

Dr. Shyam Sundar Dana
Registrar & HOD, Dept. of Fishery Extension, Faculty of Fishery Sciences (WBUAFS), Kolkata

Dr. Subhabaha Pal
Senior Faculty, MAHE South Bangalore Campus (Manipal Pro Learn – Data Science)

Prof. (Dr.) Satyabrata Pal
Formerly, Dean, Post-Graduate Studies and Professor, BCKV, W. B., Formerly Principal, Swami Vivekananda Institute of Management and Computer Science, Sonarpur Kolkata, W.B. Formerly, Honorary Visiting Professor, Indian Statistical Institute. Kolkata W.B.

Dr. K.V. Radhakrishnan
Assistant Professor (SS), Department of Fisheries Resource Management, College of Fisheries Central Agricultural University (Imphal), Lembucherra, Agartala, Tripura-799210

1

Economics of Production Technologies in Marine Fisheries in India

Aswathy N. and R. Narayana Kumar

Introduction

Global fisheries produced roughly 171 million tons of fish, with an estimated sale value of US$ 362 billion, generating over US$143 billion in exports in 2016. Around 60 million people across the world are employed in fisheries and aquaculture, with the majority in the small scale capture fisheries sector of developing countries (FAO, 2018). India has an annual harvestable potential of 4.414 mt. and with 9.9 lakh active fishers, the sector provides livelihood to nearly 4.0 million people including the post-harvest sector (Sathiananthan, 2017). The estimated value of marine fish at landing centre level was Rs. 48,381 crores and at retail level was Rs. 73,289 crores during 2016 (CMFRI, 2017). Seafood exports from the country reached 1.38 million tons valued at $7.1 billion in 2017-18 (MPEDA, 2018). The fishing fleet of the country includes 1, 94,190 fishing units consisting of 72,559 (37.31%) mechanized, 71,313 (36.67%) motorized and 50,618 (26.03%) artisanal craft (CMFRI, 2010). The mechanised sector contributed 82% of catch followed by motorised (17%) and non-mechanised (1%) sectors in 2016 (CMFRI, 2017).

The development of marine fisheries sector in India occurred in three phases. In the first phase (up to 1965-66) fishing was mainly done by using non-mechanized indigenous craft and gear. The second phase (up to 1985-86) witnessed increased mechanization, improved gear materials and introduction of motorized country craft. In the last phase (after 1986) there was intensification of mechanization, motorization of the country craft and multi-day voyage fishing (Sathianandan, 2013).

The craft and gear technologies in marine fisheries sector vary based on the types of fish caught and nature of fishing ground. The economics of different production technologies in the mechanised, motorised and non-mechanised categories were analysed based on the size of boats and capacities of engines and gear. The economic indicators developed in different coastal states helps for micro level investment decisions as well as for developing macroeconomic policies for the fisheries sector.

Economic Performance

The economic efficiency of different production methods in the fishing sector in India was studied by many authors. Previous studies on the economic performance of fishing units employed different indicators like net -benefit earnings ratio, rate of return (LeRy *et al.*, 1999; Tietze *et al.*, 2001), net returns (Panikkar *et al.*, 1994; Kasim *et al.*, 2013), capital productivity and labour productivity (Narayanakumar *et al.*, 2009; Aswathy *et al.*, 2011). The comparative analysis of various fishing units in Kerala showed that the single day purse-seiners and multiday trawlers of more than six days operation are showing the highest capital productivity with lowest operating ratios of 0.60 and 0.64 respectively(Aswathy *et al.*, 2011).

Analysis of economic performance of different fishing units by Narayanakumar *et al.*, (2009) indicated that in the mechanized trawl fishing, the multi-day fishing (MDF) of 6-10 days duration earned higher returns than the other two methods in the east coast, while the MDF of 2-5 days duration performed better than the other two in the west coast. In the mechanized gill net operation also, the economic performance of multi-day gillnet fishing of 2-5 days duration was better than the other two methods of fishing in both the east and west coasts (Narayanakumar *et al.*, 2009).

Methodology of Study

Primary data on costs and revenues of various craft and gear in the state of Kerala for the year 2016 was collected from different landing centers. Secondary data on economics of various craft and gear in other coastal states and investment in fishing units were obtained from various annual reports of Central Marine Fisheries Research Institute (CMFRI). The mechanised fishing units undertook single day fishing to fishing trips of 2-5 days to more than 6 days duration. The motorised and non-mechanised fishing units generally undertook single day fishing only. Data on operational costs, fixed cost and prices and quantities of resources landed per fishing trip were collected by following multistage random sampling method covering all the fishing seasons. The operating costs in marine fishing consists of expenses on fuel (diesel/kerosene), ice, food for the crew, crew bata, crew share, repair and maintenance of craft, gear and engine, water charges, auction charges, landing charges and other miscellaneous expenses. The input costs consist of operational costs excluding crew wages. The crew wages are paid either as fixed monthly payments or as a share of the net income earned per fishing trip. The crew share varied from 35-50% of the net income for mechanized fishing units.

The operational costs and returns, prices and quantities of the major species caught and the capital investment in various mechanized units were collected at weekly intervals from the selected landing centres in Kerala, covering all the seasons. The different economic indicators are calculated as follows:

Operating cost, OC/trip = (Fuel charges + Crew wage/ bata + Food expenses +Auction charges + Repair charges + Other charges).

The major operational cost components consist of fuel (diesel/ kerosene), food for the crew, crew share or crew wages, auction charges, repair and maintenance of fishing craft, gear and engines. Other charges consists of mobile charges, water charges etc. The crew wage varies with fishing units and regions. The fish workers are paid either as monthly salary or as a percentage of the net returns in addition to daily bata.

The gross revenue per trip is calculated from the species composition of the catch and price per species. The gross revenue per trip is estimated as follows:

$$\text{GR per trip} = \sum_{i=1}^{n} q_i p_i$$

Where pi and qi are the price and quantity of ith variety of fish caught

Net operating income = Gross revenue minus operational cost

Capital productivity (Operating ratio) = Operating costs/ Gross revenue. The smaller the ratio, the greater the firm's ability to generate profit

Labour productivity = Total catch/Man days

Input-output ratio= Input cost/ Gross revenue

Gross value added= Net operating profit + Labour wages. Gross value added is an important indicator in the GDP calculations of an economy

Capital Investment in Marine Fisheries Sector

The private capital investment in marine fisheries sector was estimated based on inventory valuation of different categories of fishing units operated in the sector. The capital investment increased from Rs. 21,023 crores in 2014 to Rs. 23,668 crores in 2016. The gross value of marine fish at landing centre level increased from Rs. 31,754 crores in 2014 to Rs. 48,381 crores in 2016. At retail level the gross value increased from Rs. 52,363 crores to Rs. 73,289 crores in 2016. The return over investment increased from 1.51 to 2.04 times at landing centre level and from 2.49 to 3.09 times at retail level during 2014 to 2016, period. The increase in return over investment was due to increase in the value of marine fish both at landing centre and retail levels (Table 1.1).

Table 1.1: Private Capital Investment in Marine Fishing Units in India

Year	Valuation of inventory investment (Rs. crore)	Valuation at LC level (Rs. crore)	Return over investment	Valuation at Retail level (Rs. crore)	Return over investment
2014	21023	31,754	1.51	52,363	2.49
2015	22662	37,317	1.65	64,593	2.85
2016	23668	48,381	2.04	73,289	3.09

Source: CMFRI, 2017

Economic Performance of Different Fishing Units

The economic performance of various categories of fishing units in selected maritime states is discussed in this chapter. Under the mechanised category trawlers, purse seiners, ring seiners, gillnetters are included. Under motorised sector, gillnetters drift gillnetters, ring seiners, minitrawls and hooks and lines are included. Since the non-mechanised sector contributed meager share of catch and is currently operating in very few states, it is not included in the economic analysis.

Economic Performance of Trawlers

Trawlers contributed 57% of the marine fish landings in 2016 (CMFRI, 2017). The trawler means any vessel that pulls a bag net through the water. A trawl is a conical net which catches fish when dragged through the water at a specific depth or along the bottom of the ocean or sea. There are many types of trawlers varying in size from open boats, powered by inboard engines to huge factory ships, which can fish in the most distant waters. The trawlers can be classified depending on type of net, number of nets, and where it's being used in the water. Otter trawlers, Side trawlers, Stern trawlers, Double rig trawlers, Shrimp trawlers, Beam trawlers, Scallop draggers, Pair trawlers, Danish seiners etc. are commonly used trawlers. Shrimp trawlers are commonly seen in the west coast, especially in Kerala and Karnataka. Pair trawlers are seen in Maharashtra and Gujarat coast. Danish seiners are used for deep sea trawling. These trawlers are commonly seen in the Cochin Fisheries harbour. Otter trawler, the common trawler seen in various maritime states in India, is the single most productive method of fishing.

Characteristics of Trawlers

The overall length of the trawlers operating in India ranges from 32-96 ft and the crew size varies from 5 to 15 (Table 1.2). A minimum of 25 different types of trawl nets are used in each highly equipped fishing boat. These are prawn nets (2-3 numbers), cuttle fish net (5 numbers), Nemipterus net (2), Saurida net (2), Roller/ Gundu net (4-5), Sardine net (2), common fish net (6). Marine diesel engines used for trawling are Ashok Leyland (India), followed by Sinotruk, Weichai Power, Yuchai (Chinese), Cummins , Caterpillar (USA) etc.

Table 1.2: Common Characteristics of Trawlers

Main features					
Type	Overall length (OAL)	Engine (HP)	Equipments used	Gear	Crew size
Trawl	32-96 ft	120-550	Eco sounder, compass, GPS, Radio transmission set	Prawn nets, Cuttle fish net, Nemipterus net, Sauridanet, Roller/ Gundunet, Sardine net, common fish net	5-15

Source: Field survey

Comparative economic analysis of the trawlers showed that multiday trawlers operating in Maharashtra had the highest operational efficiency with the lowest operating ratio of 0.35, followed by multiday trawlers of Gujarat (Operating ratio -0.36). The trawlers operating in Kerala and Tamil Nadu had the lowest operational efficiency. The highest gross revenue was earned by multiday trawlers operating in Maharashtra followed by multiday trawlers in Gujarat and Kerala (Table 1.3).

Economic Performance of Purse Seiners

Purse seining is called an active fishing method that uses a huge encircling net with a small mesh size capable of catching tonnes of shoals of fish such as mackerels and sardines. Purse seining is an active fishing method for harvesting of shoaling fishes and propulsion is required for reaching and returning from fishing ground and for operation of the gear (CMFRI, 2012) (Table 1.4).

Table 1.3: Economic Performance of Trawlers in Different Coastal States (2016)

Particulars	Kerala		Maharashtra	Gujarat	AP		TN	Odisha
	Trawl MD	Trawl MD	Trawl MD	Trawl MD	Trawl MD	Trawl	Trawl	Trawl MD
	(2-5D)	(>6D)	(>6D)	(>6D)	(2-5D)	(SD)	(SD)	(>6D)
Total Operational cost (in Rs.)	78903	280913	223201	215872	66348	18891	28896	181496
Gross Revenue (in Rs.)	125132	490586	629756	598500	137313	39718	48764	404912
Net operating income (in Rs.)	46228	209672	406555	382718	70965	20827	19891	223416
Capital productivity	0.63	0.57	0.35	0.36	0.48	0.48	0.62	0.54
Labour productivity(kg/crew)	99	252	392	399	270	151	240	304
Input-output ratio	0.29	0.25	0.30		0.18	0.18	0.42	0.25
Gross value added (in Rs.)	88962	366748	442393		112158	32741	28,304	304410

Source: CMFRI, 2017

Table 1.4: Main Features of Seiners

Type	Overall Length	Engine (HP)	Equipment's used	Gear	Crew size
Ring seine/ Purse seine	32-90 ft	9.9-350 hp	VHF, compass, GPS, Eco sounder	Ring seine/ purse seine	8-55

Source: Field survey

Purse seiners are mainly operated in Maharashtra, Karnataka and Goa. In case of purse seiners also, the highest gross revenue and operational efficiency were shown by multiday purse seiners in Maharashtra. The gross revenue realised was Rs. 5.73 lakhs for the multiday purse seiners in Maharashtra whereas for the single day purse seiners, the gross revenue varied from Rs. 1.54 lakhs in Maharashtra to Rs. 2.93 lakhs in Odisha (Table 1.5).

Table 1.5: Economic Performance of Purse Seiners

Particulars	Maharashtra		Odisha
	Purse seine(SD)	Purse seine(2-5 D)	Purse seine(SD)
Total Operational cost (in Rs.)	42635	108649	172885
Gross Revenue (in Rs.)	154590	573696	292724
Net operating income (in Rs.)	111955	465047	119839
Capital productivity	0.28	0.19	0.59
Labour productivity(kg/crew)	71	249	327
Input-output ratio	0.20	0.12	0.33
Gross value added (in Rs.)	124205	503447	292725

Source: CMFRI, 20017

Economic Performance of Inboard Ring Seiners

Ring Seine, an encircling n*et al*most similar to a mini-purse seine was first introduced along the Kerala coast in the Alappuzha region during 1985. It is simply a technologically improved version of the traditional boat seine (*thanguvala*) of Kerala coast. Contribution of ringseine to total marine fish landings of Kerala increased from 21.4 % in the nineties to 36.7 % during 2000-2004, period. In recent years, about 90% of the oil sardine and about 60 % of the mackerel landed, in Kerala, were caught in ring seines (Abdussamad *et al.*, 2015).The inboard ring seiners operating in Kerala had an OAL of up to 70ft with gear size ranging from 600-1000m. It is mainly owned by traditional fishermen groups.

Economic performance of inboard ring seiners indicated that the gross revenue was Rs. 1.01 lakhs with net operating income of Rs. 24,604 (Table 1.6). The capital productivity was low due to the high operational cost.

Table 1.6: Economic Performance of Inboard Ring Seiners in Kerala (2016)

Economic indicators	Inboard ring seiners
Total Operating Cost (Rs.)	76437
Gross revenue (Rs.)	101041
Net operating income (Rs.)	24604
Capital productivity (OR)	0.76
Labour productivity	32.85
Input-Output ratio	0.30
Gross value added (Rs.)	71041

Source: Computed by authors

Economic Performance of Gillnetters

Gillnetting is a passive method and propulsion is used for reaching the fishing ground, deployment of the gear and return to the base. About 460 gillnetters are operating from Kerala (CMFRI, 2012). Gill nets catch fish that attempt to swim through the net, which are caught if their head passes through the mesh but not the rest of the body. The mesh size used depends upon the species and size range being targeted. The main features of Gill Netter are mentioned in Table 1.7.

Table 1.7: Main Features of Gill Netter

	Main features of Gill netter				
Type	Overall Length	Engine (HP)	Equipment's used	Gear	Crew size
Gill net	32-55 ft	120-225 hp	Compass, GPS, Eco sounder	gill nets	8-10

Source: Field survey

The multiday gillnetters operating in Kerala undertook fishing trips of duration 10-20 days. The gross revenue earned was Rs. 6.32 lakhs for multiday gillnetters and Rs.1.97 lakhs for single day gillnetters (Table 1.8). The main resources caught are tunnies, sharks, rays, sailfish and groupers.

Table 1.8: Economic Performance of Gill Netters in Kerala, 2016

Kerala	Mechanised Gillnet (SD)	Mechanised Gillnet (MDF>6D)
Total Operational cost (Rs.)	107941	351658
Gross Revenue (Rs.)	197500	632092
Net operating income	89559	280434
Labour productivity	146	240
Operating ratio	0.55	0.56
Input-output ratio	0.20	0.31
Gross value added (in Rs.)	158632	437510

Source: CMFRI, 2017

Among the mechanised fishing units in India, the highest operational efficiency was shown by multiday purse seiners followed by multiday trawlers in Maharashtra. Mechanised gillnetters in Kerala and Purse seiners in Odisha had the lowest operational efficiency.

Economic Performance of Motorised Units

Traditional craft using inboard or outboard motor for propulsion are known as motorized fishing units. Mini trawlers, dugout canoes, stolephorus boats are included in this category (Table 1.9).

Table 1.9: Main Features of Motorised Fishing Craft

Type	Overall length	Engine (HP)	Equipments used	Gear	Crew size
Motorised fishing craft	25-45 ft	9.9-45 hp	VHF, Compass, GPS, Eco sounder	Small ring seines, stolephorus nets, small gill nets, minitrawl net and Hook and lines	2-55

The coastal states of Kerala, Tamil Nadu and AP have the highest number of motorised units. Among the motorised units, the highest gross revenue was obtained for motorised gillnetters and trawlers operated in Tamil Nadu. The highest operational efficiency was shown by motorised gillnetters and drift gillnetters operating in AP. Motorised units operated in Kerala showed the lowest operational efficiency. The decline in catches of most of the motorised units in Kerala may be attributed for the lower operational efficiency.

Conclusion

Economics of production technologies has greater implications for micro levels investment decisions as well as macro level policy decisions on subsidies, taxes, incentives etc. in the marine sector. Marine fish catch shows very high year to year fluctuations depending on variations in environmental parameters, stock status and production technologies. Time series data of catch, revenues and costs are necessary for arriving at meaningful policy decisions. Mechanized fishing is an economically viable enterprise in almost all the maritime states in India with a significant share in the total catch and remunerative prices for the resources caught. All the mechanized categories generated sufficient funds for reinvestment in the fishery. High speed engines and other technological advancements in fishing gear aided the mechanized fishing units to venture into distant waters and species diversification in catch. However, there is a growing concern over the rising fishing costs associated with these technological advancements. In addition, the decline in the catch share and lower economic efficiency of the traditional sector necessitates policy interventions in terms of subsidies or incentives for the small scale fishing units for ensuring sustainable production systems.

References

Abdussamad E. M., U. Ganga, K. P. Said Koya, D. Prakasan and R. Gireesh, (2015). Ring seine fishery of Kerala: An overview, Mar. Fish. Infor. Serv., T & E Ser., No. 225, 2015

Annual report, (2017-18). Ministry of Commerce & Industry, Department of Commerce, Government of India

Aswathy, N., R. Narayankumar, N. K. Harshan and C. Ulvekar, (2017). Techno-Economic Performance of Mechanized Fishing in Karwar, Karnataka, *Indian J. Fish.*, 64(1): 61-65.

Aswathy, N., Shanmugam, T. R. and Sathiadhas, R. (2011). Economic viability of mechanised fishing units and socio-economics of fishing ban in Kerala. Indian J. Fish., 58(2): 115-120.

CMFRI (2012). Annual report 2011-12, Central Marine Fisheries Research Institute, Kochi.

FAO, (2018). The state of world fisheries and aquaculture, Meeting the sustainable development goals. Food and Agricultural organisation of the United Nations, Rome, 2018.

Geetha, R. and Narayanakumar, R. and Shyam, S. Salim and Aswathy, N. and Chandrasekhar, S and Srinivasa Raghavan, V and Divipala, Indira (2014). Economic efficiency of mechanised fishing in Tamil Nadu – a case study in Chennai. *Indian Journal of Fisheries*, 61 (1). pp. 31-35.

Kasim, H. M., Syda Rao, G., Rajagopalan, M., Vivekanandan, E. Mohanraj, G., Kandasami, D., Muthiah, P., Jagdis, I., Gopakumar, G and Mohan, S. (2013). Economic performance of artificial reefs deployed along Tamil Nadu coast, South India. Indian J. Fish., 60(1): 1-8

LeryRy, J. M., Pardao, L. J. M and Tietze, U. (1999). Economic viability of marine capture fisheries, Findings of a global study and an interregional workshop, FAO Fisheries Technical Paper No. 377, FAO, Rome, 130 pp.

Marine Fisheries Information Service T&E Ser., No. 200, 2009

Narayanakumar, R., Sathiadhas, R. and Aswathy, N. (2009). Economic performance of marine fishing methods in India. Mar. Fish. Inf. Serv. T&E Ser., 200: 3-16.

Panikkar, K. K. P., Sehara, D. B. S. and Kanakkan, A. (1994). An economic evaluation of purseseine fishery along Goa coast, *Mar. Fish. Inf. Serv. T&E Ser.*, 127: 4-8.

Pillai, N. G. K. and Srinath, M. (2006). Marine fisheries of India: An approach to responsible fisheries management. *Fishing Chimes*, 26 (4):23-28.

Sathianandan, T.V. (2017). *Marine fish production in India - Present Status.* In: Course Manual Summer School on Advanced Methods for Fish Stock Assessment and Fisheries Management. Lecture Note Series No. 2/2017. CMFRI; Kochi, Kochi, pp. 23-27.

Sathiananthan, (2013). Summer school, advanced fish stock assessment techniques.

Tietze, U., Prado, J., Le Ry, J. M. and Lasch, R. 2001. Techno-economic performance of marine capture fisheries. FAO Fisheries Technical Paper, 421, FAO, Rome, 79 pp.

www.mpeda.gov.in.

2

Analysis of Aquaculture Farm Changes through Partial Budgeting Technique

Ajit Kumar Roy

Introduction

A recent study investigated the feasibility of sustaining current and increased per capita fish consumption rates in 2050, based on extensive data with respect to the changes in global and regional climate, marine ecosystem and fisheries production estimates, human population estimates, fishmeal and oil price estimations and projections of the technological advancement in aquaculture technology. The study concludes that meeting current and larger consumption rates is feasible only if fish resources are managed sustainably and fisheries management are effective (1). Aquaculture Farm Managers facing uncertainty has to make decisions every day. Some decisions have vital consequences for farm business, while others are not as crucial. These decisions mainly relates to seed, feed, fertilisers and regular water intake as management tools of farm. The choices made today may have an immediate impact on the aquaculture business or may take much longer to have an effect. These decisions may involve any facet of the farm business, including production, personnel or financing. The bottom line is that irrespective of the size or scope of any single decision, nearly all decisions may have important implications for the immediate and future success of the farm business. Because many decisions have such important impacts, farm managers need to analyze alternatives in a methodical fashion. Some alternatives can easily be analyzed, and a decision can be made quickly. In other cases, farm managers must take more time to recognize and evaluate all potential effects of that decision. To do this, farm managers need a framework for analyzing the relevant trade-offs. Agricultural economists are usually members of multidisciplinary teams, engaged in solving problems of fisheries. The economist applies economic concepts and tools to estimate economic profitability of a technology. This input can be made during any phase of the research process, technology development, testing and evaluation on farm. This paper will focus on partial budgeting as a tool that can be used by non-economists. This paper discusses on the concept of partial budgeting, a useful and easily implementable framework for such analysis.

Basic Concept of Partial Budgeting

Partial budgeting is a planning and decision-making framework used to compare the costs and benefits of alternatives faced by a farm business. It focuses only on the changes in income and expenses that would result from implementing a specific alternative. Thus, all aspects of farm profits that are unchanged by the decision can be safely ignored. In a nutshell, partial budgeting allows getting a better handle on how a decision will affect the profitability of the enterprise, and ultimately the profitability of the farm itself. However, the value of a partial budget analysis is highly dependent upon the quality of the information used in the analysis.

Basic Principles of Partial Budgeting

Partial budgets are based on the principle that small business changes have effects in one or more of the following areas:

1. Increase in income
2. Reduction or elimination of costs
3. Increase in costs
4. Reduction or elimination of income

The net impact of the above effects will be positive financial changes minus the negative financial changes. A positive net profit indicates that farm income will increase due to the change, while a negative net indicates the profit change will reduce farm income.

Components of Partial Budget

A partial budget consists of two columns, a subtotal for each column and a grand total. The left hand column has the items that increase income while the right hand column notes those that reduce income for a farm business. The budget can be divided into four parts:

1. Added Income

This area is usually an estimate of a new enterprise is to be added. Use realistic yields, product quality and prices. Over estimation may lead to incorrect decisions and possibly reduced financial performance when the change is meant to improve it. While deciding on a projected price, use average prices from the markets where produce is most likely to be sold. Also use average quality unless the change under consideration is meant to improve product quality. Income increase may come from expansion of an enterprise. If the expansion is minor, current production quantities, quality and average prices are reasonable approximations to use. But if the expansion is large, during the early production periods lower yields and quality may result due to start-up difficulties to be kept in mind while estimating changes in income.

2. Added Costs

List all increased expenses due to the change being considered. Most of these will be costs of production for new enterprise. This list may also include non-cash costs such as labor and depreciation. For example, it might be appropriate to include unpaid labor to be certain that the operator is equitably paid for his/ her labor and management input. A depreciation charge, if included, will help analyze whether there is a return on the investment that the operator makes.

3. Reduced Costs

Items for inclusion in the section would be fish/ livestock expenses which are no longer likely to incur. These costs could be reductions or total elimination of certain expenses. Examples include cost of seed, custom work, repairs, interest and paid labor etc. Inclusion of non-cash costs such as unpaid labor and depreciation would provide a full economic analysis instead of just changes in cash costs.

4. Reduced Income

Items to include here might be reductions in product sales, such as fish, vegetables, rental income, and custom work income. Another consideration here may be reductions in yields due to reduced stocking or harvest timeliness.

Partial Budget calculation

Summarization of the above four partial budget components is the last step in partial budgeting. Total each of the two factors in column 1 and write this result on the column 1 subtotal line. Repeat the process for column 2. Then take column 1 (added income/reduced cost) and subtract column 2 (increased costs/ reduced income) to arrive at a projected net return from adoption of the change under consideration. A negative number indicates the change as considered will reduce whole farm income. A positive number indicates that the change will be profitable.

When and How to Use Partial Budgets?

Partial budget framework can be used to analyze a number of important farm decisions, including:

i. Adopting a new technology

ii. Changing enterprises

iii. Choosing to specialize

iv. Hiring custom work

v. Leasing instead of buying machinery

vi. Modifying production practices

vii. Making capital improvements

The structure of analysis depends upon the nature of the decision being analyzed. For example, if you want to analyze the installation of a new hatchery for fish breeding, it would be wise to perform a partial budget analysis on the hatchery by analyzing costs and returns on per hatchery basis. On the other hand, a farmer choosing to purchase improvised aeration system to see the effects on the hatchery, in terms of total income and costs. The partial budgeting framework is flexible enough to allow for these modifications. It is worth mentioning that partial budgeting analyzes the impacts of some change on profit. Prior analysis should be performed to assure that the hatchery operation is profitable. If it is not profitable, one may have to rethink about the alternatives. There are several steps to the successful use of partial budget analysis as a decision-making tool. Each step serves a specific, unique purpose and is vital to an accurate, meaningful analysis. It may be remembered that a partial budget helps farm owners/ managers to evaluate the financial effect of incremental changes. A partial budget only includes resources that will be changed. It does not consider the resources in the business those are left unchanged. Only the change under consideration is evaluated for its ability to increase or decrease income in the farm business.

Impact Assessment is the process of identifying anticipated or actual impacts of a development intervention, on those social, economic and environmental factors which the intervention is designed to affect or may inadvertently affect. It may take place before approval of an intervention (ex ante), after completion (ex post), or at any stage in between. Ex ante assessment forecasts potential impacts as part of the planning, design and approval of an intervention. Ex post assessment identifies actual impacts during and after implementation, to enable corrective action to be taken if necessary, and to provide information for improving the design of future interventions.

Partial Budget focuses only on the changes in income and expenses that would result from implementing a specific alternative. The construction of representative budgets for specific crops is time-consuming, even when secondary data are available to provide most of the quantity and price information. Once the budget is constructed, the marginal costs of further use and modifications of budget data are relatively small. Numerous variations of a representative budget can be generated easily by alteration of a subset of input and output data. This exercise is termed partial budgeting.

Partial budgeting is most often used in the Policy Analysis Matrix (PAM) methodology as a means of assessing the effects of new technologies on farm profitability. A new seed-fertilizer package for rice, for example, would be modeled by alteration of a traditional technology budget for changes in seeds, fertilizer, and yield. If the new technology increases yields, the budget might have to be modified further to recognize additional labor requirements for tending and harvesting the crop. Although such procedures seem mechanical, they are often useful portraits of the actual process of technological change. Farmers rarely jump from one set of practices to a new technology that uses entirely different inputs and practices. Instead, they modify current practices to incorporate a particular innovation. The input and output data required for partial budgeting ideally are drawn from observations of the actual practice of farmers. Even if producers do not know specific quantities of inputs used and outputs derived, estimates can be obtained with comparative questionnaire techniques, in which producers provide information about the performance of the new technology relative to the old one. If information about actual practice is not available, the analyst is forced to rely on experiment station results or to modify them to reflect expected farm practice. Comparisons of old and new budgets give the analyst information about the economic incentives for technological change. Consideration of both profitability and changes in the structure of costs is necessary for this assessment. Even if the new technology proves more profitable than the old one, potential constraints to adoption could appear. Cash-flow problems sometimes arise when new technologies entail a greater use of purchased inputs. The lack of marketing services can also limit adaptation. Marketing boards must handle the increment in production induced by technological change, physical and financial facilities might have to be expanded. By aggregating the representative budgets to a regional or national level, the analyst can generate the total impact of technological changes.

Partial budgeting techniques can also be used in formulation of the agenda for future research and development. New technologies, such as high-yielding seed-fertilizer packages, improved means of pest control, substitutions among machinery and labor, and better water control and management, are often a direct consequence of the pattern of investment in research. Working with technical experts, agricultural economists can use partial budgeting techniques to simulate the impact of hypothetical technological changes on profitability. If potential changes do not create positive private and social profits and improvements over traditional techniques, alternative investment paths or changes in policies need investigation.

A lot of literature is available on impact assessment study on various interventions reported worldwide covering methodology as well as field implementations from FAO, UNDP, UN, World Bank, ADB, NAIP and different researchers on various crops, technologies, intervention and scenario (2-32). It is well known that *Partial budget analysis* is a simple but effective technique for assessing the profitability of new technology for an existing enterprise. It also provides the foundation for comparing the relative profitability of alternative treatments, evaluating their riskiness, and testing how robust profits are in the event of changing product or input prices. The method developed by *International Wheat and Maize Improvement Center* (CIMMYT), is extensively used for estimating the financial impact of implementing a new technology, in dairy research and plant protection research and on various crops (33). Besides a lot of reports on Partial Budgeting, economic analysis, Partial Budget to Analyze Farm Change, to estimate the cost and benefit of adaptation of a new technology, partial budgeting technique for assessment of new technology that can be evaluated in terms of its impact on the productivity, profitability, acceptability and sustainability of farming systems are available (34-42).

Partial Budget Analysis

Partial budget analysis is the tabulation of expected gains and losses due to a relatively minor change (marginal) in farming method or technology. The new technology can be technically feasible but this is not a necessary condition for adoption by farmers, the new technology must be profitable. Therefore, it is important for scientists developing a new technology or improving an existing one to determine the profitability of the technology. However, most national research systems do not have enough economists and biological scientists often lack the ability to conduct economic analysis. This chapter therefore, aims at describing a simple technique to determine the profitability of technologies at farmer level through the partial budget analysis. Scientists need to be equipped with basic economic tools to evaluate technologies.

Partial budgets list only those items of income and expenses that change. They (i) measure change in income and returns to limited resources, (ii) provide a limited assessment of risk and (iii) suggest a range of prices or costs at which a technology is profitable. After the changes are determined, the relationship is shown by the following equation:

[Added returns + Reduced costs] – [Added costs + Reduced returns] = [Profit or Loss]

The farmer's goal is to maximise profit and not necessarily only productivity. Determining the farmer's goal is a good start. For example the household may set as a goal, 'to increase income during drought periods' and achieve this goal by introducing cattle in addition to cropping. The advantages and disadvantages of partial budgeting are listed as follows:

Merits and Demerits and Right Situations for Partial Budget Analysis Technique

Merits

- The technique is simple (it can be performed with a hand calculator) and easy to learn for non-economists.
- It examines only net changes in costs and benefits. Therefore, it is effective for assessing economic viability of single intervention technologies.
- It requires less data than whole farm budgeting since fixed costs are not examined.
- The data required for partial budget analysis are collected for almost all other economic analyses.
- It allows early conclusion about the adaptability of the new technology.

Demerits

- There is a danger of forgetting that farmer's resources are limited and sometimes knowledge about the resource base may be lacking; this happens as most new technologies require an increase in purchased inputs and additional labour.
- Often scientists do not understand the farmer's objectives; the partial view of a farming system might obscure the secondary character of a given farm component.
- Lack of a time analysis; researchers do not understand the time needed for various farming activities which may overlap with or contradict a new schedule.
- Linearity of factors is assumed by using small-scale input/output data for a large-scale operation. The assumption that an increase in a unit of resource will increase the profit proportionally is questionable. The increase in management skills by many technologies often is not considered.

Limitations of Partial Budgeting

Although partial budgeting can be applied in a variety of situations it does have limitations to its use.

i. The first limitation of partial budgeting is that it is restricted to evaluating only two alternatives.

ii. The second limitation is that the results obtained from a partial budget are only estimates, and are only as good as the original data that is entered. If you enter inaccurate information in the budget, you receive inaccurate results.

iii. A third limitation is that partial budgeting does not account for the time value of money. That is, the difference in the value of cash received and/ or expended now, versus its value at some future date.

iv. Another limitation is that partial budgeting only provides an estimate of the profitability of an alternative relative to current operations. It does not provide an estimate of the absolute profitability of the business.

v. Finally, costs and returns that are not affected by an intended change are not included in the partial budget. In other words, you can only use the partial budget to consider the costs and returns of a specific action. If you cannot determine all the areas that will be affected by the intended change, it might be better to use a whole-farm budget to evaluate the impacts of the change.

Right Situations for Partial Budget Analysis

Partial budget analysis is appropriate in trials where:

i. A single component must be analysed.

ii. Inputs and outputs are measurable and easy to price.

iii. Animal yields vary little between farms.

iv. Profitability is the major concern rather than issues such as equity and income distribution.

v. Fixed cost do not change economic evaluation is needed for new technologies which are not yet well developed.

Why Partial Budget Analysis?

The farmer's goal is to maximise returns. The following equation describes this relationship:

$$NI = TR - TC$$

Where NI is the net income (e.g. from sale of cattle), i.e. the amount of money left when total costs (TC) are subtracted from total returns (TR). Total cost (TC) is composed of the cost of all inputs: variable costs (VC) and fixed costs (FC). Since FC is the same for comparing the new technology and the technology practised by the farmer, then:

$$TC = VC$$

The farmer wants to know if the new technology will increase his income while the researcher wants to know which technology is potentially more attractive economically. The change in income can be expressed as follows:

$$NI = TR - TC$$

$$NI = TR - VC$$

This relationship can be used as a rule of thumb for partial budgeting, since FC = 0 because by definition fixed costs do not vary.

If capital is not a constraint, choose the highest NI. Since higher benefits require higher costs, it is necessary to compare the extra or marginal costs with the extra or marginal benefits.

Marginal rate of return (MRR) is another way of taking the cost factor into account. It measures the NI which is generated by each additional unit of expenditure (VC):

$$MRR = NI/VC$$

MRR is also a measure of additional capital invested in a new technology and its effect on net return.

Researchers make recommendations to farmers about what to adopt. To make these recommendations, the following economic criteria should be observed provided the technology is in line with the farmer's goal, viable, and sustainable.

- If net income remains the same or decreases, the new technology should not be recommended as it is not more profitable than the farmers' current technology.
- If net income increases and variable costs remain the same or decrease, the new technology should be recommended as it is more profitable than the farmers' current technology.
- If both net income and variable cost increase being the most common case, the MRR should be considered. The greater the increase in NI and the higher the MRR, the more economically attractive the new technology is.

Partial budgets for fish/animal enterprises are essential because

- Fish/animals have a longer life cycle than food crops
- Fish/animals use non-market inputs (natural productivity of fish pond/water body, labour and land) and market prices often do not exist
- Fish/animals have multiple outputs like fish, vegetables, banana and coconuts in case of integrated farming.
- The life cycle of fish/animals is not synchronised which makes it difficult to measure output.

Data Required for Partial Budget Analysis

Collection of Input and Output Data: Agricultural production is characterized by large numbers of farms at dispersed locations. In most cases, farms lack formal records of input use, particularly with regard to individual crops. Output records are somewhat more common, but usually this information is not expressed in the terms of yield needed for economic analysis. As a result, primary farm surveys are expensive and time-consuming and place heavy demands on skilled manpower for monitoring and evaluating the survey data. In PAM-related work, the constraints of time and financial support for research usually mean that primary farm surveys are not possible. Instead, the analyst relies on secondary data in the preparation of representative farm budgets. Fieldwork remains critical to the construction of the PAM, but efforts focus on the verification of secondary data, the collection of information about current prices, and the introduction of modifications of input-output relations to account for technological change. In most circumstances, prior data on farm budgets are available. The ministry of agriculture, producer organizations, and university researchers in agricultural economics often produce farm budgets, and their surveys can provide estimates of input and output quantities. Agricultural investment project proposals require economic feasibility analyses; estimates of farm-level costs and returns are usually included in this work. Extension service personnel might also have useful information about the input and output quantity requirements for particular commodity systems. Finally, studies of comparable technologies in neighboring countries sometimes provide useful farm budgets. Whatever the source of budget information, fieldwork usually begins with interviews of the employees who originally prepared the budgets. Such interviews are useful to disaggregate information about costs and returns beyond the level provided in published documents, to assess the extent of heterogeneity of production practices and the need for multiple budgets, and to gain initial impressions about the price and quantity effects of particular policy distortions or market failures faced by producers. Field informants might also arrange interviews with farmers and other informed observers of the local agricultural economy, such as providers of input or marketing services. Interviews provide supplementary information about prices and the efficiency of various input and output markets. They can also cover the particular policy issues that motivate the research. Because the selection of expert informants is not random, care must be exercised in using responses to characterize the various representative systems. But this approach has the advantage of confining fieldwork to several weeks rather than many months.

Preparation of a good partial budget requires

1. A detailed understanding of farmer's objectives, practices, resource use pattern, constraints etc.
2. Some understanding of the fundamental concepts and principles of economic theory e.g. profitability, risk, opportunity cost, scarcity of resources, marginal concepts etc.
3. Good judgment and common sense. Educated guesses are always better than ignoring a cost or a benefit. It should be remembered that:
   a. All sources of benefits and costs to farmers are included in the analysis.
   b. The realism of the costs and yield assumptions are as important as the type of analysis chosen.
   c. A partial budget does not give the net effect of a technology/production process. It only gives the change in NET BENEFITS. It does not show the profitability of the enterprise or the farm. In order to get this one has to do an enterprise/whole farm budget.

In order to construct a Partial Budget one should calculate

A. benefits of different treatments

B. costs that vary across treatments

A. Estimation of Benefits

1. Identify the sources of benefits
   (a) Direct Benefits - main products

   - byproducts

   (b) Indirect benefits - often difficult to identify and quantify if the systems interaction is not known.

2. Quantify the benefits derived from the technology or treatment
   (a) Estimate the yield
   (b) Estimate the yield adjustment coefficient
   (c) Calculate the adjusted yield
   (d) Determine the field price of the product (and by-products).
3. Calculate the Gross Field Benefit (GFB)

   GFB = Adjusted Yield X Field Price

B. Estimation of Costs

Here the total costs that vary across treatments are estimated:

a. Identify and list the input items that vary across treatments.
b. Quantify the level of input in each treatment.
c. Estimate the unit value of the input. Once again we use the field price for input.
d. Estimate the Field Costs of the inputs
e. Field Cost = Field Price X Quantity of the Input
f. Calculate the total costs that vary for each treatment.

C. Calculation of Net Benefit (NB)

Net Benefit = Gross Field Benefit – Total Cost(s) that vary across treatments.

Steps for Partial Budgeting

The most important step in performing partial budget analysis is the proper identification of data on the costs and benefits associated with the alternative technologies. Generally the following are essential data that must be collected.

- quantities of inputs which vary between alternative technologies
- prices of these variable inputs
- yields or productivity levels resulting from the alternative technologies
- Prices of the outputs valuing non-market inputs or products opportunity cost (the value of the resource or product in its next best alternative use, e.g. family labour vs. market labour wages).

Important products of fish culture include reproductive capacity (offspring), flesh yield (weight gain), vegetables and horticultural crops. Inputs may depend on the technology being used. Input costs should include cash costs (e.g. seed, fertiliser, feed, aeration) and non-cash costs (family labour, capital costs, depreciation costs, crop residues and household wastes).

Care should be taken in determining the value to assign to inputs and outputs while calculating partial budgets. In some cases opportunity cost needs to be used. For example, manure provides a low cost fertiliser for crops and in this case it can be used as the cost of reduction in fertiliser use. Family labour is another example. Rural wages for hired labour can be used to evaluate family labour. When labour is not an option for household members, it is difficult to evaluate. Fish that are consumed by the household can be valued against the purchase value in the market.

Evaluation of on-farm trials is done to determine the impact of the technology/ intervention. It may be ex-ante, on-going as well as ex-post. Traditionally experimental results are assessed using statistical techniques only. Over the years it has been demonstrated that the criteria used by farmers to evaluate and adopt technologies may be totally different to that of researchers. Very often socio-economic considerations play a major role in accepting or rejecting a technology. Assessment of trials are therefore, one of the important stages of an on-farm research programme. The main elements of such an assessment are farmer assessment, agronomic evaluation, statistical analysis and economic analysis.

In this section various socio-economic criteria and simple techniques used to assess economic performance of a technology are discussed. The economic analysis helps researchers:

- to look at the results from the farmers' view point,
- to decide which treatments merit further investigation,
- to decide on the recommendations to make to farmers.

On-farm data upon which the recommendations are based must be relevant to the farmers' own agro-ecological conditions, and the evaluation of those data must be consistent with the farmers' goals and socio-economic circumstances.

The Need for Socio-Economic Evaluation

Traditionally, aquaculture trials are evaluated by biological scientists, using statistical techniques and the dominant evaluation criterion used is yield per unit area. The primary objective is the maximum exploitation of the biological potential and the test used often is statistical significance. Normally, the experiments/ treatments are tested at 0.01, 0.05, 0.10 per cent level. If they do not pass this test they are not considered any more. This approach has following limitations in deriving recommendations in a systems context:

(a) Evaluation criteria that a farmer may use are varied, location specific and depend greatly on the degree of market orientation. The appropriateness of any technology should always be evaluated in relation to their priority objectives and resource use pattern. The relevant criteria for the target group should be identified and used in the design and interpretation of the experiment. A number of physical, biological and socio-economic variables influence the farmers' choice of a technology. Very often the socio-economic factors such as lack of marketing, shortage of labour, clash with food priority etc., determine these choices.

(b) Agronomic data only establishes the technical relationships that can be used to determine the technical optimum. It should be complemented with market information in order to establish the economic optimum. The 'economic optimum' for any input is always lower than the 'technical optimum'. If the treatment means are not significantly different but an economic analysis shows that one treatment is a better recommendation than the others then we need to have a more careful analysis.

Examples of situations where farmers choose technologies even without significant differences in yield may be:

- varietal means may not be different but one could observe differences in palatability (preference), prices, storability, early maturing etc;
- technology may not increase the yield/profit, but may reduce the labour requirement at the critical period;
- A new technology may provide possibility of introducing a second crop i.e. increasing the total production of the systems;
- The intervention may reduce malnutrition in the target group.

It should be remembered that these variable attributes are also valid even in situations where there is significant difference in yield between treatments. It is important to remember that a farmer is managing a system and he/she is interested in improving the total production of the system without creating serious contradictions to his/her priority objectives and resource use patterns.

c) Farm level decisions are made and actions are taken in an attempt to reach goals in a world of uncertainty and scarcity of resources. These two elements are therefore, crucial in assessing the impact of any technology. Thus it is necessary for the biological scientist to conduct economic analysis in a similar manner, as they are responsible for the statistical analysis of their trials. The usefulness of the result of many bio-physical research experiments can be greatly enhanced if relevant economic analysis can be applied to the results. In this regard, it makes sense for biological scientists and agricultural economists to jointly evaluate experiments to establish both biological and economic viability.

The Majority of Economic Analysis with Reference to On-Farm Trial Work is in Three Categories

1. ***Average Returns Analysis***. This analysis consists of a listing of the average costs of producing a particular product and the average value of the product for each technology being compared. The cost-and-returns analysis requires information on both variable and fixed inputs. A more limited average variable cost-and-returns analysis is often used to compare different technologies that used the same fixed inputs. This information can be used to compare the average returns above variable costs (RAVC) (i.e., sometimes called gross margin analysis) for different technologies and the returns to other production factors such as total labour or labour for a single operation e.g. weeding. The process of valuing inputs and commodities is of particular importance in making realistic cost-and-return analyses.

The average returns (gross margin) analysis allows a comparison between various technologies being tested based on the inputs the farmer must provide. The procedure is as follows:

- Calculate an average yield or an average amount of product for each separate technology.
- Calculate average inputs often with particular emphasis on labour inputs for each technology separately.
- Calculate the gross return (i.e., gross total value product), which is the adjusted yield times the appropriate field price for the product, or products, if there are more than one.
- Calculate the variable costs associated with each technology. The variable costs for a crop trial usually include labour, seed, draught hire (i.e., if hired draught is used), and a charge for equipment depreciation.
- The average return above variable cost (RAVC) also called gross margin is then calculated by subtracting the variable costs from the gross total value product.
- These average net returns can then be compared between technologies. Some scientists believe that the return for a new technology must be at

least 30 percent higher than that for the traditional technology before farmers will be willing to consider adopting the new technology.

This analysis is generally on a per hectare basis, and the return calculated is a return to management, assuming that land is fixed and that labour has been valued at the price of its best alternative use (i.e., opportunity cost). In order to maximise profits, it is necessary to maximise returns to the most limiting resource. For example, the most limiting resource may labour for deweeding of a pond. When the most limiting resource is known, it is possible to calculate an average net return to that resource for the different technologies being compared and choose the most favourable technology.

Measures of return are calculated as follows

(a) Return above variable cost (RAV) i.e. Gross Margin (GM)

= Gross Return (GR) - Total Variable Costs (TVC)

(b) Returns to Factor A.

The general equation for the rate of return to Factor A is:

$$\text{Rate of Return to A} = \frac{\text{GR - all costs other than Costs of A}}{\text{Costs of A}}$$

So:

(i) Return to TVC = GR/TVC

(ii) $$\text{Return to labour and power Cost} = \frac{\text{GR - Material Cost}}{\text{Labour and power cost}}$$

(iii) $$\text{Return to material Cost} = \frac{\text{GR - Labour and power cost}}{\text{Material Cost}}$$

2. *Budget Analysis*. There are several types of budget analysis. The ***enterprise budget*** is statement of costs-and-returns (i.e., both variable and fixed) for a particular enterprise or technology. This type of budget can be used as a building block in making whole farm budgets. The ***partial budget***, on the other hand, is a direct comparison of the elements within enterprise budgets that change between technologies. This type of budget requires fewer data than the enterprise budget and offers the advantage of direct comparison. Finally, whole farm budgets can be used to look at allocation of resources between enterprises and at the impact of a new technology on the allocation of resources to other enterprises on a farm. The partial budget technique is used most frequently in FSR.

3. *Risk Analysis*. When a farmer undertakes a fishery enterprise she/he always faces the risk of failure and loss of their time, cash or other inputs invested in the enterprise. When farmers consider a new technology, they are concerned about the risk involved in the new technology compared to the risk of their present technology. Measuring risk is difficult and is of somewhat limited value because different farmers look at risk differently. Risk analysis needs to be kept as simple as possible. Some indications of risk can be obtained from doing sensitivity analysis with partial budgets.

Steps in Partial Budget Analysis

The first step in doing an economic analysis of on-farm experiments is to calculate the costs that vary for each treatment. In developing a partial budget all outputs and inputs are measured in currency units, as a common denominator. This is necessary because otherwise it would be impossible, for example, to add hours of labour to litres of herbicide and compare these with kilograms of grain. The use of currency units does not, however, necessarily imply that farmers spend money on inputs or receive money for the outputs. Neither does it imply that farmers are concerned only with money. It is simply a device to represent the process that farmers go through when comparing the value of the things gained and the value of the things given up. In a partial budget not all costs (e.g. family labour) necessarily represent the exchange of cash. An important concept used in the calculation of such cost items is that of opportunity cost. This cost is defined as the value of any resource in its best alternative use. In partial budgeting one has to be concerned with the differences in costs and benefits between experimental treatments.

Example of Partial Budget Analysis for *Assessment of Technological Change*

Example of Partial Budget Analysis of an Intervention: 'Brackish water Aquaculture Technology' compared to the control farms selected from the neighboring areas engaged in conventional culture practices at the District South 24 Parganas in the state of West Bengal. (Adopted from *NAIP Final Report: Impact Assessment Technologies from East and North East India*)

Cost of Cultivation: An account of cost of cultivation being a very important factor is being compiled and presented for both intervened and control groups and displayed in Table 2.1 for comprehension of changes that has occurred as a result of adoption technology compared to the control groups.

Table 2.1: Cost of Cultivation (in Rs. / Acre): Intervention: Brackish water Aquaculture: (Treated VS Control)

Field	Treated	Control	Comparison
Average Labour cost (in Rs.)	3678.50	3722.48	-43.97
Average Farm power cost (in Rs.)	947.23	1685.39	-738.16
Material Inputs cost (in Rs.)	4068.23	1235.54	2832.68
Other associated cost (in Rs.)	9.99	15.11	-5.12
Total capital/long term investment per year (in Rs.)	4445.44	1877.43	2568.00
Other cost if any (in Rs.)	0	575.79	-575.79
Total cost of cultivation (in Rs./ acre) without support from NAIP	13149.40	9111.75	4037.64
Support provided by the project (in Rs.)		N.A.	
Average support provided in Labour cost (in Rs.)	0	N.A.	
Average support provided in farm power cost (in Rs.)	0	N.A.	
Average support provided in input cost (in Rs.)	0	N.A.	
Average support provided in associated cost (in Rs.)	0	N.A.	
Average support provided in Capital cost/long term investment (in Rs.)	4445.44	N.A.	
Total support provided from project (in Rs.)	4445.44	N.A.	
Actual cost of cultivation borne by farmer (in Rs/Acre)	8703.96	9111.75	-407.79

Income: A complete picture of Income (in Rs. / Acre) (Intervention: Brackish water Aquaculture): Treated VS Control is presented in Table-2.2.

Table 2.2: Income (in Rs. / Acre) (Intervention: Brackish water Aquaculture): Treated vs. Control

Field	Treated	Control	Comparison
Income from crop (in Rs./Acre)	0	0	0
Income from vegetable (in Rs./Acre)	0	0	0
Income from straw (in Rs./ Acre)	0	223.54	-223.54
Income from Fishery (in Rs./Acre)	26363.84	4280.32	22083.52
Income from poultry (in Rs./Acre)	0	0	0
Income from Livestock (in Rs./Acre)	0	0	0
Gross Income generated (in Rs./acre)	26363.84	4503.86	21859.98
Subtract total cost of cultivation without support from NAIP (in Rs./Acre)	13149.40	9111.75	
Net Income (in Rs./acre) without support from NAIP	13214.45	-4607.89	17822.34
Add support provided from NAIP (in Rs./Acre)	4445.44	N.A.	
Net Income (in Rs./acre) with support from NAIP	17659.89	-4607.89	22267.78
Benefit cost ratio *	1.34	-0.51	
Profit from competing crop/agro-enterprise (in Rs./Acre)	80304.08	1370.47	

* *Net Income (in Rs. / Acre) / Total cost of cultivation (in Rs. / Acre)*

The ultimate interest is income for any enterprise. Therefore, details of gross income, net income profit and cost benefit ratio is presented side by side for comparison of treated and control groups of farmers table. The analysis section of the partial budget containing both net change in profits and benefit/cost ratio analysis. It is clear from above table that in case of Treated group average income was Rs.26364/acre as against an average income of Rs 4504/acre from Control group i.e. almost 6 times more. This is reflected in cost benefit ratio(Treated:1.34 & Control -0.51).It is well known that in theory, any project with a B/C ratio exceeding 1 is worthwhile, most public agencies have recognized that there is some uncertainty associated with both the benefit and the cost estimates. Accordingly, it is not uncommon for agencies to desire a threshold of B/C exceeding 1.5 for large new projects, and 1.3 for incremental projects in which uncertainty is less. The present case B/C ratio is at threshold level. Moreover, added returns and reduced costs fall into the benefits section of the partial budget and are the positive effects of a proposed change in the business as can be evidenced from the present analysis. Here the net change between positive and negative economic effects is an estimate of the net effect of making the proposed change in the total farm budget. A positive net change indicates a potential increase in income and a negative net change indicates a potential reduction in income due to the proposed change.

Conclusion

Partial budgeting can be useful in the decision process farm owners and managers used to decide on alternative uses of resources they have in their businesses. Partial budgeting is a systematic approach that can assist the manager in making informed decisions. But this budgeting process can only estimate possible financial impacts, not assure them. Management decisions and chance can change the projections. These may result in better or poorer than expected performance. Repeating the analysis using different assumptions about key variables will give some idea about the degree of risk involved in making the proposed change. In general, partial budget analysis provides a useful structure for analyzing potentially complex decisions. A computer spreadsheet package provides a simple method for performing this type of analysis. This is especially helpful in situations where many soft numbers might be used, because the computer can easily recalculate numbers when others change. Regardless of whether the analysis is done on paper or on the computer, progressive producers should use partial budget analysis to examine alternative choices and make better decisions. Partial and breakeven budgets are valuable aids to the primary producer by virtue of their simplicity. They are very easy to construct and understand, and provide a sound preliminary basis for management decisions. Proper budgeting by producers will lead to a better understanding of the farm's financial status, and, hopefully, more efficient use of the resources available particularly for aquaculture practices.

Acknowledgement

The author is thankful to ICAR for granting the project to carryout the impact Assessment of NAIP intervention on farm change.

References

Asian Development Bank (2006). Impact evaluation: methodological and operational issues. Economics and Research Department, Asian Development Bank, Manila.

Baker, J. (2000). Evaluating the impact of development projects on poverty: A Handbook for Practitioners. Washington, DC: World Bank.

Baker, J.L. (2000) Evaluating the Impact of Development Projects on Poverty, The World Bank, Washington, D.C.

Bamberger, M. (2006) Conducting Quality Impact Evaluations under Budget, Time and Data Constraints, World Bank, Washington, D.C.

Bamberger, M. (2006). Conducting quality impact evaluations under budget, timc and data constraints. Washington. DC: World Bank.

Bamberger, M., and H. White (2007) "Using strong evaluation designs in developing Countries: Experience and challenges", *Journal of Multidisciplinary Evaluation* 4(8), 58–73.

Bamberger, M., J. Rugh and L. Mabry (2006), Real-World Evaluation Working Under Budget, Time, Data, and Political Constraints, Sage Publications, Thousand Oaks, CA.

Bird, K. (2002). Impact assessment: An Overview. London, UK: ODI.

Bosc, P.M. (1988b). Evaluation economique de l'expérimentationagronomique: Approchebibliographique. (France: CIRAD)

Caroline Ashley and Karim Hussein. (2000). Developing methodologies for livelihood impact assessment: Experience of the African Wildlife Foundation in East Africa, Overseas Development Institute, Portland House, Stag Place, London, SW1E 5DP, UK

CIMMYT. (1988). From agronomic data to farmer recommendations: An economics training manual. CIMMYT, Mexico

Collinson, M. P. (1982). Farming systems research in Eastern Africa: The experience of CIMMYT and some national agricultural research services 1976-81. MSU International Development Paper No. 2. East Lansing, MI: Dept. of Agricultural Economics, Michigan State University

Crawford, E. and M. Kamuanga. (1988). "Economic analysis of agronomic trials for the formulation of farmer recommendations." MSU International Development Reprint Paper No. 6. East Lansing, MI: Dept. of Agricultural Economics, Michigan State University.

CSSRI, NAIP. 2013. Progress Report of NAIP sub-project on: Strategies for Sustainable Management of Degraded Coastal Land and Water for Enhancing Livelihood Security of Farming Communities (Component 3, GEF funded). Central Soil Salinity Research Institute, Regional Research Station (CSSRI, RRS), Canning Town - 743 329, South 24 Parganas, West Bengal. p. 49

Dayohimi Rymbai, Sheikh Mohammad Feroze, Ram Singh and Lala. P. Ray (2017). Application of Partial Budgeting for the Rice Growers of Eastern Himalaya in India. *British Journal of Applied Science & Technology* 20(5): 1-8, 2017; Article no.BJAST.33216 ISSN: 2231-0843, NLM ID: 101664541.

Devillet, R., J. Degand, and E. Dardenne. "The economic benefits of maize cultivation." Guide du maisiculteur et rapport C.I.P.F. (1981): 112-127.

Duflo, E. and M. Kremer (2003). Use of randomization in the evaluation of development effectiveness. Paper presented at the World Bank OED Conference on Evaluation and Development Effectiveness, Washington, DC, 15-16 July.

Elbers, C., Gunning, J., & Hoop, K., (2008). Assessing sector-wide programs with statistical impact evaluation: A Methodological Proposal' World Development, Vol. **20**, No. 10.

European Evaluation Society (EES) (2007). The importance of a methodologically diverse approach to impact evaluation – specifically with respect to development aid and development interventions. EES Statement. Nijkerk, Netherlands: EES secretariat

Fiszbein, A. (2006). Development impact evaluation: New Trends and Challenges'. Evidence and Policy, 2(3): 385-393.

Food Policy Research Institute (IFPRI) and the Mexican Progress Anti-Poverty and Human Resource Investment Conditional Cash Transfer Program. Impact Assessment Discussion Paper27. Washington, DC: IFPRI

Herdt, R. (1987). Whither farming systems" In: How Systems Work: Proceedings of farming systems research symposium 1987, 3-7. Fayetteville, Arkansas: Winrock International Institute for Agricultural Development. University of Arkansas

Hildebrand, P.E., and F. Poey. (1985). On-farm agronomic trials in farming systems research and extension. Boulder, CO: Lynne Rienner.

Independent Evaluation Group of the World Bank (IEG) (2006). Impact Evaluation: An overview and some issues for discussion. OECD DAC

Lessley, B. V., D. M. Johnson, and J. C. Hanson (1991) Using the Partial Budget to Analyze Farm Change, 1990-1991 Edition, University of Maryland System, pp. 7.

Mandal, Subhasis. Sarangi, S. K., Burman, D., Bandyopadhyay, B.K., Maji, B., Mandal, U.K. and Sharma, D. K. (2013). Land shaping models for enhancing agricultural productivity in salt affected coastal areas of West Bengal – An economic analysis. Indian Journal of Agricultural Economics, 68(3):389-401.

Merino, G. et al. (2012). Can marine fisheries and aquaculture meet fish demand from a growing human population in a changing climate? Glob. *Environ. Change* 22(4): 795–806.

NAIP (2012). Annual Report 2011-12, National Agricultural Innovative Project, ICAR, New Delhi

NAIP, (2012).Annual report, 2012-13. NAIP, ICAR, Krishi Bhawan, New Delhi-110012

NAIP, (2014). Monitoring and Evaluation Report on National Agricultural Innovation Project. Indian Council of Agricultural Research, New Delhi, India.

National Household Survey Capability Programme: Sampling Frames and Sample Designs for Integrated Household Survey Programmes, Preliminary Version (DP/UN/INT-84-014/5E), New York, 1986

Norman, D. W. (1987). "Farmer groups for technology development: Experiences from Botswana." Paper presented at a workshop on Farmers and Agricultural Research: Complementary Methods, IDS, Sussex, UK.

Ravallion, M. (2008). Evaluation in the practice of development. Policy Research Working Paper 4547.Washington, DC: World Bank.

Roy, A. K. (2010). Evaluation and Impact Assessment of Technologies and Developmental Activities in Agriculture, Fisheries and Allied Fields. New India Publishing Agency.101 Vikas Surya Plaza, Pitampura, New Delhi.xiv+510p (ISBN: 978-93-8023540-0)

Roy, A.K. (2010). Methods, Tools and Techniques for Evaluation, Monitoring and Impact Assessment of Development Programme.In:.Evaluation and Impact Assessment of Technologies and Developmental Activities in Agriculture, Fisheries and Allied Fields, NIPA, New Delhi(eds. A.K Roy).pp:1-76

Roy, A.K. (2018). Impact Assessment of an NAIP Intervention "Brackish Water Aquaculture Technology' Through Partial Budgeting Analysis *JFLS*, Vol 3(2):58-66.

Roy,A.K. (2018): Partial Budgeting Analysis of an Intervention of 'Land Shaping Technology' at South 24 Parganas, West Bengal, India. *Open Acc J Envi Soi Sci.*, Volume-1.issue-4: 94-102

Science Council (2006c). Impact assessment of policy-oriented research in the CGIAR: A Scoping Study Report'. CGIAR Science Council.

Science Council (2006d) 'Spill over Increases Returns to Sorghum Genetic Enhancement'. CGIAR, Science Council Brief 4.

Singer, U. (2013). Livelihoods and resource management survey on the Mekong between LouangPhabang and Vientiane Cities, Lao PDR. Vientiane, Lao PDR: IUCN. 122 pp.Sustainable Livelihoods Guidance Sheets. FID, UK.

Sita Devi K, Ponnarasi T. (2009). An economic analysis of modern rice production technology and its adoption behaviour in Tamil Nadu. *Agricultural Economics Research Review.* 2009; 22:341-34.

The World Economic and Social Survey (2013). Sustainable development challenges department of economic and social affairs of the United Nations Secretariat (UN/DESA), E/2013/50/ Rev1 ST/ESA/344.

United Nations (2011a). The millennium development goals report 2011. Sales No. E.11.I.10.

3

Impact Assessment Tools for Evaluating Fisheries Research Programmes

Neha W. Qureshi; Sivaramane N. and M. Krishnan

Introduction

Many a times research programs appear potentially promising before implementation yet fail to generate expected impacts or benefits. Impact Assessment (IA) has become an integral part of any research study. It, not only justifies the investment made, but also helps in prioritizing projects. The ultimate objective is to make an impact from the output of the projects carried out. Researchers and extension functionaries involved in the upliftment of farmers/fishers through various research projects, training activities and interventions are bound to justify their resource utilization through unbiased and objective based impact assessments.

What is Impact Assessment?

Impact assessment (IA) is a structured process for considering the implications, for people and their environment, of proposed actions while there is still an opportunity to modify or abandon the proposals. It is applied at all levels of decision-making, from policies to specific projects. Impact Assessment is simply defined as the process of identifying the future consequences of a current or proposed action. The difference between input, output, outcomes and how they encompass the monitoring and evaluation programmes of any organisation later transforming into the study of impacts has been shown in Figure 3.1.

The process involves the identification and characterization of the most likely impacts of the proposed actions (impact prediction/forecasting), and an assessment of the social significance of those impacts (impact evaluation) (Khandker *et al.*, 2009)

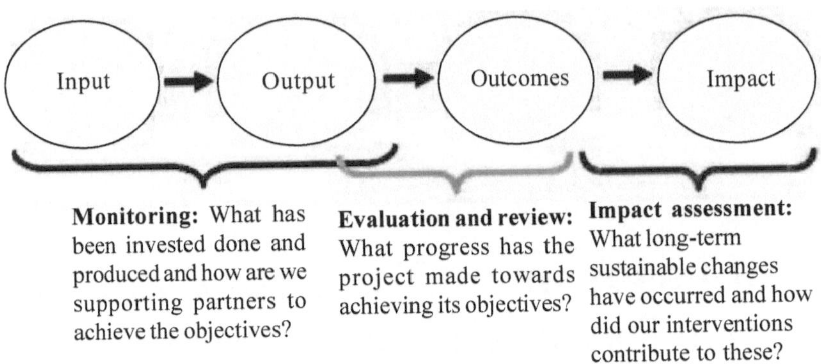

Fig. 3.1: Pathway Leading to Impact Assessment

Why is the need of IA?

The obvious need for impact evaluation is to help policy makers decide whether programs are generating intended effects; to promote accountability in the allocation of resources across public programs; and to fill gaps in understanding what works, what does not, and how measured changes in well-being are attributable to a particular project or policy intervention. Effective impact evaluation should therefore, be able to assess precisely the mechanisms by which beneficiaries are responding to the intervention (Chen *et al.*, 2008). These mechanisms can include links through markets or improved social networks as well as tie-ins with other existing policies. The last link is particularly important because an impact evaluation that helps policy makers understand the effects of one intervention can guide concurrent and future impact evaluations of related interventions. The benefits of a well-designed impact evaluation are therefore, long term and can have substantial spill-over effects (Figure 3.2).

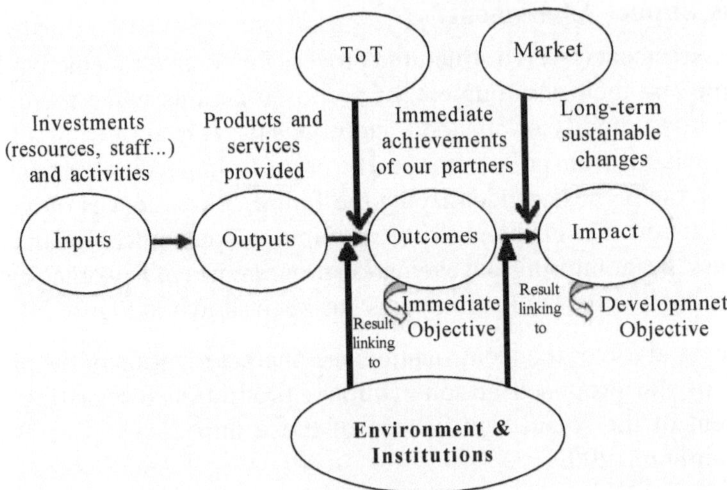

Fig. 3.2: Benefits of Well-Designed Impact Assessment Programme

Methods to Study Impact

Impact evaluation includes qualitative and quantitative methods, as well as ex ante and ex post methods. Qualitative analysis, as compared with the quantitative approach, seeks to gauge potential impacts that the program may generate the mechanisms of such impacts, and the extent of benefits to recipients from in-depth and group-based interviews (Figure 3.3). Whereas quantitative results can be generalized, the qualitative results cannot be. Nonetheless, qualitative methods generate information that may be critical for understanding the mechanisms through which the program helps beneficiaries.

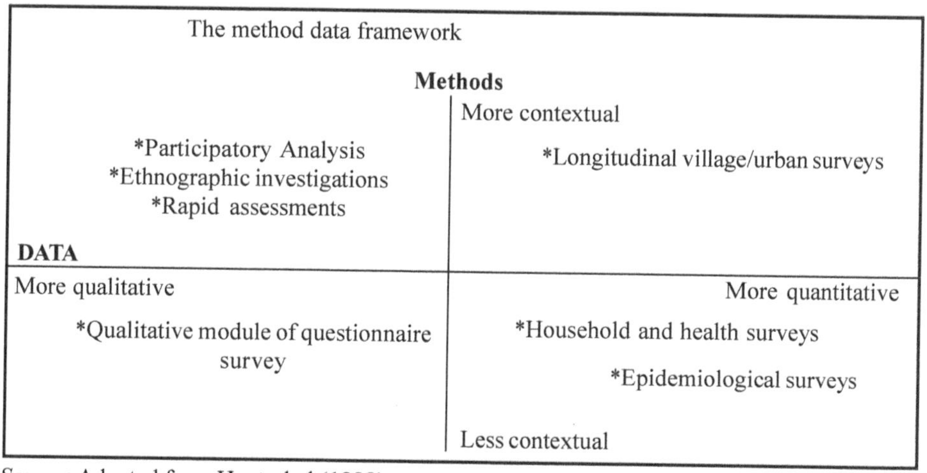

The method data framework

Methods

	More contextual
*Participatory Analysis *Ethnographic investigations *Rapid assessments **DATA**	*Longitudinal village/urban surveys
More qualitative *Qualitative module of questionnaire survey	More quantitative *Household and health surveys *Epidemiological surveys Less contextual

Source: Adapted from Hentschel (1999)

Fig. 3.3: Qualitative v/s Quantitative Techniques

Data Collection Designs and their Characteristics

The data collection methods for impact assessment studies vary with regard to the context in which they are being used. Some of them are enlisted in the Table 3.1 below:

Experiments with Control

Such kinds of experiments which mostly happen in laboratories have a control group and impact can be easily assessed. Control group is a group which is not subjected to or not influenced by intended treatment. For example the impact of a nutraceutical on the growth of a fish in terms of weight and length or the impact of mid-day meal on the nutrition of kids measured in terms of physical characteristics such as height, weight and no. of days of absence to the classes due to sickness.

- Problem: Demonstration effect
- Tools of analysis:

Table 3.1: Characteristics of various impact assessment methodologies

Characteristics evaluation design	Cost	Reliability	Technical expertise	Types of evaluation (primarily adoptive to the design)	Ability to measure what is happening	Ability to exclude rival hypothesis
Case study: one measurement (actual vs. planned)	low	very low	low	reporting	very low	Non-existent
Case study: two measurements (before and after)	medium	low	low	process evaluation	good	Low
Time series design (prior trend vs. actual)	relatively low, if feasible	medium	medium	impact evaluation	very good	medium
Case study with one measurement and a control group (with and without)	medium	low	low	formative evaluation	low	Low
Quasi-experimental design	relatively high (variable)	relatively high (variable)	relatively high	impact evaluation	very good	good (variable)
Experimental design	expensive			evaluation research	very good	very good

- t-test of the differences
- ANOVA

Different Evaluation Approaches to Ex Post Impact Evaluation

A number of different methods can be used in impact evaluation theory to address the fundamental question of the missing counterfactual. Each of these methods carries its own assumptions about the nature of potential selection bias in program targeting and participation, and the assumptions are crucial to developing the appropriate model to determine program impacts. (King *et al.*, 2009). In IA, even though the scale of measurement, criteria and indicators varies widely across the programmes, over years, the development in methodologies and associated analytical techniques have drifted impact study more towards objective assessment. IA is complex and viewed through various facets like social, cultural, institutional, technological, economic, environmental, etc.

These methods include:

1. Randomized evaluations
2. Matching methods, specifically propensity score matching (PSM)
3. Double-difference (DD) methods
4. Instrumental variable (IV) methods
5. Regression discontinuity (RD) design and pipeline methods
6. Structural and other modeling approaches

1. Randomized evaluations

Allocating a program or intervention randomly across a sample of observations is one solution to avoiding selection bias, provided that program impacts are examined at the level of randomization (Angrist *et al.*, 2002). Careful selection of control areas (or the counterfactual) is also important in ensuring comparability with participant areas and ultimately calculating the treatment effect (or difference in outcomes) between the two groups. The treatment effect can be distinguished as the average treatment effect (ATE) between participants and control units, or the treatment effect on the treated (TOT), a narrower measure that compares participant and control units, conditional on participants being in a treated area. Randomization could be conducted purely randomly (where treated and control units have the same expected outcome in absence of the program); this method requires ensuring external and internal validity of the targeting design. In actuality, however, researchers have worked in partial randomization settings, where treatment and control samples are chosen randomly, conditional on some observable characteristics (for example, landholding or income) (Figure 3.4). If these programs are exogenously placed, conditional on these observed characteristics, an unbiased program estimate can be made. Despite the clarity of a randomized approach, a number of factors still need to be addressed in practice (Banerjee *et al.,* 2007). They include resolving ethical issues in excluding areas that share similar characteristics with the targeted sample, accounting for spill-over to non-targeted areas as well as for selective attrition, and ensuring heterogeneity in participation and ultimate outcomes, even if the program is randomized.

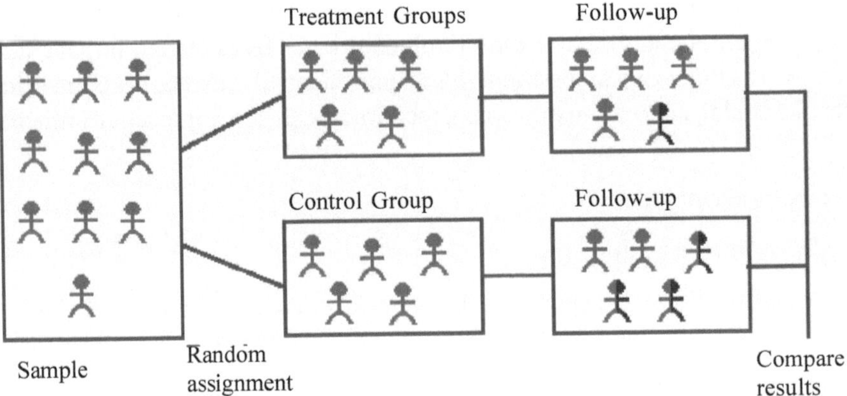

Fig. 3.4: Randomised control trials

2. Propensity Score Matching

Matching techniques aim at comparing treatment and control units, which are similar with respect to some observable characters. Matching is a simple affair when the dimensionality is small, i.e., with small number of characteristics. When the dimensionality increases, as in many cases of its application, it becomes difficult to decide which dimensions to match. When dimension becomes large, in fact, difficulties arise in finding suitable match for comparison. In literature, this is often termed as 'curse of dimensionality'. Propensity Score Matching (PSM) provide a natural waiting scheme that overcome previously mentioned problems and yields reliable estimator of treatment effect (Bryson *et al.*, 2002). In PSM, we estimate the probability of a unit being in the treatment group based on all relevant observable characteristics, which is called as propensity score. Participants are then matched on the basis of this to nonparticipants. The average treatment effect of the program is then calculated as the mean difference in outcomes across these two groups (Caliendo *et al.*, 2008). The validity of PSM depends on two conditions: (a) conditional independence (namely, that unobserved factors do not affect participation) and (b) sizable common support or overlap in propensity scores across the participant and nonparticipant samples.

Steps for PSM

- PSM constructs a statistical comparison group that is based on a model of the probability of participating in the treatment, using observed characteristics.
- Participants are then matched on the basis of this probability, or *propensity score*, to nonparticipants.
- The average effect of the treatment (*ATT*) of the program is then calculated as the mean difference in outcomes across these two groups.

Necessary conditions:

- Conditional independence (namely, that unobserved factors do not affect participation)
- Sizeable common support or overlap in propensity scores across the participant and nonparticipant samples.

Matching Methods

- Nearest neighbor matching (Fig. 3.5)
- Radius matching
- Kernel Matching (Fig. 3.ig 5)
- Stratification Matching

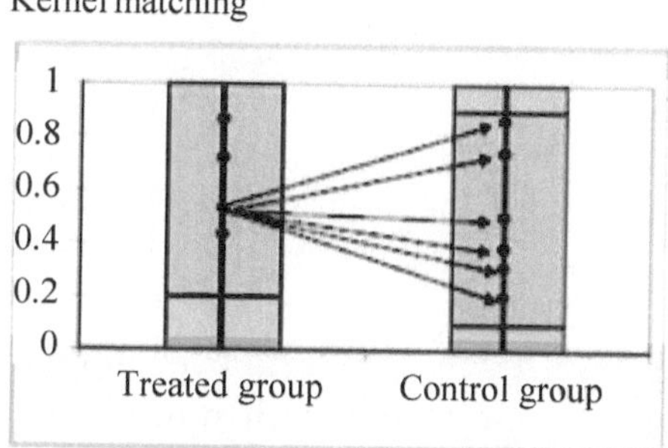

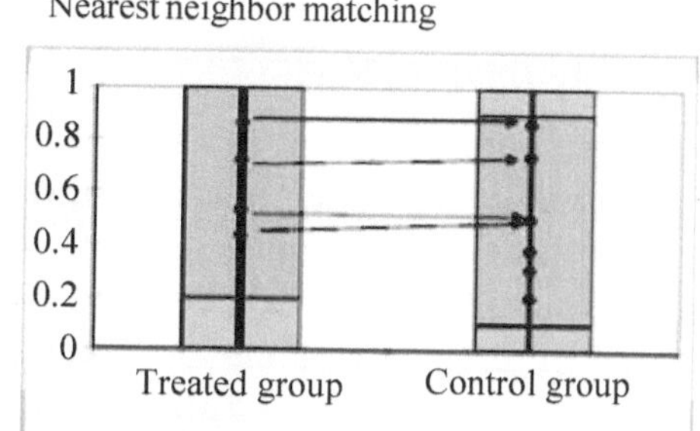

Fig. 3.5: Kernel matching and nearest neighbour matching

3. Double Difference

Double-difference (DD) methods, compared with propensity score matching (PSM), assume that unobserved heterogeneity in participation is present—but that such factors are time invariant. With data on project and control observations before and after the program intervention, therefore, this fixed component can be differenced out. (Figure 3.6) Some variants of the DD approach have been introduced to account for potential sources of selection bias. Combining PSM with DD methods can help resolve this problem, by matching units in the common support. Controlling for initial area conditions can also resolve non-random program placement that might bias the program effect. Where a baseline might not be available, using a triple-difference method with an entirely separate control experiment after program intervention (that is, a separate set of untreated observations) offers an alternate calculation of the program's impact.

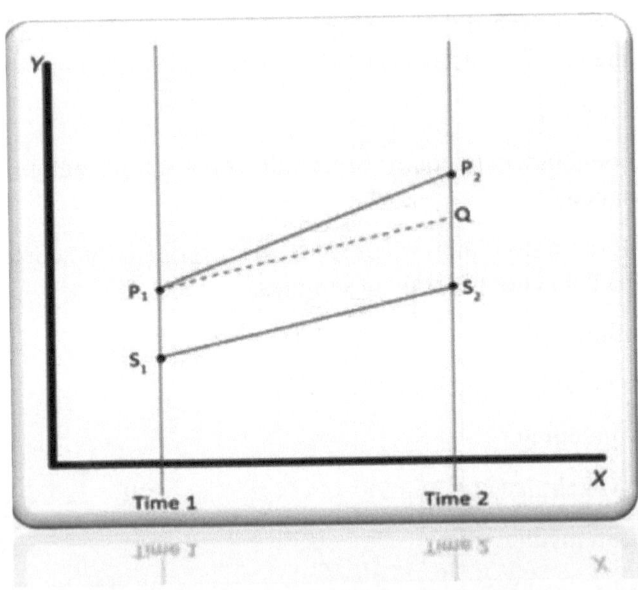

Fig. 3.6: Difference in difference technique

4. Instrumental Variable

Instrumental variable (IV) methods allow for endogenity in individual participation, program placement, or both. With panel data, IV methods can allow for time-varying selection bias. Measurement error that results in attenuation bias can also be resolved through this procedure. The IV approach involves finding a variable (or instrument) that is highly correlated with program placement or participation but that is not correlated with unobserved characteristics affecting outcomes (Abadie *et al.*, 2002). Instruments can be constructed from program design (for example, if the program of interest was randomized or if exogenous rules were used in determining eligibility for the program). Instruments should be selected carefully. Weak instruments can potentially worsen the bias even more than when estimated by ordinary least squares (OLS) if those instruments are correlated with unobserved characteristics or omitted variables affecting the outcome. Testing for weak instruments can help avoid this problem. Another problem can arise if the instrument still correlates with unobserved anticipated gains from the program that affect participation; local average treatment effects (LATEs) based on the instruments can help address this issue.

5. Regression Discontinuity and Pipeline Methods

In a non-experimental setting, program eligibility rules can sometimes be used as instruments for exogenously identifying participants and nonparticipants. To establish comparability, one can use participants and nonparticipants within a certain neighborhood of the eligibility threshold as the relevant sample for estimating the treatment impact known as regression discontinuity (RD). This method allows observed as well as unobserved heterogeneity to be accounted for (RD). (Koenker, 1978). Although the cut-off or eligibility threshold can be defined non-parametrically, the cut-off has in practice traditionally been defined through an instrument. Concerns with the RD approach include the possibility that eligibility rules will not be adhered to consistently, as well as the potential for eligibility rules to change over time. Robustness checks can be conducted to examine the validity of the discontinuity design, including sudden changes in other control variables at the cut-off point. Examining the pattern in the variable determining eligibility can also be useful whether, for example, the average outcome exhibits jumps at values of the variable other than the eligibility Figure 3.7.

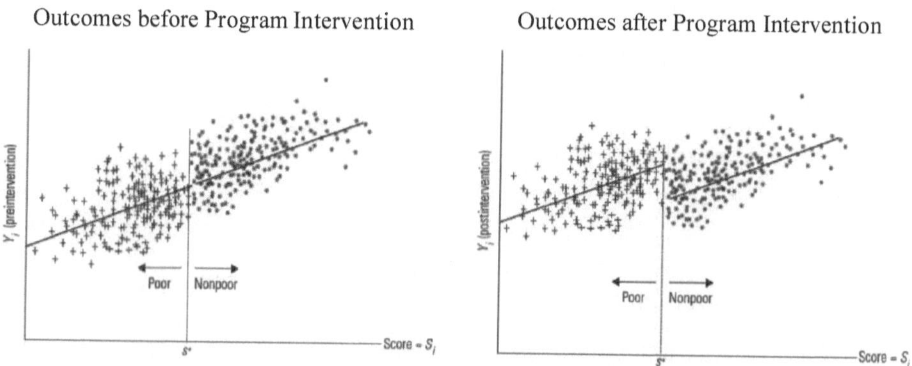

Fig. 3.7: Regression discontinuity before and after an intervention

Pipeline comparisons exploit variation in the timing of program implementation, using as a comparison group eligible participants who have not yet received the program. One additional empirical strategy considered by program evaluators is to exploit data on program expansion along a given route (for example, an infrastructure project such as water, transport, or communication networks) to compare outcomes for eligible participants at different sides of the project boundary as the program is phased in. This method involves a combination of pipeline and RD approaches that could yield interesting comparisons over time.

6. Economic Models to Evaluate Policies

Economic models can help in understanding the potential interactions—and interdependence—of a program with other existing policies and individual behavior. Unlike reduced-form estimations, which focus on a one-way, direct relationship between a program intervention and ultimate outcomes for the targeted population, structural estimation approaches explicitly specify interrelationships between endogenous variables (such as household outcomes) and exogenous variables or factors. Structural approaches can help create a schematic for interpreting policy effects from regressions, particularly when multiple factors are at work.

Ex ante evaluations also build economic models that predict program impacts amid other factors. Such evaluations can help reduce costs as well by focusing policy makers' attention on areas where impacts are potentially greater. The evaluations can also provide a framework for understanding how the program or policy might operate in a different economic environment if some parameters (such as rates of return on capital or other prices) were changed. Some economic impact assessment methods are enlisted below:

- NPV / BCR / IRR
- Partial budgeting technique
- Total factor productivity
- Trend forecasting
- Economic surplus
- Case studies
- Simulation

Other important tools are:

- Adoption rate/index/quotient
- Factors influencing the adoption of a technology (logit/probit)
- Consumer surplus model
- Dummy variable regression models
- Structural change in time series models
- Input-output model/SAM
- Competitiveness index (for exports)

Intricacies in Measurement of Impact

Impact assessment studies are not that simple as they seem to be. Lot of externalities arises while doing it and some of them are:

- Impact assessment should start before the commencement of the project
- May be highly complex
- May have both negative and positive externalities
- May be lasting very long or forever (timeframe should be clearly defined)
- Externalities may be difficult to measure or quantify
- Boundary or stakeholders should be clearly defined
- Delineating effects of different interventions may sometimes be impossible
- Highly subjective / qualitative and arriving at consensus may be difficult
- Success of IA lies in selection of appropriate indicators
- Authenticity or Reliability of IA (Third party evaluation)

Training Impact Assessment Tools

There is lot of training assessment tools used to evaluate the impact of different training programmes and most commonly used among them is Kirkpatrick model

The Kirkpatrick Model is the worldwide standard for evaluating the effectiveness of training. It considers the value of any type of training, formal or informal, across four levels. It was created by Dr. Don Kirkpatrick in the 1950s, the model is applied before, during and after training to both maximize and demonstrate training's value to the organization.

Level 1: Reaction

The degree to which participants find the training favorable, engaging and relevant to their jobs

Level 2: Learning

The degree to which participants acquire the intended knowledge, skills, attitude, confidence and commitment based on their participation in the training

Level 3: Behavior

The degree to which participants apply what they learned during training when they are back on the job

Level 4: Results

The degree to which targeted outcomes occur as a result of the training and the support and accountability package

Impact Assessment of Fisheries Research and Education

Fisheries and aquaculture is one of the fastest growing food production sector and industry globally and in India. Recognised as the sunrise sector, it contributes 5.15% to India's agricultural GDP and 1.1% to total GDP value. There are 26 colleges providing higher education in fisheries, ICAR-Central Institute of Fisheries Education, being the only deemed university at national level. In addition, there are 7 ICAR institutes contributing to fisheries research in the country. The impact of fisheries research on fish production has not been studied so far and there are no exhaustive works in such domain but ICAR-CIFE (Sahoo *et al*, 2018) has attempted first of its kind Research in India to estimate the impact of ICAR-CIFE on fisheries higher education in India and on the overall Indian economy. The investment being made by ICAR-CIFE for generating human capital and their subsequent benefits/impact in the form of economic value to fisheries sector has been quantified. Mapping the careers of more than 1000 CIFEians (MFSc & PhD) over the years (1995-2017), the study provides the first estimate of economic worth of human resource generated by a University. Besides descriptive statistics, a modified Flamholtz's human cost accounting method, Lev and Schwartz model (for estimating present value of future earnings) and the Divisia Index method of Total Factor Productivity were employed as methods in the study.

CIFE has produced 9.5% of total professional human capital pool available in the country during last two decades. CIFE's contribution to national economy, in the form of human capital helping increase the overall factor productivity of fisheries sector and thus the growth rate, was found to be as high as Rs. 51.11 crores during 2016-17 (if contribution of professional fisheries HR to total fisheries HR in the country is taken at a minimum value of 50%). The value would be Rs. 76.66 crores if the contribution is taken as 75%. Thus, the nation reaped a benefit of Rs. 51 to Rs. 77 crores through the contribution of CIFE's professional human capital while it invested Rs. 14 crores in 2016-17. For every single rupee spent on generating professional Fisheries human resource, it yields a very high return of 3.58 to Rs. 5.38.

It was an initiation and many more such studies can be carried out in terms of impact of fisheries research as well impact of human resource generated in fisheries sector on Indian economy.

Conclusion

In reality, no single assignment or evaluation method may be perfect, and verifying the findings with alternative methods is wise. Different ex ante and ex post evaluation methods can be combined, as can quantitative and qualitative approaches. The main lesson from the practice of impact evaluation is that an application of particular methods to evaluate a program depends critically on understanding the design and implementation of an intervention, the goals and mechanisms by which program objectives can be achieved, and the detailed characteristics of targeted and non-targeted areas. By conducting good impact assessment over the course of the program and by beginning early in the design and implementation of the project, one can also judge whether certain aspects of the program can be changed to make it better. Impact of fisheries research and education on blue economy of the country has to be studied in detail to delve deeper into the dynamics of variables that drive the engines of the economy.

References

Abadie, Alberto, Joshua D. Angrist, and Guido W. Imbens (2002). Instrumental variables estimates of the effect of subsidized training on the quantiles of trainee earnings. *Econometrica,* 70 (1): 91–117.

Angrist, Joshua, Eric Bettinger, Erik Bloom, Elizabeth King, and Michael Kremer. (2002). Vouchers for private schooling in colombia: evidence from a randomized natural experiment. *American Economic Review*, 92 (5):1535–58.

Banerjee, Abhijit, Shawn Cole, Esther Dufl o, and Leigh Linden. (2007). Remedying education: evidence from two randomized experiments in India. *Quarterly Journal of Economics,* 122 (3): 1235–64.

Bryson, Alex, Richard Dorsett, and Susan Purdon. (2002). The use of propensity score matching in the evaluation of active labour market policies. Working Paper 4, Department for Work and Pensions, London.

Caliendo, Marco, and Sabine Kopeinig. (2008). Some practical guidance for the implementation of propensity score matching. *Journal of Economic Surveys*, 22 (1): 31–72.

Chen, Shaohua, Ren Mu, and Martin Ravallion. (2008). Are there lasting impacts of aid to poor areas? Evidence for rural china. Policy Research Working Paper 4084, World Bank, Washington, DC.

Khandker, Shahidur R., Zaid Bakht, and Gayatri B. Koolwal. (2009). The poverty impact of rural roads: evidence from Bangladesh. *Economic Development and Cultural Change,* 57 (4): 685–722.

Khandker, S., B. Koolwal, G. and Samad, H., (2009). Handbook on impact evaluation: quantitative methods and practices. The World Bank.

King, Elizabeth M., and Jere R. Behrman. (2009). Timing and duration of exposure in evaluations of social programs. *World Bank Research Observer*, 24 (1): 55–82.

Koenker, Roger, and Gilbert Bassett. (1978). Regression quantiles. *Econometrica*, 46 (1) 33-50

Sahoo Jyotimanjari (2018). Human resource accounting of ICAR-CIFE's Higher Education. Unpublished M.F.Sc Dissertation work.

4

Economic Significance of Resource Use and Species Diversification in Aquaculture

A.D. Upadhyay; A.B Patel and D.K. Pandey

Introduction

Tripura is a small state blessed with rich flora and fauna. The small scale aquaculture with other agricultural enterprises and allied economic activities are common practice followed by majority of the farmers in rural areas. The small scale aquaculture is not only common in rural areas but is also practiced in the backyard ponds in urban which have multiple uses. The homestead fish production is common in Tripura because fish is highly preferred food item in the diet of about 95% people of Tripura. But due to the subsistence level of aquaculture practices in the state, productivity and production of fish is quite insufficient to meet out market demand for fish. Though the state Government and other stakeholders of fish production systems are making relentless effort to enhance the production to the level of 74 MT to meet the current market demand, the average fish productivity in the state is comparatively lower as compared to many states of the country. The lower productivity is an important issue and one of the major impediments in enhancing fish production in the state. The experts and research finding have identified several factors such as short culture period, acidic nature of soil and water, more iron content, poor quality of seed, improper feeding, lack of technical knowhow etc are the main reasons behind the lower productivity. Still, there are many other issues uncovered and unaddressed which may be the lower growth of fishes and consequent lower fish productivity. For instance, factors like pond size and number of species in the culture system at farmers' level may also have impact on the farm productivity. It is necessary to adopt region-specific models for furthering freshwater aquaculture in view of the available aquatic biodiversity as well as the consumer preferences and economics of operations in different parts of the country (Ayappan et al, 2009). This is particularly important considering polyculture is the major form of aquaculture in the state. Datta 2015, has opined that due to consumer's preference, high market value as well as conservation-revitalization of various cultivable species, importance of diversification in culture of various species has been found of immense importance in the present aquaculture scenario. Among the species, endemic minor carps, catfishes in monoculture, perches, murrels live species are important. In this study, an attempt has been made to analyse level of species diversification at the farm level and their socio-economic and scientific significance at farm level.

Sampling Method

In whole north eastern region of country, fish production is very common and it is integral part of the farming system. The fish production in the region not only supplements nutritional security of the people of the region but it also provide better income than many other agricultural enterprises. Fish production also provides gainful employment to the gainful employment to the farming community. This study was taken up to analyze farmers' strategies in fish culture with respect to stocking ratio and species diversification to enhance the production and farm income. Total area under culture fisheries was 26538.18 ha with total fish production 71981.28 t during 2016-17 where as the total demand for fish in the state was 92532.45 t (Economic Review, 2017). Tripura is second smallest state in the NEH states, where as the state has highest area and production of fish among seven NEH states. The study was taken up in Tripura. The multistage stratified random sampling technique was applied to for selection of the districts, blocks, villages and the respondents. First of all out of eight districts in Tripura Two districts namely, Sepahijala and Gomati districts were selected on the basis of the area under fish culture followed by two blocks Bishalgarh and Melagarh from Sepahijala district and Matabari and Kokraban blocks from Gomati districts were selected based on area under culture fisheries. Thereafter, two villages from each block and altogether total 12 villages were chosen for selection of respondents. At last stage the fish farmers according to number of fish farmers in each village and sample comprises 226 respondents selected by using simple random sampling without replacement (SRSWR).

This study was based on primary data. The personal interview method of data collection along with pretested semi structured survey schedule including both open ended and closed ended questions was adopted for collection of data. The primary data included the details of socioeconomic information, land holding, pond area, fish production system, details of culture practice such as pre-stocking and post stocking management, stocking density, details of species cultured, species selection for culture factors considered for selection of species, input and output quantities and prices, technical knowhow of species diversification and fish culture harvesting of fish and input and output marketing etc. The survey was conducted during May-December 2014.

There are quite few methods, which explain either concentration (i.e. specialization) or diversification of activities in a given time and space by a single indicator (Joshi et al. 2003). These measures include (i) Index of maximum proportion, (ii) Herfindal Index, (iii) Simpson Index, (iv) Ogive Index, (v) Entropy Index, (vi) Modified Entropy Index, and (vii) Composite Entropy Index (Kelley, Ryan and Patel 1995; Pandey and Sharma 1996; Ramesh Chand 1996; Joshi et al. 2003). The selection of method was done based on objective of the study, we used Simpson Index. The Simpson Index provides a clear picture on dispersion of the events (stocking ratio and species combinations) across the selected fish farms in two districts of Tripura. The index ranges between 0 and 1. If there complete specialization (monoculture), the index moves towards 0 and on the other hand if number of species increases or proportion of a species reduces in total stock then index move towards 1. This index is easy to compute and interpret, as follows:

$$SID = 1 - \sum_{i=1}^{n} Pi^2$$

Where, SID is the Simpson Index of Diversity, *i* is fish species *n* total number of species stocked in a pond and *Pi* is the proportionate of ith fish species in total fish stocked in a pond.

For analysis of time series data of average fish product and bevaviour of variables with respect to diversitification index the trend equations were fitted by using different growth models. Growth models are nothing but the models that describe the behavior of a variable overtime. Linear, logarithmic, Inverse, Quadratic, Cubic, Power, Sigmoid, Growth, and Exponential models have been used in the present study.

Further, for identification of the factors responsible for diversification of species at farm level, the Henry Garrett's Ranking Technique was applied (Zalkuwi *et al.* 2015). The responses of the farmers were converted into ranks using following formula-

$$Percent\ position = \frac{100(R_{ij} - 0.5)}{N_j}$$

Where,

R_{ij} = Rank given to i^{th} factor by j^{th} farmer

N_j = Number of factors ranked by j^{th} farmer

The percentage position of each rank obtained through above formula was converted into score using referring to the table given by Garrett and Woodworth (1969). Then for the each factor, score of individual respondents were added and then divided by the total number of respondents. Then the ranks were assigned to each factors based on the value of mean score.

The farmers' perception on ranks of relative importance of factors motivating them to adopt species diversification at their farms was assessed. The Henry Garrett's Ranking Technique was applied to rank the factors. The mean score for each factor was calculated and factors were ranked based on their mean score

Challenges of Small Scale Aquaculture

There are number constraints is aquaculture, particularly when semi-intensive fish production is the key income generator for a farmer. In Tripura most of the fish culture ponds are very small in size. This survey indicated that among fish farmers, 37 % of total ponds were in the categories 0.5 to 1.0 canie (0.08 to 0.16 ha) and 82% ponds were of less than 1.5 canie (0.24 ha) and only 7% ponds are >=2.5 canie (Figure 4.1). The average land holding and pond size were recorded as 3.15 canie (0.504 ha) and 1.03 canie (0.165ha), respectively. It was interesting to note that the distribution pattern of both land and water area across the fish farmers were observed to be skewed but distribution of water area was less skewed as compared to that of the land area (Figure 4.2). This may lead to better income distribution in case of fish farmers. In addition to smaller size of pond, short culture period, acidic nature of soil and water, higher suboptimal iron content, poor quality seed, improper feeding, lack of technical knowhow. Poorly developed infrastructure and limited access to transportation or urban centers, are major challenge to small scale aquaculture. Further, poor transportation, communication services, limited market access, financial and social assistance restricts farmer from generating more income through small scale aquaculture.

Prices for fish are higher in urban market and the ability of farmers to access and sell fish in these markets can greatly influence potential income generation. Many farmers rely on local markets to sell their produce; however, with on increase in supply of fish reduces the market price and decrease potential benefits. Middlemen and traders are often engaged to sell fish further afield, where prices for fish are higher. Farmers without access to urban markets are restricted have little option but to sell fish at a set price to these middlemen, who then sell the same at a higher price to the urban buyers. Governing bodies need to ensure access to markets for local farmers in priority manner so as to increase profit margins.

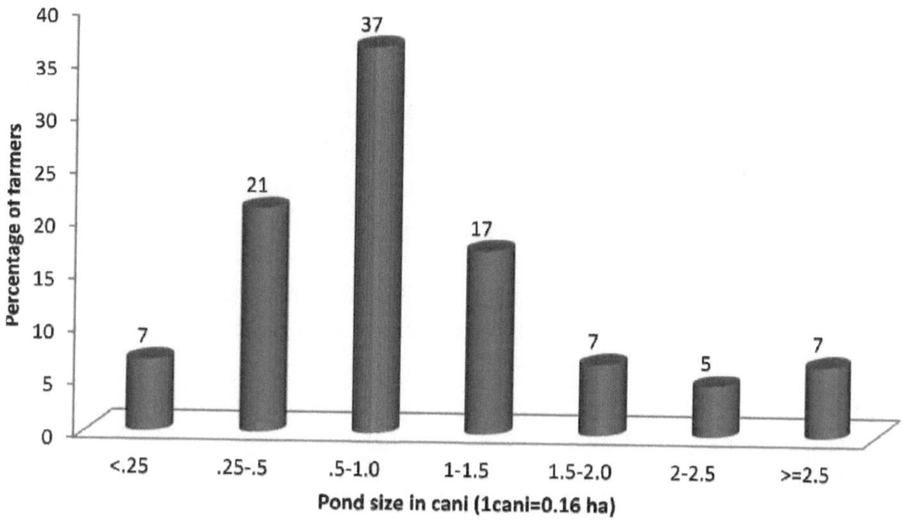

Fig. 4.1: Pond Size of the Fish Farmers of Tripura

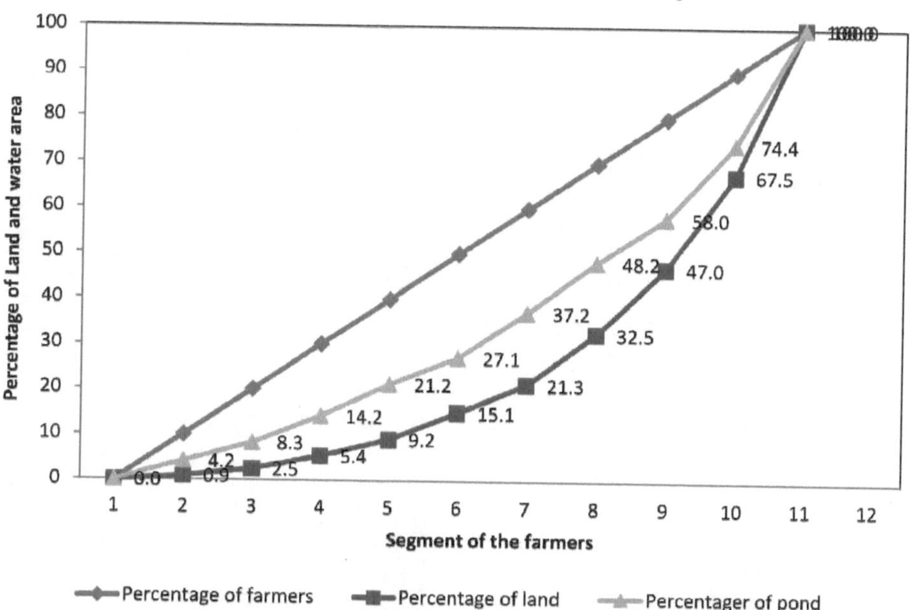

Fig. 4.2: Distribution of Land And Water Area in the Fish Farming Community

Trend of Fish Productivity in Tripura

The State Fisheries Department has devised a strategic plan for enhancing fish production of the state to make the state self reliant. This strategy targets to increase fish production in state through horizontal and vertical expansion. But considering that scope for horizontal expansion is limited in the state, vertical expansion i.e. productivity enhancement was only way forward to significantly enhance the level of fish production. The effect of joint efforts of stake holders is reflected by progressive increase in the fish productivity in the state (Figure 4.3). In fact state has been declared as self sufficient according set target which was to produce sufficient quantum of fish required to meet the state's requirement of existing fish eating population as per recommended rate of percapita fish consumption (11kg/capita/year). Though the present level of per capita consumption in the state is already much higher than the recommended level, still large quantity of fish is imported into the state from other states as well as from nearby country Bangladesh on daily basis. This indicates that market demand for fish is remarkebly higher than the FAO recommended level of fish consumption.

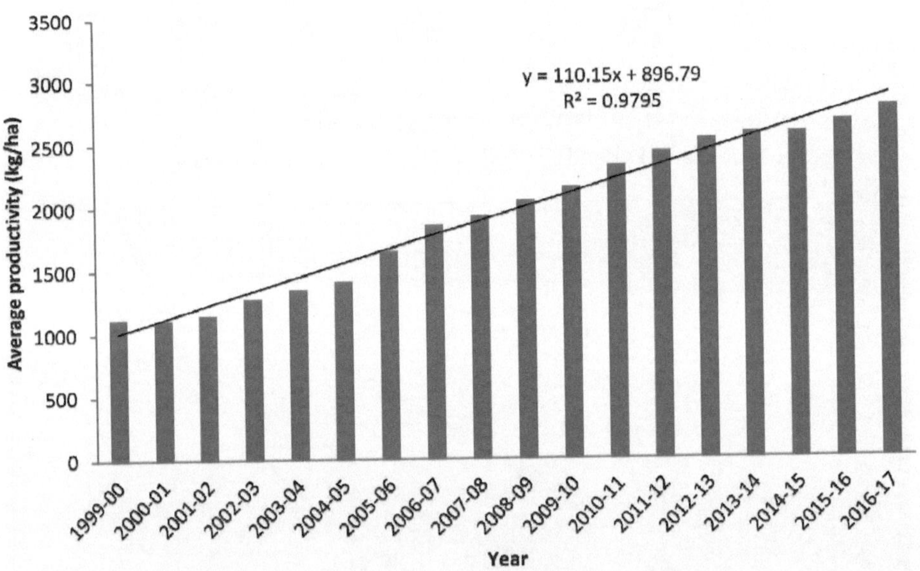

Fig. 4.3: Increase in Average Productivity (in kg) of Fish in Culture Fisheries of Tripura

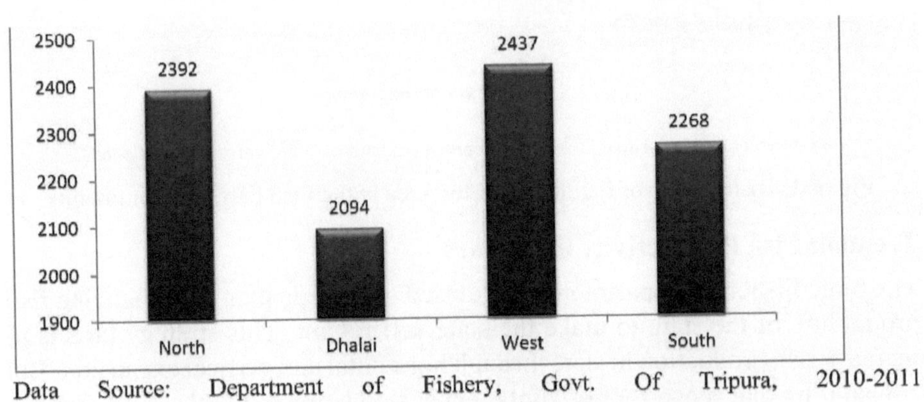

Data Source: Department of Fishery, Govt. Of Tripura, 2010-2011

Fig. 4.4: District-wise Productivity (Kg /ha./year)

Yield Gap: The secondary data on average fish productivity of different districts of Tripura is shown in Figure 4.4. The average productivity of fish in the state (2010-11) varied from one district to the other. The fish productivity was highest in West Tripura district (2437 kg/ha/y) and it was lowest in Dhalai district (2094 kg/ha/y). The marked variation in the productivity may be due many reasons like pond size, water depth, stocking, feeding, culture period, water quality management, etc.

Water level in the Surveyed ponds: The culture period was recorded to be months (June to February) in seasonal ponds. Mostly as expected, water level showed highly pronounced variation during monsoon and post monsoon period. While the maximum water depth was observed to be 3.5-5 m during mansoon and it was 0.25 - 0.5 m during dry period (Table 4.1). It must be emphasized that both maximum and minimum recorded water depths are suboptimal for aquaculture and potentially may reduce the fish productivity.

Table 4.1: Water level in the culture ponds of Tripura

Water level	Monsoon	Post monsoon
Maximum	5	0.5
Minimum	3.5	0.25
Average	4.5	0.275

Stocking: The size of seed, stocking density, species composition and time of stocking are some of the important determinants of fish growth and the total biomass at the time of harvest. In the study area staggered multiple stocking was reported by most of the respondents. Farmers are adopting this practice due unavailability of all the required seeds at one time and seeds are purchased from the vendors as and when available. There were large variations in the stocking density and it varies from 6250/ha to 31250/ha (Figure 4.5). However, the required optimum stocking density (9375 to 12500/ha) was followed only by 15% farmers and about 50% farmers had practiced higher stocking densities. In addition to imbalanced stocking density, majority of the farmers purchased seed from the venders which were of poor quality. It was also observed that the farmers were not maintaining optimum recommended proportion/ratio of each species in the ponds. The farmers were stocking their fish ponds with species whichever and whenever made available to them by the village vendors. These all added to inefficiencies and hindered the increase in growth of fishes and ultimately lead to lower fish production. Sarkar and Lakra, (2010) reported that small indigenous fishes traditionally occupied an unenviable position and an inseparable link in the life, livelihood, health and the general well being of the rural mass, especially the poor in the NE region. Therefore, farmers utilize varieties of indigenous fishes. For that matter, a total of 17 fish species were recorded from the culture systems at farm level, used by the fish farmers of the Tripura.

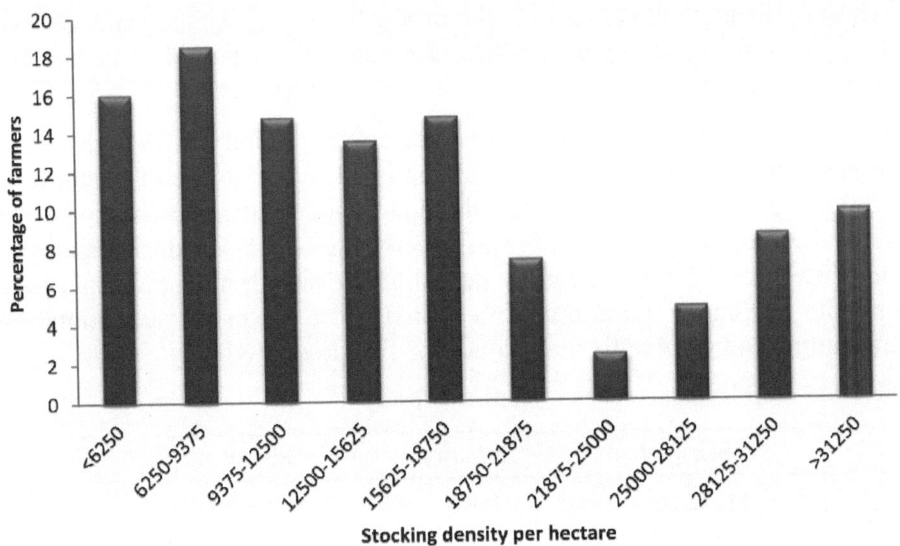

Fig. 4.5: Stocking Density maintained by Fish farmers of Tripura

Stocking Ratio

In order to achieve higher production of fishes in unit water area, fast growing compatible species of different feeding habits, different growth rate and different weight class are cultured together, so that, all ecological niches can be utilized by stocked species. According to Sinha *et al.* 1985, a pond having average water depth of 2.0-3.0 m may be stocked at the rate of 5,000 fingerlings/ha. However, Datta, 2014, advocated 6,000-12,000 fingerlings/ha in pond having an average water depth of 2.5 m. In any case the principle behind determining the stocking ratio is to fulfill the habitat and feeding niches operating in the upper, column and bottom layer of the pond with species combinations in such a away that it minimizes overlap. In this study number of fingerling of each species stocked into the pond and proportion of individual species into total fish stock were calculated and thereafter average of proportion of each species in case all 226 respondents were calculated and given in Table 4.2. The stocking ratio of major eight fish species were 7.03% Catla (*Catla catla*), 7.38% Silver carp (*Hypophthalmichthys molitrix)*, 1.51% Grass carp (*Ctenopharyngodon idella)*, 28.28% Rohu *(Labeo rohita*, 26.49% Mrigal (*Cirrhinus mrigala)*l, 13.92% Common/Mirror carp (*Cyprinus spp.),* 13.92% Japani putti (*Puntius gonionotus*) and 15.09% bata (*Labeo bata)*. Whereas, the recommended level

Table 4.2: Stocking Ratio Maintained by the Farmers in the Study Area

Sl.N.	Species Combination	Number of respondents	Mean Stocking Ratio	Recommended Level of stocking ratio in 6 species polyculture	Spatio-Tropic habits
2	Catla (*Catla catla*)	216	7.03	10-15	Surface feeder - Zooplankton form the major diet
5	Silver carp (*Hypophthalmichthys molitrix*)	128	7.38	30-35	Surface feeder - Phytoplanktophagous
6	Grass carp (*Ctenopharyngodon idella*)	104	1.51	5-10	Surface/column feeder - Macrophyte feeder
1	Rohu (*Labeo rohita*)	218	28.28	20-25	Predominantly column feeder - plankton and organic debris from the major diet
3	Mrigal(*Cirrhinus mrigala*)	208	26.49	15-20	Bottom feeder - Detritivore
4	Common/Mirror carp (*Cyprinus* spp.)	195	13.92	15-20	Bottom feeder - Omnivore
7	Japani puti (Puntius spp.) (Puntius gonionotus	173	12.38	-	Column/bottom feeder - Plankton ...
8	Bata (Labeo bata)	162	15.09	-	Bottom feeder

of stocking ratio of six species were 10-15% catla, 30-35% silver carp, 5-10% grass carp, 20-25% rohu, 15-20% mrigal and 15-20% common carp. These results shows that the stocking ratio of silver carp and grass carp were below to the recommended level.

Species Diversification

Generally, in six species combination, the upper layer of the pond is stocked with 30% of the total stock, the column and bottom layer holds 40% and 30% of the stocks respectively. Again, in each layer one IMC is stocked with one exotic carp with non-competing habits e.g. *catla* with silver carp (upper layer), rohu with grass carp (column layer) and *mrigal* with common carp (bottom layer). Species combination of SC2.5: C 1.0: R 2.50 GC 1.0: M 1.0: Cc 2.0 at 7500 per hectare stocking density was reported to be ideal (Chaudhuri *et al.*, 1975). Since in this ratio, all carp species attained the marketable size of one kilogram with perhaps least inter and intra-specific competition. Keeping, this stocking ratio as a benchmark, Simpson Index of Diversity which is determined by proportion of ith fish species in a given stock was calculated for 226 selected farms. The obtained results are given in Table 4.3.

The results indicated that in case of Gomati District; mean diversification index of 31.58% fish farms was 0.805 which was close to Bench mark DI where majority of the farms (66.32%) were found to be less diversified as their DI ranging 0.7 to 0.8. In case of Sepahijala district majority of the farms (83.75%) were more diversified and their DI were found to be above 0.85. Only 5.69% farms of the Sepahijala were found to be less diversified. Further, out of total 17 fish species were recorded from the study area of Tripura, eight species namely Rohu, Catla, Mrigal, Grass carp, and Silver carp, Common carp, Japani putti (*Puntius gonionotus*), and bata (*Labeo bata*) were found to be more dominated and preferred by the farmers to include in the culture system. A radar diagram for these species were developed which shows percentage of farmers of the two selected districts which are included these species in their stocking of fishes (Figure 4.6). The figure is also indicated strategies related to trade off among the species adopted by the fish farmers.

It evident from the figure that the fish species Gross carp recommended under the polyculture system was adopted by 40 and 60% of farmers in Gomati and Sepehijala districts, respectively. Similarly the Silver carp was adopted by 51 and 65% of farmers in Sepehijala and Gomati districts, respectively. It is clearly indicated from the figure that these two species were heavily replaced by Japaniputti (*Puntius gonionotus*), and bata (*Labeo bata*) in the study area. These species become popular and widely adopted by the fish farmers of Tripura because they are compatible with other species, higher

Table 4.3: Distribution Farms Based on Species Diversification Index

DI Range	Gomati			Sepahijala		
	Mean DI	Fish production (kg/ha)	Percentage of Farm	Mean DI	Fish production (kg/ha)	Percentage of Farm
<0.70	0.687	2999.504	15.79	-		-
>0.70-0.75	0.727	2791.209	20.00	0.738		0.81
>0.75-0.80	0.782	3946.280	30.53	0.782	2357.94	4.88
>0.80-0.85	0.815	3896.150	31.58	0.820	2542.69	10.57
>0.85-0.90	0.860	2476.563	2.11	0.883	3682.51	9.76
>0.90-0.95	-		-	0.925	2704.14	37.40
>0.95-1.0	-		-	0.974	2079.06	36.59
Average DI	0.751	3529.918	100.00	0.919	2569.98	100.00
Benchmark DI	0.805		-	0.805		

demand in the market and comparable growth in the culture system. Though the grass carp and silver carp had better growth but due to less preference to these species by the consumers majority of fish farmers does not grow this species at farm. Alternatively, several other species *like calbasu, bighead (Aristichthys nobilis,)* chingri, rupchanda, bhagna, tilapia, pabda (*Ompok bimaculatous*), *chital* (*Notopterus chitala*), singhi and magur were stocked in limited numbers in view of better returns and to meet out highly diversified fish demand in the local markets. However, it is noticed that the income of the fish farmers can be further enhanced through standardization stocking of IMCs with these local species, stocking management and better input supports for the fish farmer of Tripura.

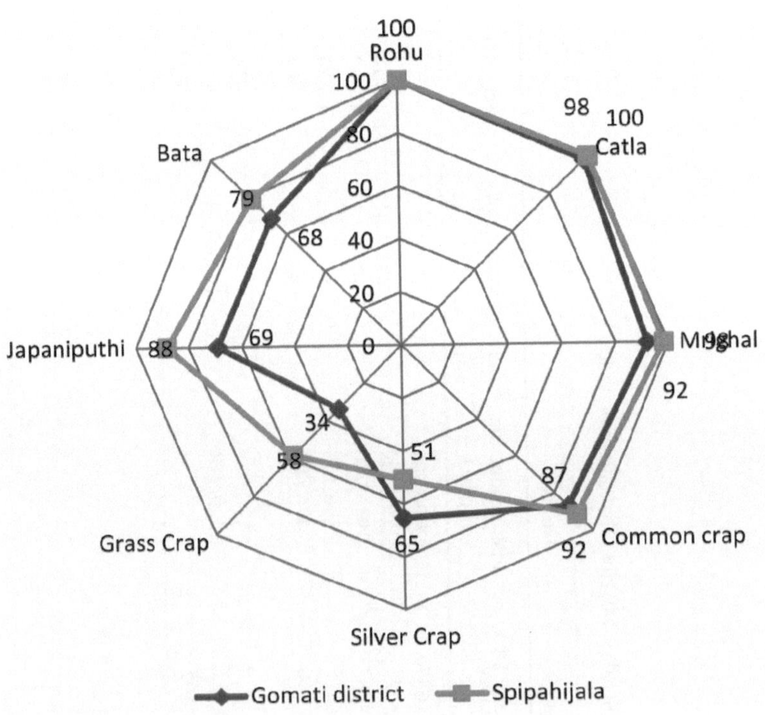

Fig. 4.6: Adoption major species in selected districts of Tripura

Impact of Diversification Strategy on Fish Productivity and Gross Return

Further, to explore impact of species diversity at farm level on fish productivity and gross return, relationship between DI and productivity and DI and gross return were worked out for both districts and it represented in Figure 4.7 to Figure 4.10. Diversification indices were considered as independent variable and fish productivity and gross return were separately treated as independent variable. It was found that in all the cases cubic function were found best fit with R^2 0.736 and 0.845 in case of DI and Fish productivity in Gomati and Sepahijala Districts, respectively. Whereas, it was 0.812 and 0.998 in case of DI and Gross return in Gomati and Sepahijala Districts, respectively. It is noticed that with increase in level of diversification upto certain level of productivity as well as gross return increased but later on both productivity as well as gross return start declining because of overcrowding and species turned out to be competitive for available feed.

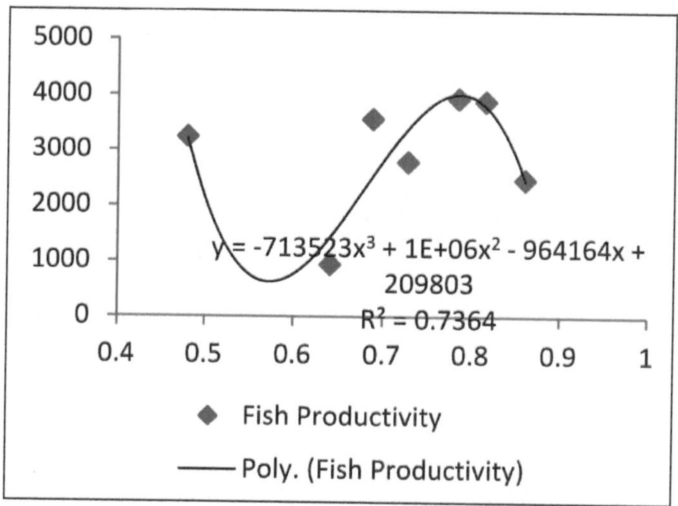

Fig. 4.7: Relationship in DI and Fish productivity in Goamti district (thousand)

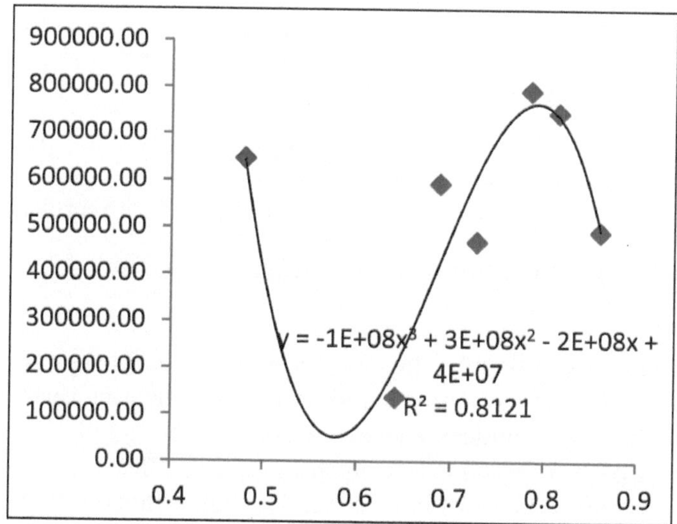

Fig. 4.8: Relationship in DI and Gross Return Rs/ha in Gomati (thousand)

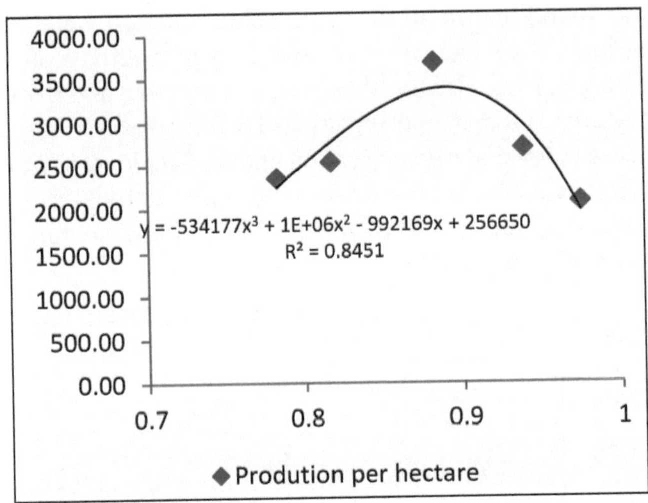

Fig. 4.9: Relationship in DI and Fish productivity in Sepahijala

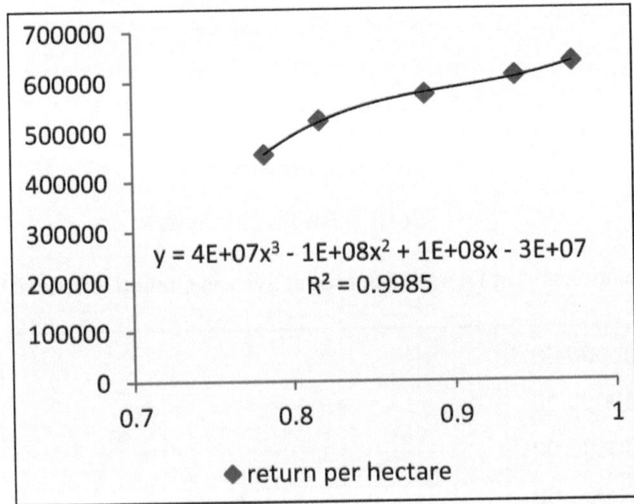

Fig. 4.10: Relationship in DI and Gross Return R/ha in Sepahijala

Factors Influencing the Species Diversification at Farm Level

i. Pond size

The fish production system of the Tripura is dominated by small pond size owned by the fish farmers. In this exploratory study the relationship between average diversification indices and pond size were analyzed using scattered plot and fitting of the regression line on observed data. The results obtained are presented in Figure 4.11 and Figure 4.12. The cubic function with R square values 0.4779 and 0.545 were found to be best fit for the observed data in case of Gomati and Sepahijala districts, respectively. These findings in the fish production system of Tripura not supports general analogy exists in agriculture as small farm are more diversified as compared to large farm. The observation of larger size of ponds was relatively more diversified as compared to the smaller size of ponds may be because of small scale of fisheries, high purchasing power of farmers for seed and feed and also the general belief of the farmers that higher stocking of seed will results more harvest. These results reveal that larger farmers whose seed purchasing capacity is more were inclined towards stocking of more number of species on their farm in view of getting additional benefit.

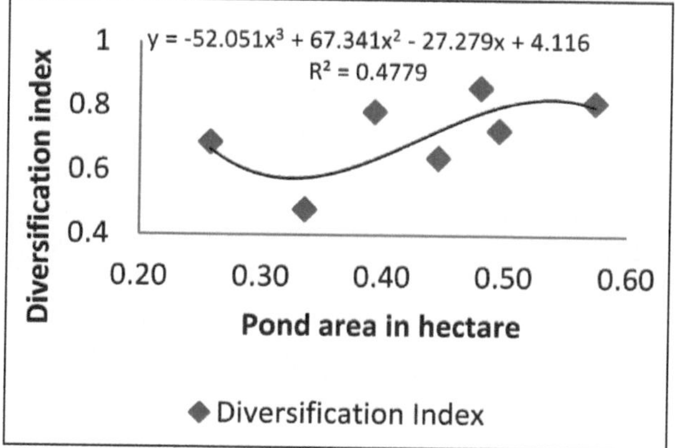

Fig. 4.11: Relationship between DI and water area in Gomati district

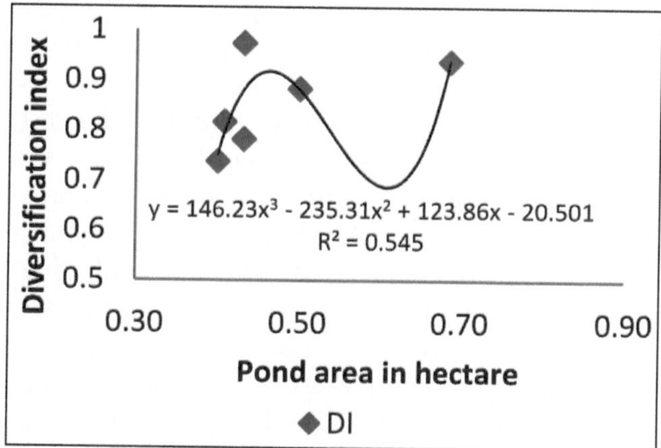

Fig. 4.12: Relationship between DI and water area in Sepahijala district

ii. Input Supply and Technical Knowhow

The farmers' perceptions toward motivational factors for selection of species for stocking at their farms were recorded through personal interview. The Henry Garrett's Ranking Technique was applied to workout mean score for each factor and factors were ranked on the basis of their mean score. The result obtained is presented in Table 4.4. The factors like market demand for fish species, species specific market price, less incidence of fish disease, easy availability of seeds and better growth were ranked as I, II, III, IV and V, respectively. However, factors such as familiarity with culture practice of species, fish attain marketable size before they attain maturity and growth performance in winter season was ranked VI, VII and VIII, respectively. These results revealed that the selection of fish species for stocking and hence, species diversification at farm level were found to be more economic motive, growth performance, technical knowhow and input availability.

Table 4.4: Ranking of the factors which affected the species diversification

Sl.No.	Factors	Mean Score	Ranks
1.	High market demand	72.08	II
2.	Easy availability of seed	52.04	IV
3.	Better Growth Performance at farm	49.39	V
4.	Fetched high market price	74.06	I
5.	Familiar with culture practice of species	35.67	VI
6.	Fish attain marketable size before they attain maturity	37.33	VII
7.	Can grow even during winter	22.33	VIII
8.	No disease problem	60.19	III

Conclusions

The above results indicated that in case of Gomati District, mean diversification index of 31.58% fish farms was 0.815 which was close to Bench mark DI where, majority of the farms (66.32%) were found to be less diversified as their DI ranging between 0.7 to 0.8. On the other hand in Sepahijala district majority of the farms (83.75%) were more diversified and their DI were found to be above 0.85. Only 5.69% farms of the Sepahijala were found to be less diversified. The two species namely grass carp and silver carp were found to be partial to complete replacement by the Japaniputti (*Puntius gonionotus*) and bata (*Labeo bata*) in the study area. This is mainly due to consumer preferences and market price. It was further noticed that increase in level of diversification up to certain level leads to support increment in productivity as well as gross return but after a certain level both productivity as well as gross return start declining. It is mainly because of the diversification without scientific basis farmers could not manage the stock of the mixed species. The factors like market demand for fish species, species specific market price, less incidence of fish disease, easy availability of seeds and better growth were ranked as I, II, III, IV and V, respectively. However, factors such as familiarity with culture practice of species, fish attain marketable size before they attain maturity and growth performance in winter season was ranked VI, VII and VIII, respectively.

These results revealed that the selection of fish species for stocking and hence, species diversification at farm level were found to be more economic motive, growth performance, technical knowhow and input availability. However, the farmers' strategies for species diversification at farm level required sound technical support for the selection of the species as well as the feed management and other culture practices.

References

Anon (2011). Diversification in aquaculture: A tool for sustainability Published by Spanish Ministry of Environmental, Rural and Marine Affairs General Technical Secretariat Publications Centre

Anon (2012). State Fisheries Department Report, 2011-12.

Ayyappan S. Gopal Krishna A, and Ganesh Kumar B. (2009). Species Diversification. *In* aquaculture and domestic fish marketing in India pp 13-22 in Indaqua 2009 Souvenir 21-23 January at CIFA Bhunmeshwar, 105pp, Organized MPEDA, GOI.

Bhat Mudasir Hassan and Salam Md. Abdus (2016). An Assessment of Nature and Extent of Crop Diversification across Agro-Climatic Zones of Jammu and Kashmir: Spatial and Temporal Analysis. *Journal of Agriculture and Veterinary Science*, 9(11) 33-40.

Biswas B., Das S.K., Mondal I., and Mandal A. (2018). Composite fish farming in West Bengal, India: redesigning management practices during the course of last five decades. *International Journal of Aquaculture*, 8(12): 90-97 (doi: 10.5376/ija.2018.08.0012).

Chaudhuri, H., R.D. Chakrabarty, P.R. Sen, N.G.S. Rao and S. Jena (1975). A new high in fish production in India with record yields by composite fish culture in freshwater ponds. *Aquaculture*, 6: 343–355.

Datta M.K. (2014). Avalanches in composite fish farming of carps in tropical state of Tripura, India http://aquafind.com/articles/Farming-Carp-in-India.php

Datta, M.K. (2015). Diversification of fish species in breeding in India. Online http:/aquafind.com/article/diversification-Of-Fish-Species_breeding-In-India.php.

Devi, S. Z., Singh, N. R. Singh, Anandkumar, N; and Th. Laxmi (2014). Fish production in Manipur- an economic analysis. *Journal of Crop and Weed*, 10(2):19-23.

Sinha V.R.P., Chakraborty R.D., Tripathi S.D., Das P., and Sinha M., (1985). Carp culture package of practices for increasing production aquaculture extension manual new series, No 2, CIFRI, Barrackpore, 4: 6-26.

Upadhyay, A.D. Patel, A.B. and Pandey D.K. (2015). Species diversification at farm level and its economic significance in fish production system of Tripura. *Journal of India Fisheries Association* 42: 39-48.

Zalkuwi, Jimjel; Singh, Rakesh; Bhattarai, Madhusudan; O.P Singh and Rao Dayakar (2015). Analysis of constraints influencing sorghum farmers using Garrett's ranking technique; A Comparative Study of India and Nigeria International Journal of scientific research and management, 3(3)2435-2440.

5

Advances in Modeling in Fisheries and Aquaculture in Big Data Era

Ajit Kumar Roy

Introduction

It is already established that mathematical modeling aims to describe the different aspects of the real world, their interaction, and their dynamics through mathematics. It constitutes the third pillar of science and engineering, achieving the fulfillment of the two more traditional disciplines, which are theoretical analysis and experimentation. Nowadays, mathematical modeling has a key role also in aquaculture. The problem is the complex modeling other than counting. Modeling and reasoning with data of different kinds can get extremely complex. Data streams are common source of Big Data. Now the task is to collect, prepare, represent, model, reason, visualize and additional issues like usage, quality, context, streaming, scalability. A few publications are available on modeling of fisheries and aquaculture in Indian context with live examples of modeling (Roy, *et al.* 1998, Roy, *et al.*, 2000, Bhar, *et al.*, 2002, Roy *et al.*, 2002, Ghosh *et al.*, 2017, Roy, *et al.*, 2006, Roy *et al.*, 2008, Upadhyay *et al.*, 2012). A book on exclusively with various types of model applicable in fisheries and aquaculture has been published (Roy *et al.*, 2009). In all these earlier publications conventional statistical and econometrics techniques are applied. But today there is sea change in this field with the advancement of technologies like artificial intelligence, machine learning, and big data analytics. There are attempts to apply these cutting edge technologies aiming sustainable productivity through efficient management of resources. In the following sections an attempt has been made to present an overview of the advances in analytics along with a briefing on what was developed in the past, and what is available today. Before proceeding to the analytics, I am presenting the scenario of basic characteristics of data along with different types of data generated today with automation of aquaculture industry.

Types of Data in Aquaculture

The input and output variables of the data arising out of aquaculture may include numerical and categorical variables. Numerical variables can be:

Continuous – measured quantities expressed as a float (e.g. 'av. weight');

Discrete – count expressed as an integer (e.g. 'number of fish');

While categorical variables may be:

Regular categorical- data including non-ordered classes (e.g. species Bream/ Bass);

Ordinal – classes that can be ordered in levels (e.g. estimations poor/fair/good).

From the variables that can be measured it is important to distinguish between:

i. Variables that do not change over time, often identifying population attributes (e.g. identifications such as 'year' or 'hatchery');

ii. Variables that can change over time but do not change within a sampling period (e.g. 'batch');

iii. Variables that change daily, taken into account when samplings occur (e.g. 'average weight')

Generally we have *three types of input data* according to the impact they assure:

1. ***Identification data***. This is the data that permits the fish farmer to manage the production and correctly identify the fish;
2. ***Daily data***. This is the data that is provided by the fish farmers resulting from their everyday data input (e.g. 'date', 'av. wt.', 'actual feed', etc.);
3. ***Sampling data***. At predetermined points of the fish growth timeline, a sample of the fish is done to confirm the model values and make the appropriate adjustments
4. ***Life to Date*** [LTD]. This is cumulative data that is calculated from the time when the fish enters the net as a fry to the date of data collection, and will last until the date of the harvest.

Accuracy of the data

There is the probability of error from human observation that can be expressed in the data and consequently in the output on the fish production optimization. Sometimes that error can be measured and controlled by imposing limits to that same error. That brings us to three categories of data:

i. *Inaccurate data*. Data that can be measured and evaluated but may contain errors (e.g. 'av. weight');

ii. *Accurate data*. Data that can be measured and evaluated but cannot contain errors (e.g. 'food');

iii. *Unclassified data*. Data that cannot be measured, quantified or evaluated (e.g. 'feed').

Sampling methods and data

The sampling at the fish farms is a common procedure that permits an evaluation of the production process. It consists of the collection of a sample of the fish in a certain unit during a period of time. Moreover, the average weight is corrected providing a more precise assessment to FCR and SFR (Pullin *et al.,* 2007). In particular, the aquaculture sample is perceived to gain the following significant benefits:

i. Costs savings in frame updating and maintenance;

ii. Aquaculture indices i.e. Indices for sustainable aquaculture: e.g. biological, environmental, etc. can be selected more quickly and economically.

iii. Improvement of the reliability of aquaculture indices from different aspects;

iv. Possibility of integration and substantive understanding of the global production.

New Sources of Data Generation from Aquaculture

In the past we used to collect data manually from different experimental set up, surveys, secondary Govt. and private sources. Now with technological changes particularly ICT, there is a paradigm shift in data generation, collection, analytics and being mostly automatic. The sources of Big Data in aquaculture are often originated from:

- The ability to track much more information than we used to;
- New and pervasive sensors, measuring water temperature, oxygen level, etc.;
- The ability to collect all recorded data, whether or not we need it or can extract relevant information from it.
- Freely availability of data relating to environmental statistics

The following points of the report of FAO will give us the latest trend of technologies used for efficient management of fisheries. On board vessels, cameras and other sensors are used to improve the monitoring of the catch, including the deployment of gear and processing equipment. Images and videos are used to identify species. The use of image recognition software to detect and classify caught species automatically, which is already being tested and used in selected fisheries, results in disruptive improvement of on-board observations and catch reporting and much better understanding of stocks and fisheries. Added with the above, sensors placed on board vessels such as acoustic sounders and in the open waters for example, on buoys or as autonomous drones, fish are now easier to detect and study. The information they provide, when combined with catch reports, can radically change the number and quality of environmental and stock assessments (FAO 2018).

Uncertainties and Challenges in Aquaculture

As per FAO report, a major challenge in implementation of the 2030 Agenda is the sustainability. It is well known that the production in aquaculture depends on so many factors. Most important to the production in aquaculture come from weather prediction, oxygen levels and water temperature, which are very specific to this activity. The tasks in fish farming carry several uncertainties often expressed by measurements or even evaluations that permit further, optimization. A classical example is the aim for a better control on the food loss and food quality. A contribution of data mining in this context would be of interest to the aqua farming industry with a view to saving or relocating resources. Further, the aquaculture industry is facing challenges such as high cost of inputs and climatic changes. Climate effect such as increase in temperature is observed to increase disease transmission, deplete oxygen, increase incidence of harmful algal blooms in ponds to mention a few important variables that remains undetermined during the complete production pipeline is the exact number of fish. This represents a big lack of knowledge about production. In fact, the unknown number of the fish until the end of the production is important for the amount of food given and, consequently, for the resources spent. Modern research and commercial aquaculture operations have begun to adopt new technologies, including computer control systems. Aqua farmers realize that by controlling the environmental conditions and system inputs like water, oxygen, temperature, feed rate and stocking density, physiological rates of cultured species and final process outputs e.g. ammonia, pH and growth. These are exactly the kinds of practical measurements that will allow commercial aquaculture facilities to optimize their efficiency by reducing labour and utility costs as per AquaSmart Consortium, The aqua Smart Project. The benefits for aquaculture process control and artificial intelligence systems are:

1. Increased process efficiency;
2. Reduced energy and water losses;

3. Reduced labour costs;
4. Reduced stress and disease;
5. Improved accounting;
6. Improved understanding of the process.

Opportunities for Big data in Aquaculture and Fisheries

In the following paragraphs, we turn our focus to the specific characteristics of aquaculture and discuss what are the likely advantages of application of Big Data methods in that context. Here Aquasmart approach to Big Data analytics is notable. According to the latest available statistics collected globally by FAO (2018), the 2030 Agenda for Sustainable Development offers a vision of a fairer, more peaceful world in which no one is left behind. The 2030 Agenda also sets aims for the contribution and conduct of fisheries and aquaculture towards food security and nutrition, and the sector's use of natural resources, in a way that ensures sustainable development in economic, social and environmental terms, within the context of the FAO Code of Conduct for Responsible Fisheries. A major challenge to implementation of the 2030 Agenda is the sustainability. *Global fish production* peaked at about 171 million tons in 2016, with aquaculture representing 47 percent of the total and 53 percent if non-food uses (including reduction to fishmeal and fish oil) are excluded. The total first sale value of fisheries and aquaculture production in 2016 was estimated at USD 362 billion, of which USD 232 billion was from aquaculture production. With *capture fishery production* relatively static since the late 1980s, aquaculture has been responsible for the continuing impressive growth in the supply of fish for human consumption. Between 1961 and 2016, the average annual increase in global food fish consumption (3.2 per cent) outpaced population growth (1.6 per cent) and exceeded that of meat from all terrestrial animals combined (2.8 per cent). In per capita terms, food fish consumption grew from 9.0 kg in 1961 to 20.2 kg in 2015, at an average rate of about 1.5 per cent per year. Preliminary estimates for 2016 and 2017 point to further growth to about 20.3 and 20.5 kg, respectively. The expansion in consumption has been driven not only by increased production, but also by other factors, including reduced wastage. In 2015, fish accounted for about 17 percent of animal protein consumed by the global population. Moreover, fish provided about 3.2 billion people with almost 20 percent of their average per capita intake of animal protein. Despite their relatively low levels of fish consumption, people in developing countries have a higher share of fish protein in their diets than those in developed countries. The highest per capita fish consumption, over 50 kg, is found in several small islands developing States (SIDS), particularly in Oceania, while the lowest levels, just above 2 kg, are in Central Asia and some landlocked countries. Global capture fisheries production was 90.9 million tons in 2016, a small decrease in comparison to the two previous years. Fisheries in marine and inland waters provided 87.2 and 12.8 per cent of the global total, respectively. The next decade is likely to see major changes in the environment, resources, macroeconomic conditions, international trade rules and tariffs, market characteristics and social conduct, which may affect production and fish markets in the medium term. Influences include climate change, climate variability and extreme weather events, environmental degradation and habitat destruction, overfishing, IUU fishing, poor governance, diseases and escapes, invasion of non-native species; issues associated with accessibility and availability of sites and water resources and access to credit; as well as improved fisheries management, efficient aquaculture growth and improvement in technology and research. In addition, issues related to food safety and traceability, including the need to demonstrate that products are not derived from illegal and proscribed fishing operations, can have a relevant impact in terms of market access (FAO 2018). In Europe, the largest market for fish in the world with increasing consumption, aquaculture accounts for about 20% of fish production with 65% of the seafood consumed in the EU being imported (European Commission Aquaculture). The future demand for fish is expected to increase due to increasing population and income and health benefits associated with fish consumption. Increased demand for high protein food and apparent declines in capture fisheries together have helped drive rapid expansion of the aquaculture industry in the past two decades. With it, the increase of aquaculture is expected to provide the opportunity for the access to Big Data in the sector.

Big Data in Fisheries/Aquaculture

Analysis of the ocean of data provided by sensors involves a complex workflow which extends beyond traditional fisheries data centers. Cloud-based services are required to cope with much larger data storage needs at the point of creation. The prime examples of such "big data" are the huge datasets from satellites that monitor the environment, but video and data from mobile phones also require a software solution that can easily be adapted to an increasing volume of data or users. The big-data approach will change the understanding of natural and human processes, such as the growth and distribution of species or the spatial planning of fisheries and aquaculture. Through big data, new opportunities arise for tracing how and where vessels operate and for tracking products all the way to shops and consumers (FAO 2018).

Big Data in aquaculture is focused towards unlocking the economic potential of improved management decisions. The benefits of precision in the *aquaculture* production pipeline provide a relevant improvement in the optimization of resources. The influence of the transformations made by precision *aquaculture* also known as site specific farm management is starting to have impact on aquaculture and on the seafood sector in general. This is boosted by the recent developments on sensors, robotics, computer vision, satellite imaging and Big Data analytics. Nowadays, the available technology enables the usage of drones and driverless boats, or interconnected devices powered by advanced data mining in the context of the Internet of Things adapted to the needs of the sector. It has been announced by European Commission Aquaculture that aquaculture is the next target for the advances boosted by Big Data technology much as what happened to agriculture. Examples of that technology are the forecast of specific weather features such as the prediction of storms.

Modeling and Prediction in Aquaculture

Mathematical modeling aims to describe the different aspects of the real world, their interaction, and their dynamics through mathematics. It constitutes the third pillar of science and engineering, achieving the fulfillment of the two more traditional disciplines, which are theoretical analysis and experimentation. Nowadays, mathematical modeling has a key role also in aquaculture. In the following section, I present an overview of that. Available literature suggests that a good number of works are attempted in modeling arena. *Growth and reproductive modeling* of wild and captive species is essential to understand how much of food resources an organism must consume, and how changes to the resources in an ecosystem alter the population sizes. The study of growth means basically the determination of the body size as a function of age. Therefore, all stock assessment methods work essentially with the age composition data. This has been an important topic in the aquaculture research and development. Several numerical methods have been developed, which allow the conversion of length-frequency data into age composition and the von Bertalanffy growth model of body length as a function of age has become one of the cornerstones in fish biology because it is used as a sub-model in more complex models describing the dynamics of fish populations. A model for the growth of the eel in aquaculture is available (Houvenaghel *et al.,* 1989). This growth model is a predictive model for aquaculture, constituted by adding algometric functions related to physiological features. The methodology for length-age data of several species of fish can consider models fitted to each dataset other than the von Bertalanffy growth model (VBGM). These are generalized VBGM, Gompertz growth model, Schnute-Richards growth model, and Logistic (Katsanevakis *et al.,* 2006). A Model developed to forecast fish growth, with uncertainty, providing good predictions of future biomass of Norwegian farmed salmon. In addition, the number of new fish stocked, fish lost, slaughtered and wasted, as well as the sea temperature related to the growth, were modeled. Optimal management in a single-farm model for sea bass, based on weight classes and biological sub-models is reported (Loland *et al.*, 2011)

Models for Aquaculture (Temperature and dissolved oxygen stratification)

Accurate characterization of temperature and dissolved oxygen stratification in units used for aquaculture is of critical importance in understanding how these units may be constructed, oriented, or otherwise managed biophysically when one wishes to provide optimal environmental conditions for the organisms being cultured. While field studies can provide characterizations of water quality stratification at a single locale, to date there have been few attempts at developing reliable models which can be used at a variety of sites after initialization with appropriate local geographic and atmospheric data. Advances in model structure and reduction of data requirements relative to previous models reflect the desire to provide for culturists the ability to predict stratification events with commonly available data, obtained either by hand or from a simple weather station located at or near the unit site. A series of simulation runs was performed in to assess the quantitative effects on temperature and dissolved oxygen concentration generated by varying pond depth and phytoplankton density input values.

Artificial Intelligence (AI) Systems in Aquaculture

Today's artificial intelligence (AI) systems i.e. expert systems and neural networks offer the aqua culturist a proven methodology for implementing management systems that are both intuitive and inferential (Balakrishnan *et al.*, 2007). There have been many successful commercial applications of AI e.g. expert systems in cameras and automobiles. The major factors to consider in the design and purchase of process control and artificial intelligence software are functionality/intuitiveness, compatibility, flexibility, upgrade path, hardware requirements and cost. Of these, intuitiveness and compatibility are the most important. The software must be intuitive to the user or they will not use the system. Regarding compatibility, the manufacturer should be congruent with open architecture designs so that the chosen software is interchangeable with other software products. Reviews of the technologies and implementation of the technologies necessary for the development of computer intelligent management systems for enhanced commercial aquaculture production is available: (Lee *et al.*, 2000). AI can be used for solving a wide spectrum of problems:

- Optimization (e.g. diet, temperature, light, etc.)
- Pattern recognition (e.g. anomalies, size, etc.)
- Prediction (e.g. growth, diseases, etc.)

- Automation (e.g. feeding, water quality, etc.) sampling on big data enables off-line data analysis, enabling performing expensive operations. The use of machine learning tools in aquaculture is itself not new, but it has not been explored in its full potential

ANN to Forecast Water Quality and Temperature

Comparison of Artificial Neural Networks over traditional forecasting methods artificial neural networks has long been used for weather forecast (Maqsood *et al.,* 2004) and to generate probabilities of precipitation and quantitative precipitation forecasts (Hall *et al.,* 1999). Neural network forecasts exceeded other traditional forecasting methods such as linear or logistic regression systems boosting controversy surrounding the value of a priori knowledge in determining predictor variables. However, in many cases, incorporating a priori weather knowledge is not feasible because it is very difficult to quantify prior knowledge of weather processes as input to a neural network (Fabbian *et al.,* 2007). Additionally, research revealed that neural networks outperformed logistic regression, discriminant analysis, and rule-based prediction systems in the classification of tornado events (Marzban *et al.,* 1996). Similarly, ensemble-based neural networks combining the outputs of neural network subcomponents have been shown to outperform each individual neural network subcomponent in the prediction of wind-speed, temperature, and humidity (Clark *et al.,* 1985). Artificial neural networks are massively parallel processors that have the ability to learn patterns through a training experience. Because of this feature, they are often well suited for modeling complex and non-linear processes such as those commonly found in the heating system. They have been used to solve complicated practical problems. The simulation results indicate that, the control unit success in keeping water temperature constant at the desired temperature by controlling the hot water flow rate in closed aquaculture systems (Atia *et al.,* 2011). This methodology is also useful in estimating the water temperatures in small river streams (Government, U.S., 2011).

In neural network control is used for dissolved oxygen of aquaculture pond aeration system. In particular, for controlling the speed of air flow rate from the blower to air piping connected to the pond through control blower speed. This is also the approach in that analyses the important factors for predicting dissolved oxygen of Hyriopsis Cumingii ponds, and finally chooses solar radiation (SR), water temperature (WT), wind speed (WS), PH and dissolved oxygen (DO) as six input parameters (Liu *et al.,* 2011). Alternatively, the authors in study the oxygen dissolved in water using machine learning techniques based on Bayesian inference that can be used to enhance the computer simulation of molecular materials, focusing here on water (Bartok *et al.,* 2013). As a decision system,

ANNs are an important tool for forecast in aquacultureparticularly to forecast the freshwater fish caught (Benzer *et al.*, 2015) and for detecting fish object inside a digital image in the context of computer vision (Man *et al.*, 2011). The neural network is used to recognize the fish object from a small window that scans for the object all over the image. The neural network will examine the windows of the image and decides whether each window contains a fish.

Remote sensing is another example where ANNs are used to manage the available data, complemented with other artificial intelligence methods and techniques. Local empirical neural network algorithms can be used to evaluate the potential impact of aquaculture assessed using remote sensing data (Bengil *et al.*, 2014). ANNs have powerful pattern classification and pattern recognition capabilities, being able to learn from and generalize from experience. Recent studies have shown the classification and prediction power of the Neural Networks. ANNs can approximate any continuous function and have been successfully used for forecasting of data series in aquaculture. While ARIMA assumes that there is a linear relationship between inputs and outputs, ANN have the advantage that can approximate nonlinear functions. An ANN can be used to solve problems involving complex relationships between variables. It provides a practical and, in some situations, the only feasible way to solve real-world problems.

Feed composition has a large impact on the growth of animals, particularly marine fish. A quantitative dynamic model to predict the growth and body composition of marine fish for a given feed composition over a time span of several months is reported. The model takes into consideration the effects of environmental factors, particularly temperature on growth, and it incorporates detailed kinetics describing the main metabolic processes (protein, lipid, and central metabolism) known to play major roles in growth and body composition. That showed that multiscale models in biology can yield reasonable and useful results. The model predictions are reliable over several timescales and in the presence of strong temperature fluctuations, which are crucial factors for modeling marine organism growth. The reported model provides important improvements over existing models (Bar *et al.*, 2009).

Determining Influential Factors

Under this subsection, it will be dealt how to determine the factors that influence aquaculture production. These widely used most important factors like e.g. temperature, season, size, av. weight, oxygen, pH, and local features are very important areas of concern for aqua culturists. In such situation, generalized linear regression used for that aiming to identifying influential factors.

Key Performance Indicators (KPI)

As we know KPIs are very important for answering to stakeholders. But most of the times we donot know what are the KPI for aquaculture under different environmental conditions. Along with this, I also described their associated metrics for KPI modeling and business analytics. The formulae to determine and quantify the measurements of these KPIs in the context of the available data are expressed in this section.

The following paragraphs identify the core business KPIs to be tackled in this deliverable, and describe the relation between those KPIs and the business questions to be answered.

Core set of business KPIs

The following KPIs will be used:

i. FCR

ii. SGR

iii. GPD

iv. Mortality %

v. Production time

vi. Protein Efficiency Ratio

In the following list I present some of the user stories to be considered. Here, I represent the user stories that directly relate with the business questions to be answered. User stories that were considered of high priority by the end users are as follows:

- Evaluate feeding policy
- Evaluate feed performance
- Evaluate fry quality
- Evaluation of production practices
- Evaluate the influence of the environment
- Estimate average weight
- Estimate fish count
- Evaluate feeding process
- Define extra parameters

In the following I briefly describe the business questions to be answered in the context of this deliverable with the help of the *Big Data analytics*.

i. Evaluation of feed performance. This is done vs. model or other feeds. I want to evaluate feed suppliers, feed types and feed composition, taking into account also the time dimension (winter, summer).

ii. Evaluation of feeding process. I want to evaluate people (feeders), feeding rates and feeding times.

iii. Evaluation of fry quality. I have to evaluate different hatcheries, brood stock origins and hatchery qualities.

iv. Evaluation of production strategies. Stocking month, feeding policies (e.g. time to change feed size), protocols for grading, unit type and size, vaccinations.

v. Evaluation of the influence of the environment, Farm, anchor, unit, other environmental data like oxygen level, water currents, weather.

vi. Estimation of fish number and average weight based on the feed consumption.

Formulation of KPIs for data analytics

Now the following paragraphs I present and formalize the entities and KPIs in present in this deliverable associated to data analytics in aquaculture. This inventory of KPI formulations was made based on the expert information of the aqua farmers and on the online resources (Aquatext, Fish farming).

BWG – Body Weight Gain *(BWG)* = Final body weigh (g)-Initial body weight (g)

CV – Coefficient of Variation (or weight Coefficient of Variance)

CV = Standard deviation/Mean*100

DGR – Daily Growth Rate

DGR= Final body weigh (g) –Initial weight (g)/ (Period (days)* Initial weight (g)*100

FCR – Feed Conversion Rate

Net growth = Biomass at the end of the period – (Initial Biomass + Biomass of harvests + Biomass transferred to other units) – Biomass transferred from other units

Gross growth = Net growth + Biomass of mortalities + Biomass of culling + Biomass of adjustments

$Economical\ FCR = Total\ dry\ feed\ given/\ Net\ growth$

$Biological\ FCR = Total\ dry\ feed\ given/\ Gross\ growth$

FE =1/ $FCR * 100$

SFR – Suggested Feeding Rate

$SFR = (Daily\ feed/\ Biomass) * 100$

Daily feed could also be period feed → period SFR

SGR – Specific Growth Rate

SGR (%/day) = $(lnWf - lnWi)/t) * 100$

LnWi = the natural logarithm of the initial average body weight (g)

LnWf = the natural logarithm of the final average body weight (g)

t = time (days) between lnWi and lnWf

If the Food Conversion Rate is known, the SGR can also be calculated by dividing the Percentage body weight fed per day by the food conversion rate. This calculation can be turned round to predict growth if the SGR is known.

GPD – Growth Per Day,

ADG – Average Daily weight Gain

Average weight gain = Average body weight at the end of the period – Average body weight at the start of the period

ADG (g fish/day) = Average weight gain (g)/period(days)

FR – Feed Rate

The amount of food given to fish over a specified period of time. The most common way of expressing this is as percentage of the animal's body weight per day. For example a 1000 gram fish, being fed 20g of food per day would be on a 2% feed rate [(20 / 1000) x 100)].

FR = (Feed/Biomass)*100

Mortality bio (kg)=Mortality No. *Av. Weigh (g)/1000

SR-Survival Rate

SR= (Number of fish harvested/Initial fish number)*100

SR – Survival Rate

SR= (Number of fish harvested/Initial fish number)*100

FR=(Feed/Biomass)*100

Mortality bio (kg) = Mortality No. * Av. Weigh (g)/100

It is not correct to say that Surve = 100 – Mortality Rate. It is not correct to say that Survey because there are also other reasons that reduce the fish numbers (e.g. adjustments which are missing or extra fish and culling).

Generalized Linear Regression Models to Determine Impact Factors

In the following section we discuss how to apply generalized linear models (GLMs) to determine the influential factors in the aquaculture production. Having the features previously determined by the expert knowledge and the data collected one use automatic feature selection methods to determine those influential factors by constructing a classifier. This consists on the process of selecting a subset of relevant features (variables, predictors) for use in model construction. That will permit to:

i. reduce the noise in the output of the forecast model;
ii. simplify it to make it easier to interpret by researchers/users;
iii. enhance generalization by reducing over fitting;
iv. shorten the training times;
v. Improve its overall computational efficiency.

The main argument when using a feature selection technique is that the data contains many features which are either redundant or irrelevant, and can thus be removed without incurring much loss of information (Bermingham *et al.*, 2015). Notice that redundant or irrelevant features are two distinct notions, since one relevant feature may be redundant in the presence of another relevant feature with which it is strongly correlated (Guyon *et al.*, 2003).

Roles of two classical methods (Correlation and Regression)

Correlation refers to any of a broad class of statistical relationships involving any statistical relationship between two random variables or two sets of data. Correlation examines if there is an association between two variables, and if so to what extent.

Regression establishes an appropriate relationship between the variables. Its aim is to discover how a dependent variable Y is related to one or more independent variables X. Ordinary linear regression predicts the expected value of a given unknown quantity (the response variable, a random variable) as a linear combination of a set of observed values (predictors).

Statistical Classification

The goal of statistical classification is to use an object's characteristics to identify which class or group it belongs to. A linear classifier achieves this by making a classification decision based on the value of a linear combination of the characteristics. A training of linear classifiers includes:

- Logistic regression — maximum likelihood estimation the feature vector assuming that the observed training set was generated by a binomial model that depends on the output of the classifier;
- Support vector machine — an algorithm that maximizes the margin between the decision hyper plane and the examples in the training set.

Logistic Regression

The most common method for analyzing binary response data is logistic regression that is used to model relationships between the response variable and several explanatory variables which may be categorical or continuous. Logistic regression has been generalized to include responses with more than two nominal categories (Dobson *et al.,* 2008). We now turn our attention to two types of models where the response variable is discrete and the error terms do not follow a normal distribution, namely logistic regression and Poisson regression. Both belong to a family of regression models called generalized linear models (GLMs).In a GLM, each outcome of the dependent variables, Y, is assumed to be generated from a particular distribution in the exponential family, a large range of probability distributions that includes the normal, binomial, Poisson and gamma distributions. In linear regression, the use of the least-squares estimator is justified by the Gauss-Markov theorem, which does not assume that the distribution is normal.

Stream Story for Aquaculture

The in-house developed software stream story (freely available at http://streamstory.ijs.si) provides a qualitative representation of sensory data or any time-series data. It permits data visualization as a hierarchical state machine. It also allows the user to explore the distribution of states and transitions. Further, it offers prediction and anomaly detection services. It permits to analyse simultaneous time-series data streams corresponding to features that are being studied and determine the impact of those features against one or more selected features. This is done by a hierarchical continuous-time Markovian process, where the states, transitions and hierarchy, are automatically learned from the data.

Automatic Feeding Control

In a paper authors present an efficient visual signal processing system to continuously control the feeding process of fish in aquaculture tanks. The aim is to improve the production profit in fish farms by controlling the amount of feed at an optimal rate. The automatic feeding control includes two components: 1) a continuous decision on whether the fish are actively consuming feed, and 2) automatic detection of the amount of excess feed floating on the water surface of the tank using a two-stage approach. The amount of feed is initially detected using the correlation filer applied to an optimum local region within the video frame, and then followed by a SVM-based refinement classifier to suppress the falsely detected feed. Having both measures allows the authors to accurately control the feeding process in an automated manner. Experimental results show that their system can accurately and efficiently estimate both measures (Dunn *et al.*, 2008). Preliminary work is reported to have been done to determine if an automated feeding system can be developed through image analysis of fish feeding behaviour using a submerged surveillance camera (Blyth *et al.*, 1992). The proposed method is to obtain video images of fish feeding behaviour and subsequently applying data analysis. That data analysis entails image processing, followed by pattern recognition and machine learning methods. There are many common tasks any fish farmer would find difficult to do, such as counting, sorting, measuring and weighing of fish without having to individually handle and stress the fish. These are critical needs in fish farming, as this is essential information for financing, insurance, stock management and feeding activities. Past ongoing research has developed numerous fish feed monitoring, counting and measurement techniques without having to handle or stress the fish. In the past, the implemented techniques included acoustic equipment and signal processing and even the x-raying fish fed with spiked iron powder. Today, the conventional practice of hand-feeding is based on the use of feed tables and the experienced eye of the feeder adjusting the feed quantity to suit the needs of the fish. As units and holding units have become larger and deeper, accurate visual observations of the fish have become more difficult. The information feedback of feed consumption can be improved by implementing methods such as: - the airlift pump with pellet counters to provide an automatic feeder cut-off, and a facility for recycling the uneaten pellets (Bjordal *et al.*, 1993, Beveridge *et al.*, 2004, Kadri *et al.*, 1991); -the underwater video camera to observe the stock during feeding (Thorpe *et al.*, 1992, Phillips *et al.*, 1985, Man *et al.*, 2011). One of the fundamental challenging problems in computer vision is detecting object inside an image or video frames. The author presents a neural network-based approach for detecting fish object inside a digital image. The ANN is used to recognize the fish object from a small window that scans for the object all over the image. The ANN will examine the windows of the image and decides whether each window contains a fish. This approach is aimed to eliminate the difficult task of manually selecting the fishes, which must be chosen and localized to be analyzed (Bellemans *et al.*, 2000).

Feed Conversion Ratio (FCR) and Specific Feeding Ratio (SFR) Modeling

The Feed Conversion Ratio [FCR] is an important performance indicator to estimate the growth of the fish. It is widely used by the aquaculture fish farmers in pair with the Specific Feeding Ratio [SFR]. Its importance follows from the fact that 70% of the production costs in aquaculture are due to the food given to the fish during growth. Some of it will fall through the net and some will be spared. Optimization of the feeding process of the fish can thus confer great benefits to the economic development of the farms. Specifically, the FCR permits the aqua farmer to determine how efficiently a fish is converting feed into new tissue, defined as growth . Recall that the FCR is a ratio that does not have any units provided by the formula: FCR = dry weight of feed consumed/wet weight of gain while the feed conversion efficiency (FCE) is expressed as a percentage as follow: FCE = 1/FCR × 100.

Example: If the total amount of gain over the 14-day interval of feeding was 10,500 – 10,000 kg or 500 kg, and the total amount of feed offered was 420 – 400 kg/day × 14 days = 20 kg/day × 14 = 280 kg, then the FCR and FCE are calculated as follows:

FCR = 500 kg fed/280 kg gain = 1.78

FCE = 1/1.78 × 100 = 56%

Recall that the FCR and FCE are based on dry weight of feed and fish gain, as the water in dry pelleted feed is not considered to be significant. A typical feed pellet contains about 10% moisture that will only slightly improve the FCR and FC. Each aquaculture entity draws an appropriate FCR table to that batch of fish. Higher temperature leads to lower energy spent and faster growth, and consequently to a lower FCR. As the fish gets bigger, it requires more food to increase its biomass in percentage, and thus the FCR grows higher with the increase in average weight. The quality of the food and the size of the pellet are not considered at this point. At high temperatures (above 30 degrees in the case of bream and bass) low oxygen leads to low conversion to biomass. This is one of the hidden variables in the model, which should be considered separately at a later stage. One of the possibilities would be to penalize the FCR tables for the lack of oxygen. The other variable is the high reproduction of the fish in low temperatures and high average weight, which highly affects the growth of the fish.

Recall that the Economic FCR is the real FCR index following from the quotient between food given to the fish and the fish biomass. When the temperature is too high or too low we should ignore the data that is filled in with zeros and considered empirical data. The prediction can be done based on the model cited here. Forecasting is a mature field with a vast amount of methods and approaches. The methods used can be grouped into 4 main classes:

i. Naïve approach;

ii. Parametric approach;

iii. Non-parametric data-driven approach;

iv. Hybrid approach.

Time Series Classification and Forecasting

The information provided by periodic datasets from sampling to sampling gives the basics KPIs of a unit between two samplings. With minor normalization effort, it can provide enough feedback for the evaluation on most business questions, as it involves most KPIs. Its small size is a challenge. It needs relatively small time frames between samplings, and needs relatively consistent handling (same feeder, same food type, etc.). This dataset is ideal for classification and regression algorithms. We can use it for short term forecasting of numeric KPIs with methodologies like *Neural Networks (sensitivity analysis), SVM, additive generalized linear models etc. Clustering techniques* can be used to identify clusters of each KPI and maybe even common attributes (hatchery, food type) of these clusters.

Handling of time series data using neural networks

A times series can always be decomposed according to its behaviour describing the dynamics of the data. The components of a time series can be any combinations of the following:

i. Trend [Tt] – a linear progression with either positive or negative slope;

ii. Seasonality [St] – a pattern repeated in time;

iii. Cycle [Ct] – a large pattern in time;

iv. Randomness [Rt] – a non-controlled behaviour.

In time series analysis, identified the components that can be forecasted separately, and put them together in a comprehensive way to enable an overall forecast. The most common method to put this components together is the multiplicative model Xt = Tt*St*Ct*Rt. An important step to elimination of the seasonal aspect of the time series. That is, if the time series represents a seasonal pattern of L periods, then by taking the moving average Mt of L periods, we get the mean value for the year. This would then be free of seasonality and contain little randomness. Thus, Mt=Tt*Ct. Now, to determine the trend, we take the de-seasonalized time series and use regression to fit a suitable trend line. The choices can be linear, quadratic, exponential, etc. After the trend Tt has been estimated, one can use Ct = Mt / Tt to estimate the cycle component Ct. Finally, to isolate seasonality we simply divide the original series by the moving average Xt / Mt = Tt *St *Ct *Rt/Tt*CT=St *Rt

ANN for classification and forecasting of time series data

The technological advances lead to the dominance of artificial neural networks (ANNs) for classification and forecasting of time series data, is important in the context of aquaculture. In general, the predictive power of ANNs for time-series data is based on the idea that we can use the past to predict the future. Autoregressive Moving Average (ARMA) models are appropriate to the study of time series data influenced by cycles and seasons. It is possible to generalize ARMA using the ANN approach by a diagram.

In this section, brief description of the methods that are most applicable to aquaculture is presented.

Naïve methods are considered as unsophisticated forecast approach where no forecast model is actually built (Vijayakumar *et al.,* 2005). It can produce surprisingly good results if the data is very periodic in nature. A recursive formula enables us to incrementally update the model instantly with every new record (Johnson *et al.,* 1992). To estimate a single regression model with more than one outcome variable we shall use the more general multivariate regression. In particular, if there is more than one predictor variable in a multivariate regression model, the model is a multivariate multiple regressions (Drake *et al.,* 2006).

K nearest neighbors [kNN] is a type of instance-based learning, or lazy learning, where the function is only approximated locally and all computation is deferred until classification. The kNN algorithm is among the simplest of all machine learning algorithms. Both for classification and regression, it can be useful to assign weight to the contributions of the neighbours, so that the nearer neighbours contribute more to the average than the more distant ones. An optimal parameter k (number of neighbors) has to be found for best results.

Random forests is an ensemble learning method for both classification and regression tasks, that operates by constructing a multitude of decision trees at training time and outputting the class that is the mode of the classes (classification) or mean prediction (regression) of the individual trees.

Support vector machines (SVMs) are supervised learning models that analyse data and recognize patterns, used for classification and regression problems.

Artificial Neural Networks [ANNs] are models that are inspired by the structure and/or function of biological neural networks. They are a class of pattern matching that are commonly used for regression and classification problems but are really an enormous subfield comprised of hundreds of algorithms and variations for all manner of problem types. The most common in load forecasting is back propagation method.

Advanced Artificial Intelligence Techniques such as support vector machines are used in environmental modeling in considering the multidimensional nature of the study cases (Gelman *et al.,* 2004). The use of Bayesian hierarchical models within a state-space framework allows for separate modeling of the system process and the data process (Schulstad 1997).

Fish Count Model

The number of fish in a batch during the fish production is undetermined until harvest. Though, it is an important factor in the calculation of most of the aquaculture metrics. There is exact data available from external sources (e.g. temperature, oxygen level) and from sampling (e.g. average weight). But, there exist several non-exact measurements, such as fish count and its dependent variables (e.g. number of fish mortalities). To better estimate the number of fish during the production cycle, a model is developed a mathematical learning model of exponential nature that estimates that number built on historical data of the number of mortalities and the adjustments made. The fish count model used for the machine learning experiments is built according to the probability of the fish dying and/or disappearing for other reasons (escape etc.), *using a Poisson distribution.*

Case-Based Reasoning and Fish Disease Diagnosis

Case-Based Reasoning (CBR) is a reasoning method. CBR systems perform remarkably well on complex and poorly formalized domains. CBR classifiers use a database of problem solutions to solve new problems. CBR is intuitive, relatively simple to implement, transparent and it learns. Decision support system developers have problems with the knowledge elicitation bottleneck, the dynamics of decision support, the constant maintenance that systems require, the fact that systems must be accepted and that advice must be justified. CBR addresses each of these problems. A case-based reasoner solves new problems by using or adapting solutions that were used to solve old problems. It also offers a reasoning paradigm that is similar to the way many people routinely solve problems. Cases are several features describing a problem plus an outcome or a solution. Cases can be very rich (text, numbers, symbols, plans, multimedia, etc.), are not distilled knowledge, are records of real events and are excellent for justifying decisions. Unlike CBR, neural nets and genetic algorithms cannot justify their decisions. A large amount of work is done in fish disease diagnosis with the help of CBR. Detailed work done on a DSS for hatchery production management for Atlantic salmon in Norway and a developed decision support tool for aquaculture to assess economic and ecologic impacts of alternative decisions on aquaculture production (a system based on simulation models and enterprise budgeting) are reported (Bolte *et al.*, 2000, Li *et al.*, 2002). A web-based expert system for diagnosing fish disease in aquaculture facilities in China using short message service with great success is available (Garaas *et al.*, 2011). It considers the usage of the mobile phone instead of the computer systems, widely used in rural areas of China. They implemented a SMS platform in Java with the high accurate diagnosis rate of 93.57% for fresh-water-fish diseases validated by experts. CBR can also be used to capture the information on the sorting, and to also encapsulate the knowledge and intuition of the fish farmer with it, combining sensor data with the knowledge of the fish farmer (Tidemann *et al.*, 2012). This system can diagnose 48 kinds of fresh-water-fish of in Tian Jin area; the system has been tested with 140 fish diseases cases that were diagnosed by fish expert. Nine could not be diagnosed, the rest diagnosis results agreed with the experts', so the accurate diagnosis rate was as high as 93.57%. It was built upon the existing web-based system and the SMS platform composed of Wave com GSM/GPRS modem, SIM card, central computer and common mobile phone that can send short message, using the nearest-neighbour search model. Moreover, the main focus of the CBR application developed is to lower fish mortality during grading operations [56]. The cases in the knowledge base consist of an object (fish), the given symptoms and the treatment given. The similarity assessment between a query and the case-base is done by a clustering algorithm to find which part of the case-base the query belongs to. A simple nearest neighbor algorithm is then used to find the closest matching cases.

Conclusion

At the end to summarise, the challenges and opportunities of aquaculture for data analytics are reviewed and identified. These are to be addressed with the appropriate methodologies and technology. The mathematical models developed, reported and discussed in this paper aim to contribute to the improvement of the aquaculture procedures, providing a deeper insight on the information retained in the collected data. Using data mining help solve many problems. Moreover, the statistical analysis of the results permits a clearer visualization of the important features in the data that can boost the production and optimize the processes related to it. FAO envisaged that a major challenge to implementation of the 2030 Agenda is the sustainability as because, the next decade is likely to see major changes in the environment, resources, macroeconomic conditions, international trade rules and tariffs, market characteristics and social conduct, which may affect production and fish markets in the medium term. Influences include climate change, habitat destruction, overfishing, IUU fishing, poor governance, diseases and water resources as well as improved fisheries management, efficient aquaculture growth and improvement in technology and research. Moreover, recent developments on sensors, robotics, computer vision, satellite imaging, usage of drones, driverless boats, and interconnected devices powered by advanced data mining in the context of the Internet of Things generating huge data in petabyte scale and here lies the scope of Big Data analytics in fisheries research. The big-data approach will change the understanding of natural and human processes, such as the growth and distribution of species or the spatial planning of fisheries and aquaculture. Through big data, new opportunities arise for tracing how and where vessels operate and for tracking products all the way to shops and consumers. Modeling in the areas of *Feed Conversion Ratio (FCR) and Specific Feeding Ratio (SFR), Key* Performance Indicators (KPI) is the need of the hour for efficient management of resources for sustainable production from fisheries sector. Available report suggests that neural network forecasts exceeded other traditional forecasting methods such as linear or logistic regression systems. Application of ANN to Forecast Water Quality and Temperature that benefits aquaculture process control and artificial intelligence systems that in turn helps in increased process efficiency; reduced energy and water losses; reduced labour costs; reduced stress and disease; better understanding of the process and efficient accounting are also highlighted. The increasing adoption of Statistics, data mining and Big Data analytics in aquaculture, facilitated to bring the techniques of aquaculture to a new level of in-depth understanding and deeper insights into the aquaculture reality. Aquasmart data visualization tools that have been developed and are likely to contribute to the success of the end user businesses. (Roy and Sarangi 2009; Pullin, et al., 2017; FAO, 2012 and Aquasmart Consortium)

References

Aqua Smart Consortium, The aqua Smart Project [Online]. URL: www.aquasmartdata.eu.

AQUATEXT Dictionary http://www.aquatext.com/dicframe.htm (accessed in 28.1.2016).

Atia, D. M., Fahmy, F. H., Ahmed, N. M., and Dorrah, H. T. (2011). Solar Thermal Aquaculture System Controller Based on Artificial Neural Network. Engineering, 03(08), 815–822.

Balakrishnan, M and Meena, K and Sethi, S N and Sarangi, Aditya N (2007). Neural network and its application in aquaculture. *In* :A.K.Roy eds. Applied Bioinformatics and statistics in Fisheries Research, ICAR-CIFA, Bhubaneswar, India. 3. pp. 145-151.

Bar, N. S., and Radde, N. (2009). Long-term prediction of fish growth under varying ambient temperature using a multiscale dynamic model. BMC Systems Biology, 3(1), 1.

Bartók, A. P., Gillan, M. J., Manby, F. R., and Csányi, G. (2013). Machine-learning approach for one- and two-body corrections to density functional theory: Applications to molecular and condensed water. Physical Review B, 88(5), 054104

Bellemans, T.., De Schutter, and B. De Moor (2000) Data acquisition, interfacing and pre-processing of highway traffic data. Proceedings of Telematics. 19:0–2.

Bengil, F., and Bizsel, K. C. (2014). Assessing the impact of aquaculture farms using remote sensing: an empirical neural network algorithm for Ildýrý Bay, Turkey. Aquaculture Environment Interactions, 6(1), 67–79.

Benzer, R., and Benzer, S. (2015). Application of artificial neural network into the freshwater fish caught in Turkey. International Journal of Fisheries and Aquatic Studies, 2(5), 341–346.

Bermingham, Mairead L.; Pong-Wong, Ricardo; Spiliopoulou, Athina; Hayward, Caroline; Rudan, Igor; Campbell, Harry; Wright, Alan F.; Wilson, James F.; Agakov, Felix; Navarro, Pau; Haley, Chris S. (2015). Application of high-dimensional feature selection: evaluation for genomic prediction in man. Sci. Rep. 5

Beveridge M (2004). Cage Aquaculture. Third Edition, Oxford, UK.

Bhar Lalmohan, S. S. Walia and A. K. Roy (2002). Forecasting fish production from ponds (mimeo). Division of Forecasting Techniques, IASRI, Library Avenue, New Delhi-110012. p.30

Bjordal A, Juell JE, Lindem T, A. Ferno A (1993); Hydro acoustic monitoring and feeding control in cage rearing of Atlantic salmon, Fish Farming Technology, pp. 203-208, Balkema, Rotterdam.

Blyth PJ, Purser, GJ & Russell JF (1992). Boosting profits with adaptive feeding: letting the fish decide when they're hungry. AustAsia Aquaculture, Vol. 6:33-38

Bolte John, Nath Shree, and Ernst Doug (2000). Development of decision support tools for aquaculture: the pond experience. Aquaculture Engineering, 23:103–119.

Clark, C.W. (1985). Bio economic Modelling and Fisheries Management. Wiley, New York.

Dobson, A. J., & Barnett, A. (2008). An Introduction to Generalized Linear Models, Third Edition. CRC Press. [ASL15] Atoum, Y., Srivastava, S., & Liu, X. (2015). Automatic Feeding Control for Dense Aquaculture Fish Tanks. IEEE Signal Processing Letters, 22(8), 1089–1093.

Drake, J. M., Randin, C., & Guisan, A. (2006). Modeling ecological niches with support vector machines. Journal of Applied Ecology, 43(3), 424–432.

Dunn, Zelda (2008). Improved feed utilization in cage aquaculture by use of machine vision. Diss. Stellenbosch: Stellenbosch University.

European Commission, Aquaculture. http://ec.europa.eu/fisheries/cfp/aquaculture/index_en.htm

Fabbian, D., R. De Dear, and S. Lellyett (2007). Application or neural network forecasts to predict fog at Canberra International Airport," Weather and Forecasting, vol. 22(2) 372-381.

Fish Farming Glossary http://www.aquaculture.co.il/getting_started/glossary.html

Food and Agriculture Organisation of the United Nations. FAO. (2018). The State of World Fisheries and Aquaculture 2018. FAO, Rome. http://www.fao.org/3/i9540en/I9540EN.pdf

Garaas, M., & Stevning, G. H. (2011). Case-Based Reasoning in identifying causes of fish death in industrial fish farming.

Gelman, A., Carlin, J. B., Stern, H. S., and Rubin, D. B. (2004). Bayesian Data Analysis, 2nd Edition. Chapman & Hall/CRC, Boca Raton, Florida.

Ghosh, Ambalika;Ajit kumar Roy;and Bijay Kali Mahapatra (2017). Relationship between fish seed production and inland fish production of West Bengal, India. S.M. *Journal of Biometrics &Biostat.* 2(2):1013.

Government, U. S. (2011). Estimating Water Temperatures in Small Streams in Western Oregon Using Neural Network Models. Books LLC.

Guyon, Isabelle; Elisseeff, André (2003). An Introduction to Variable and Feature Selection". JMLR 3.

Hall, T., Brooks, H. E. and Doswell, C.A.I. (1999). Precipitation Forecasting Using a Neural Network. Weather and Forecasting, 14(3), 338–345.

Houvenaghel, T., & Huet, T. (1989). A model for eel growth in aquaculture. European Aquaculture Society.

Johnson, Richard Arnold, and Dean W. Wichern (1992). Applied multivariate statistical analysis. Vol. 4. Englewood Cliffs, NJ: Prentice hall.

Kadri, S., Metcalfe, N.B. Huntingford, F.A. & Thorpe J.E. (1991) Daily feeding rhythms in Atlantic salmon in sea cages. Aquaculture, 92:219 – 24.

Katsanevakis, S. (2006). Modeling fish growth: Model selection, multi-model inference and model selection uncertainty. Fisheries Research, 81(2-3), 229–235.

Lee P.G. (2000). Process control and artificial intelligence software for aquaculture. Aquacultural Engineering, 23, 13-36.

Li, D., Fu, Z., & Duan, Y. (2002). Fish-Expert: a web-based expert system for fish disease diagnosis. Expert Systems with Applications, 23(3):311–320.

Liu, S., Yan, M., Tai, H., Xu, L., & Li, D. (2011). Prediction of Dissolved Oxygen Content in Aquaculture of Hyriopsis Cumingii Using Elman Neural Network. In Computer and Computing Technologies in Agriculture V (Vol. 370, pp. 508–518). Berlin, Heidelberg: Springer Berlin Heidelberg.

Løland, A., Aldrin, M., Steinbakk, G. H., Huseby, R. B., Grøttum, J. A., & Quinn, T. J. (2011). Prediction of biomass in Norwegian fish farms. Canadian Journal of Fisheries and Aquatic Sciences, 68(8).

Man, Mustafa, et al. FishDTecTools: Fish Detection Solution Using Neural Network Approach. Journal of Communication and Computer 8.2 (2011): 96-102.

Maqsood, I., Khan, M.R., Abraham, A. (2004). An ensemble of neural networks for weather forecasting, Neural Comput & Applic 13: 112-122.

Marzban, C. and G.J. Stumpf (1996). A neural network for tornado prediction based on Doppler radar-derived attributes. Journal of Applied Meteorology, vol. 35(5)617-626.

Phillips, M.J., Beveridge, M.C.M & Ross L.G. (1985). The environmental impact of salmonid cage culture on inland fisheries: present status and future trend. Journal of fish Biology, 27:123–37.

Pullin, Roger SV, Rainer Froese, and Daniel Pauly (2007). Indicators for the sustainability of aquaculture. Ecological and Genetic Implications of Aquaculture Activities. Springer Netherlands. 5372

Roy, A. K. and N. Sarangi, (2009). Modeling, Forecasting, Artificial Neural Network and Expert system in Fisheries & Aquaculture, Daya Publishing House, New Delhi, xv+239p (ISBN: 1081-7035-571-0).

Roy, A. K., K.N. Sahoo, K.P. Saradhi and G.S. Saha (2002). Farm-size and aquaculture productivity relationship. *Journal of Asian Fisheries Science, Philippines*: 129-134.

Roy, A. K., M. Rout, P. K. Saha and A. K. Datta (1998). Influence of Environmental Factors on the SGR of Catla catla Grown in Sewage Ecosystem-Investigated with multiple Regression Analysis. In: A. K. Aditya & P. Halder (Eds.), Proc. of the National Seminar on Environmental Biology. Daya Publishing House, New Delhi-35, 87-93.

Roy, A. K.; P. K. Saha; A. K. Datta; P. K. Satapathy (2000). Regression model for Catla catla reared in sewage fed ponds In: B. B. Jana, R. D. Banerjee, B. Guterstam and J. Heeb (eds.) Waste Recycling and Resource Management in the Developing World, pp. 65-69. University of Kalyani, India and International Ecological Engineering Society, Switzerland held during 23-27, Nov., 1998 at Kolkata.

Roy, A. K.; P. K. Satapathy and A. Antony, (2006). Modeling in Aquaculture. *In*: A. K. Roy, A. Antony & N. Sarangi (eds.) Bioinformatics and Statistics in Fisheries Research, Vol-2. Central Institute of Freshwater Aquaculture, Kausalyaganga, Bhubaneswar-751002, India. Pp. 329-342.

Roy.A.K. N. Panda &D.P. Rath (2008). Forecasting of culture fisheries production in India. *Inland Fish Soc.India*, 40(Spl.1):54-59.

Schulstad, G. (1997). Design of a computerized decision support system for hatchery production management. Aquaculture Engineering, 16:7–25.

Thorpe, J.E. and Huntingford FA (1992). The importance of Feeding Behaviour for the Efficient Culture of Salmonoid Fishes. World Aquaculture Society, Baton Rouge, LA.

Tidemann, A., Bjørnson, F. O., & Aamodt, A. (2012). Operational Support in Fish Farming through Case-Based Reasoning. Iea/Aie, 7345(Chapter 12), 104–113.

Upadhyay, A.D., A.K. Roy and Jumli Karga (2012). Exploratory data analysis approach with fish and seed production in Arunanchal Pradesh. *Keanean Journal of Science* Vol. 1, 2012.

Vijayakumar, S., A. D'Souza, and S. Schaal (2005). Incremental online learning in high dimensions. Neural computation, vol. 17, no. 12, pp. 2602–34.

6

An Analysis of Value Chain of Processed Fish Products in North East Region of India

A.D. Upadhyay; D.K. Pandey and Bahni Dhar

Introduction

Value chain analysis has gained considerable popularity in recent years. Although many definitions are applied, value chains essentially represent enterprises in which different producers and marketing companies work within their respective businesses to pursue one or more end markets. Value chain participants sometimes cooperate to improve the overall competitiveness of the final product, but may also be completely unaware of the linkages between their operation and other upstream or downstream participants. Value chains therefore, encompass all of the factors of production including land, labor, capital, technology, and inputs as well as all economic activities including input supply, production, transformation, handling, transport, marketing, and distribution necessary to create, sell, and deliver a product to a certain destination.

Value Chain

A value chain is the "organized link across groups of producers, traders, processors and service provider, including nongovernmental organizations (NGOs), that join together to improve the productivity and value added their activities" (Asian Development Bank, 2013) A value chain is a sequence of related business activities (functions) from provision of specific inputs for a particular product to primary production, transformation and marketing, up to the final sale of a particular product to the consumer (GTZ Value Links, 2008). The two core elements are embedded in definition of a value chain: chain and value. The chain component of value chain refers to a supply chain—the processes and actors that take a product from its conception to its end use and disposal. The second component is value which implied value is being added to the product or service at each step in the supply chain.

The Casuga *et al.* (2008), has described the distinctions between the supply chain and value chain on the basis of different characteristics (Table 6.1)

Table 6.1: Difference between Characteristics of Supply Chain and Value Chain

Characteristics	Supply Chain	Value Chain
Production	Supply led bulk production	Market driven or demand driven differentiated products
Structure	Fragmented supply chain	Integrated supply chain
Marketing	Large number of intermediaries	Less number of market intermediaries or direct procurement by the lead firm or processor of marketing firm
Extension services	Local input suppliers or local extension service providing agencies	Provision of extension services and inputs in the chain
Financing	Money lenders, traders, relative or friends mostly for production	Financing within and outside chain through contract

Source: Casuga et al. (2008)

Porter's value chain Model

Michael E. Porter was the first to introduce the concept of a value chain. He developed the Five Forces Model to show businesses where they rank in competition in the current marketplace (Free Press, 1998) (Figure 6.1). Porter's Value chain model based on his opinion that a business firm cannot achieved competitive advantage in an environment it works at market place without it focuses on its performances on many discrete activities including design of product, production, and marketing, delivering and supporting its product. Porter splits a business's activities into two categories: primary activities and support activities.

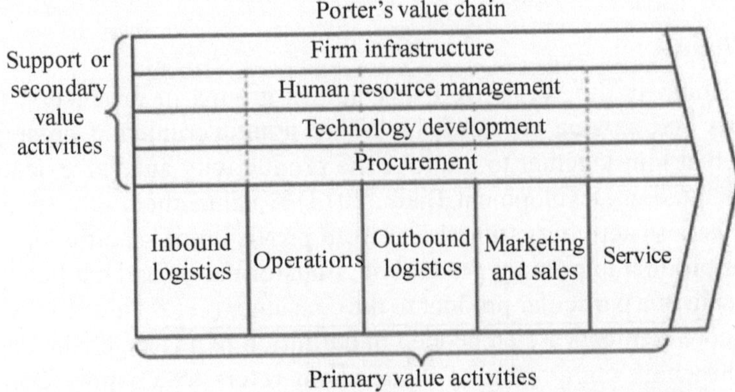

Fig. 6.1: Porter's Five Forces Model of Value Chain Analysis

Primary activities

Those activities involved in creation, transfer and sale of the products and these activities are:

1. ***Inbound logistics*** are the receiving, storing and distributing of raw materials used in the production process.
2. ***Operations*** are the stage at which the raw materials are turned into the final product.
3. ***Outbound logistics*** are the distribution of the final product to consumers.
4. ***Marketing and sales*** involve advertising, promotions, sales-force organization, distribution channels, pricing and managing the final product to ensure it is targeted to the appropriate consumer groups.
5. ***Service*** refers to the activities needed to maintain the product's performance after it has been produced, including installation, training, maintenance, repair, warranty and after-sale services.

Support activities

All those activities which support the primary activities and these are as follows:

1. ***Firm's Infrastructure***: all activities including organizational structure, control system, company's internal environment are included in this category.
2. ***Human Resources Management***: It includes recruitment, hiring, training, development and compensation etc.
2. ***Technology development***: The activities those intended to improve product and process at different level.
3. ***Procurement***: It is related with the procurement of inputs raw materials etc.

Steps of Value chain analysis

1. Identifying the all the primary activities and sub activities
2. Identifying the supporting activities and sub activities
3. Identifying the all the links of the value chain
4. Identify and workout opportunities for creating value and competitive advantage in terms cost and differentiation.

One of the advantages of the Value Chain Analysis is that it is very flexible tool that can be used to diagnose and create competitive advantage both in terms of product cost and product differentiation for the business firm.

Case Study of Value Chain Analysis of Processed Fish Products in North East Region of India

The total fish production in India reached to 11.41 million tonnes in 2016-17, the country occupies second place in the world. Out of total production about 67 per cent produce consumed in fresh form and remaining 33 per cent of total production utilized for processing and drying, fishmeal, canning etc. India's total fisheries export is about $7.08 billion in 2017-18. Fisheries sector of India provides employment for over 15 million people.

In North East region of the country, majority of population (90%) are fish eater. Further, the peoples of the region are also habitual in consumption of dry fish and several traditional processed products like Shidal, Nona Ilish, smoked fish, and canned fishes etc,. The demand for these products in the region is very high and Therefore, dry fishes in the region is imported from the coastal states like West Bengal, Orissa, Andhra Pradesh, Tamil and Gujarat and also from land locked states like Uttar Pradesh. Several Specific dry fish and fermented fish products like Shidal and Nona Ilish are also imported from the Bangladesh and Myanmar. The dry fish processing and trade involved numbers of stake holders such as fishermen, processors, input suppliers labourers, transport agencies traders, marketing agencies including wholesalers/retailers etc. This value chain is also important for nutritional security of people of NE region particularly in rural areas. Hence, it has become imperative to analyze value chain of dry fish in North Eastern region. With backdrop a study was conducted on "Socio-Economic Aspect of Value Chain Analysis of Dry Fish in North East Region of India" with specific objectives to map value chain of dry fish, estimate cost and margin of chain actors, examine the status of institutional credit in the value chain as well as to suggest up gradation strategies for the development of dry fish value chain.

Selection of respondents

The value chain of dry fish involved many actors and business operators like fish fishermen, processor, trader, wholesalers, commission agents, retailers, labourers and consumers. Therefore, for the selection of the representative sample of the respondents from different categories of value chain actors, multistage stratified random sampling was applied. Therefore, all together 12 markets including 6 wholesale and 6 retail markets and five processing centre including 2 from coastal states and 3 from NE region were selected for the survey and data collection (Table 6.2).

Table 6.2: List of Selected Processing Centers and Dry Fish Markets

S.No.	Name of the state/ Landing Centers	Wholesale Markets	Retail markets	Processing Center
1.	Assam	Jagiroad, Karimganj,	Tinsukia, Silchar,	Kusumpur
2.	Manipur	Ima market, Thoubal,	Nambol, Moirang,	Thanga
3.	Tripura	Golbajar, Teliamura,	Udaipur, Kumarghat	Gandacharra
4.	WB	-	-	Digha
5.	Gujarat	-	-	Veraval

After selection of markets and processing centres, chain actors were categorized into processor, wholesalers, retailers, labourers and consumers and from each of these categories. At last a representative sample including all categories of chain actors were selected using simple random sampling without replacement. All together 555 respondents including 60 processor from Digha and 78 processors from Veraval, 95, 102, 220 numbers of different categories of market functionaries were selected randomly from Assam Manipur and Tripura, respectively (Table 6.3).

Table 6.3: Details of Sampling of Respondents

Functionaries	Assam	Manipur	Tripura	Digha (W.B.)	Veraval (Gujarat)
Processor	5	20	43	44	24
Wholesaler	31	11	33	-	-
Retailer	22	48	53	-	-
Labor	19	11	56	16	54
Consumer	18	12	35	-	-
Total	95	102	220	60	78

Data Collection

Two methods of data collection such as Focused Group Discussion (FGD) and personal interview method were applied for the collection of primary data. The information was gathered from each category of respondents using 5 types of semi structured survey schedules. The details of information collected from the respondents are described below-

a) **Fish Processor** : data on general information of processors, family details, family asset position, expenditure pattern, fixed assets, purchase of raw materials, value addition activities and cost , sale details, inventory, risks, institutional support, quality/standards, financing related information and major constraints in fish processing business etc. have been included.

b) **Market General Information:** In order to canvas overall scenario of the market such as details of market administration and management, market regulations, type of functionaries operating in the market, type of activities undertaken in the market, marketing charges/fees, basic amenities available in the market, employment of labour in the market, market arrival, service supports available and geographical coverage and problems faced market functionaries in the dry fish trade etc have been included in this questionnaire.

c) **Traders/Wholesaler/Retailers:** the middleman involved in transactions/ distribution and flow of the products from processing centers to retail markets of the North East Region. Therefore, a common survey schedule including information on personal profile, family profile, assets details, expenditure pattern, transaction details (purchased and sale quantity and prices) value chain activities, marketing charges, agencies through which produce purchased and sold, price setting method, finance, business relationship, quality and standards, problems in dry fish business etc was developed.

d) **Labourers:** At various stages of value chain such as processing assembling trading, wholesaling, retailing, large number of labourers with varied level of skill was engaged for performing various activities. Therefore, information related to their family profile, family assets, employment details, monthly expenditure, income, other wages, monthly saving etc. were included.

e) **Consumers:** At the retail markets, respondents of this category were interviewed for gathering the information on family details; expenditure pattern, consumption pattern of processed fish products, consumers' preference, perception on quality, prices paid for different products and availability of product of his choice etc were collected. This survey work was conducted during 2015.

Value chain Mapping

The value chain of dry fish has been mapped using functional analysis. Functional analysis provide complete structure of the value chain and in this value chain was mapped by defining value chain boundaries, identifying core activities, identifying economic agents involved at different stages and related functions, interrelationship and linkages between the economic agents, depicting flow of the commodity and bottlenecks. The results related to all these have been represented using flowcharts and tables.

Economic Analysis of Activities of Value Chain

Economic analysis of value chains was performed for measuring of costs and margins of business operations at different stages of value chain. This helped in understanding how much money an actor of value chain contributed and how much money an actor received (margins). It also helped in understanding whether a value chain is pro-poor or not. Calculation of cost and returns of value chain actors: Economic benefits and costs of individual operator of the value chain will be analyzed using following formulae:

Revenue of an actor = Σ (Qi * Pi)

[Qi= total quantity of ith dry fish products sold, P= Price of ith dry fish products]

Total cost of an actor (Rs/t) = Fixed cost + Variable costs.

Net incomes earned by an operator were calculated by deducting cost from revenues.

Net income (Rs/t) = total revenue – total cost.

Net margin (Rs/kg) = Net income / Quantity sold

The average cost and margins accrued for each category of chain actors were calculated.

Institutional Credit

For effective operation and sustainable growth of dry fish trade, financing is crucial in the whole marketing system of dry fish. The Descriptive analyses were performed to assess the pattern of credit use; credit need by different business operators and the factors determining the operators' decision to use institutional credit. The financial aspects including status of bank account, status of credit in dry fish value chain, the credit need were also analyzed using qualitative and quantitative data in form of tables and graphs.

Status of dry fish production and trade in the Country

In the country annually about 3.22 lakh tonnes of unsalted dry fish and 2.93 lakh tonnes sated dried fish and smoked fish produced in the country. Through diagram it is observed that during the period of five years (2008 to 2012), there slight increase in production of both the products. Out of total production of these two categories of the products, the Gujarat alone constituted 68% of total unsalted dried fish and 46% of total salted dried fish including smoked fish produced in the country (Figure 6.2 & Figure 6.3). The other states like Andhra Pradesh, Kerala, Maharashtra, West Bengal, Orissa, Karnataka among the maritime states and Bihar, Haryana, Manipur, Nagaland in the land locked states produces small proportion of salted and unsalted dry fish products. In addition to domestic market demand for these products, there is also export demand for these processed products. Annual export of dried fish (marine) was 67901 metric tonnes in 20013-14 and the trends of the export of these products was quite encouraging during 1995-96 to 2013-14 (Figure 6.3). The major chunk of salted and unsalted dried fish produced in coastal states and landlocked states are being traded to the North Eastern Region of the country where the demands for these products are very high.

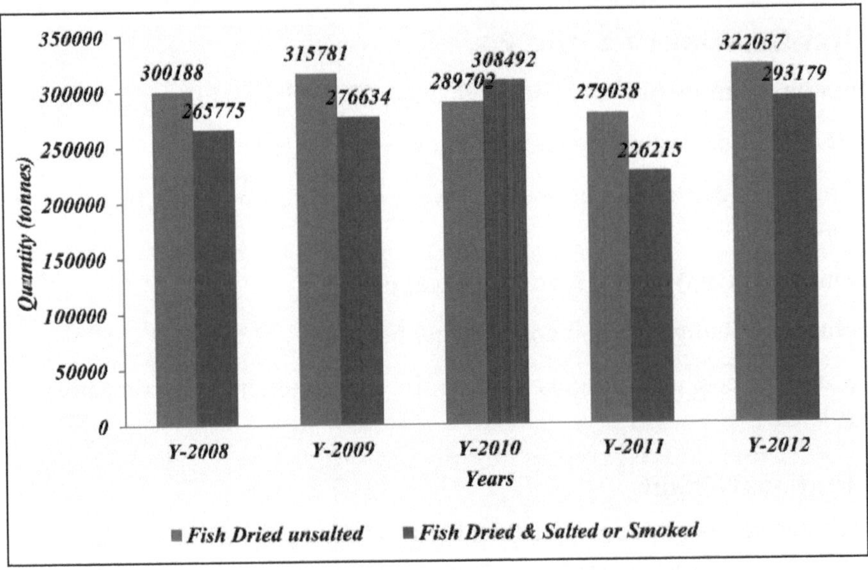

Fig. 6.2: Production of Dried Fsh in Country During 2008 To 2012
Source: Handbook of Fisheries Statistics (2014)

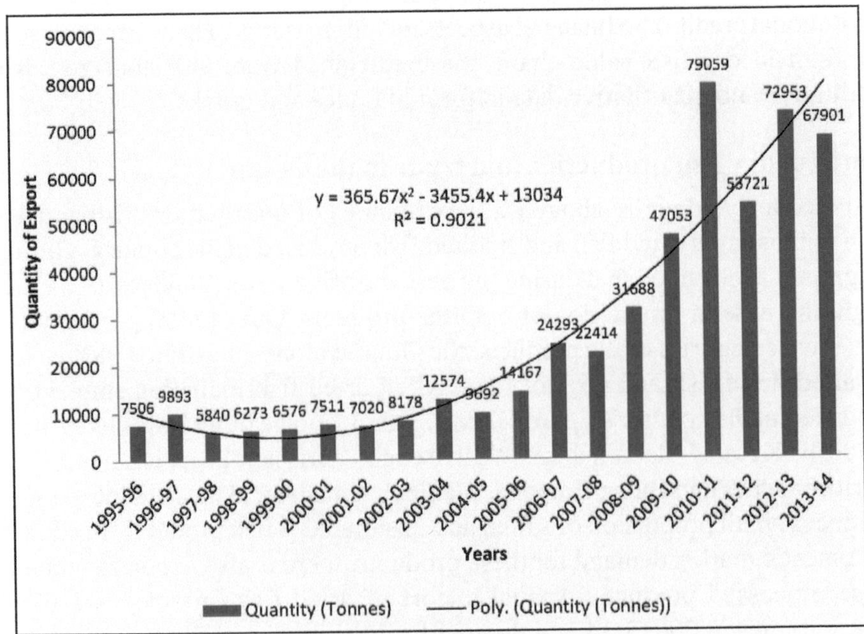

Fig. 6.3: Trend of Export of Dried Fish (in Tonnes

Core Process of Dry Fish Value Chain

Value chain of dry fish comprises a large network of distributional channels and it connect the production and processing centre that confined in coastal belt (marine fishes) and in northern States (freshwater fishes) to the consumption points distributed to whole North East Region of the country. Therefore, the nodes of the dry fish value chain are widely scattered. The core processes and activities undertaken at the different stages of value chain are represented through flow diagram in Figure 6.4.

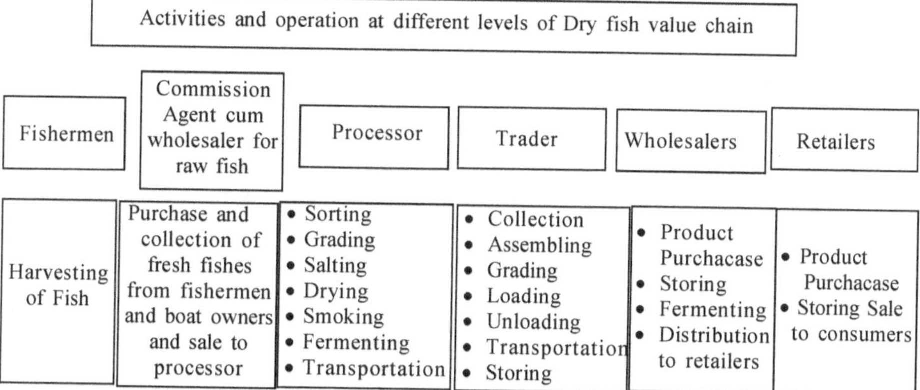

Fig. 6.4: Specific Activities Performed in Core Process of Dry Fish Value Chain

The core processes and economic agents involved in dry fish value chain were identified. The core processes were procurement of raw fish by the processors at landing centre, fish processing, assembling and trading, wholesaling and retailing. The processes are described below-

A. Core Activities

i. Fish Processing

***Drying*:** The different methods are used in value additions of fish such as drying, smoking and fermentation. The drying of fishes is an age old processing practice, which adds the value in terms of increasing its storage life and avoids spoilage of fishes. It also reduces volume and quantity of fish that makes the handling easier. Most of the medium and low value species form bulk of the landings and are marketed either in fresh condition or it is utilized for value addition.

***Fermentation*:** In the Northeast region of the country, fermentation is one of the oldest and most economical methods for producing and preserving food. In addition to preservation, fermented foods can also have the added benefits of enhanced flavour, increased digestibility, and higher nutritional and pharmacological values. The Shidal is traditional valued added fish product based on fermentation technique and highly demanded in the NE region. This product prepared by the local people of NE region by using indigenous technique. The processing of shidal involve activities like procurement of dry fish (puntius sp), sorting and cleaning of dry fishes, preparation of matka (earthen pot) with mustard oil and airtight packing and keeping it for 6 months for the fermentation. Preparation of Shidal is skilful and time consuming process. The Shidal preparation is labour intensives activity and mostly women were engaged in it.

Smoking: Fish smoking prolongs the shelf-life of the fish, enhances flavour, increases utilization of the fish, reduces waste and increases protein availability to the people (Jallow, 1995). A variety of smoked products are popular in the tropical countries (Gopakumar, 1997). Though there are several reports on smoke curing of fish in India and abroad, the smoking of fish in Manipur is unique in nature. No salting is involved in the entire smoking process (Singh *et al.*, 1990; Lilabati and Viswanath, 2000). In India, fish smoking is widely practiced in Orissa, West Bengal, Assam, Arunachal Pradesh, Manipur, and Madhya Pradesh, some pockets on the west coast, and the Godavari and Krishna deltas in Andhra Pradesh. The smoked fish are one of the popular and highly preferred products of people of Manipur, a North-Eastern state of India. Varieties of fishes were found to be smoked by fishermen families or unemployed women. The smoked fish products consumed after frying or roasting or as an ingredient to other dishes to add taste and flavour to food. These activities provide employment and income to women of the rural areas.

ii. Trading

Assembling and trading was identified as one of the most important processes in dry fish value chain. The traders are being functioning at the dry fish processing centers. Trading involves range of activities like supply of packaging material to the processors, procurement of processed products, transportation and storage of products and disposal (transportation) of dry fish produce to distant markets as per demand by the wholesalers of distant markets. In some cases it was also found that they were financed/advanced to the small scale dry fish processors for the procurement of raw fish and processing. Traders are highly specialized business concerns and they are well informed about the demand and prices of the dry fishes in the different wholesale markets of distant places. The further provide links between processors and wholesalers and disseminate information on required quantity and quality of processed products in different markets and about the prevailing prices for different type of processed fishes.

iii. Wholesaling and retailing

The wholesaling and retailing are important activities of dry fish value chain and at this stage value chain activities like transportation, sorting, grading, packaging, marking, storage and wholesale etc were performed. In addition to these activities in some cases in Tripura, it was found that wholesalers were also engaged in secondary processing of dry fish for preparation of matka shidal.

B. Supporting activities

In addition to core activities there are number of service support to the value chain of dry fish such as input suppliers at processing centre including, fishermen who supplies wet fishes to the processors, suppliers of salt, ice, packaging materials, plastics, bags, bamboo, rubbers, medicines etc. The labourers, transporters, insurance companies providing insurance during transit, financial institutions and communication services. These service supports are crucial to all stages of the value chain of dry fish.

Dry fish value chain map

A value chain map of dry fish that reflects the interrelationship and linkages among the chain actors are represented through diagram in Figure 6.5. The dry fish value chain of NE region included flows of dry fish and other processed products, produced and processed in Coastal states(outside fishes) as well as in NE region(local fishes) into value chain of NE region. The outside fish value chain further comprises two types of fishes, marine fishes that imported from costal states and inland fresh water fishes which are imported from land locked states. The important states from where the dry fish are imported in NE region are Gujarat, A.P., Tamilnadu, Orissa, W.B., U.P., and Bihar etc. In addition to this some of the processed products like matka shidal and nona elfish are imported from adjoining countries like Bangladesh and Myanmar. In case of local fresh water fishes, it produced and processed in small quantities. Though quantity of locally processed fishes is quite small but important because of consumers preference.

Value chain map of Dry fish represented in Figure 6.5, reflects that it comprises two sub value chain one outside fishes (sub value chain-I) and for locally produced and processed fishes (sub value chain II). Both chains begin with fishermen who catch the fishes and bring it to landing centers. In case of sub value chain I, processors procured the fishes from fishermen through commission agents at landing centers and transport it to processing units. At processing centre fishes are cleaned, salted, washed and sun dried. After processing it packed and sold to the traders. The traders collect the processed products from the processors and transport it to their establishments where they do sorting, grading, packaging, storing, marking etc. and based on demand in distant markets they arrange for transportation and sale to wholesalers of the distant markets. The traders were selling their products to wholesalers of Jagiraod Dry fish market which is the Asia' s biggest dry fish market located at Assam. Only in few cases it was found that traders were also directly transporting to different states and sale it to the wholesalers. From the Jagiroad, wholesalers of different states of NE region were purchase the dry fishes and sale it to the retailers at subsequent level of the value chain. Finally the retailers were responsible to cater the need of dry fish consumers. In sub value chain II fishermen he himself or small scale processors process the local fishes and sale it either to traders, wholesalers or also to retailers as per their convenience. However, major portion of the produce converse with wholesalers of the state and it move with sub value chain I.

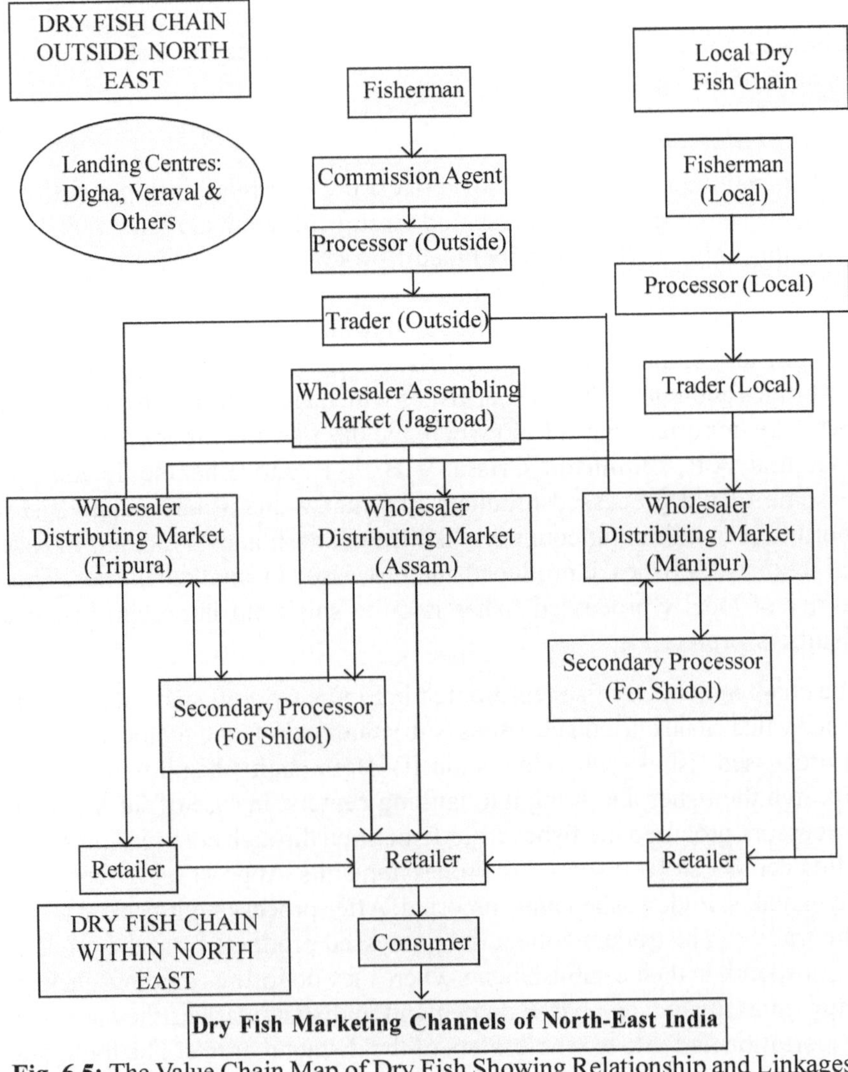

Fig. 6.5: The Value Chain Map of Dry Fish Showing Relationship and Linkages

Each of two sub value chains comprises several distribution channels or alternative routes through which the product flows from point of production to point of consumption. In sub value chain-I and category sub value chain –II, five distribution channels in each category Out of five distribution channel identified for outside dry fishes, distribution channel-1 with core vale chain actors fishermen – auctioneer (commission agent) – processer – traders – wholesalers of assembling market (Jagiroad) – wholesalers of distributing markets – retailers-consumers was found to be most common distribution channel and major proportion of dry fish produce were traded through this channel. Similarly, in case of sub value chain II out of five identified distribution channels, channel-4 with chain actors' fishermen cum processers – wholesaler – retailer – consumer was noticed to be most common for trading of local processed fish products.

Backward and forward linkages in dry fish value chain

Horizontal linkages

It appears from this survey based study that the traders at processing centers, wholesalers of different markets and even some of the retailers of dry fish markets were used to pool transportation facilities in transportation of smaller quantities of produce procured to distant markets together for economizing transportation charges. This pooling helped them even to trade the smaller quantities by individual agencies; otherwise it was not possible for them to transport the produce individually. It also benefited the small chain actors in participating to dry fish value chain.

Vertical linkages

Vertical linkages were observed at processing centre Veraval Gujarat where some of the processors were also performing the function of traders. They themselves used to procure the dry fishes of other processors also and trade it to the distant markets. It was also found that in some cases processors of Digha that they made verbal agreement with traders for supplying of all the processed dry fish to them. However, it was more common in case of small scale processors where they used to take money from traders and do the processing business. Further, it was found that some of the wholesalers of dry fish in dry fish markets of Tripura, Assam and Manipur were also involved in processing of Matka Shidal.

If we consider the processors as reference point in dry fish value chain then backward linkages are fishermen for supply for raw fish and input suppliers and labourers and the traders, wholesalers, retailers and consumers are the forward linkages in dry fish value chain.

Nature and Types of Products Flows in Dry Fish Value Chain

The value chain analysis of the dry fish indicated that the three types of processed products namely dried fishes, smoked fishes and fermented fishes are traded. More than 39 varieties of dry fish products including both the sources marine as well as inland fisheries are being in dry fish value chain. The varieties of dry fish products included fish species namely Indian Mackerel (Rastrelliger kanagurta), Gar Fish *(Xenentodon cancila*), Ribbon Fish (*Trichurus lepturus*), Bombay Duck (*Harpodon nehereus*), Crockers (Scianeids), Rainbow Sardine (*Dussumeria acuta*), Cat fish (*Mysteus and Arius species*), Glass fish (*Chanda ranga*), Tardoore (*Opisthopterus tardoore*), Chapra (*Gudusia chapra*), Large razor belly minnow (*Salmostoma bacaila*), Prawns (*Acetes species*), Big Crockers (Scianeids), Golden scad (*Carax para*), Golden Anchovey (*Coilia dussumieri*), Corica (*Corica soborna*), Oil sardine (*Sardinella longiceps*), Mola (*Amblypharyngodon mola*), Butter catfish (*Ompak species*), Puntius (*Puntius species*), lesser sardine (*Sardinella species*) and Flying fish (*Exocoetus species* etc). The fermented fish products namely Puthi Shidal, Baspati Shidal (*Puntius species*), Nona Ilish (*Tenolisa ilisha*) were found in the markets. The smoked fishes which recorded from Manipur were Tank Gobi (*Glossogobius garius*), Singhi (Heteropneustes fossilis), Glass fish (*Chanda ranga*),Feather back (*Notoperus notopterus*), Snake head (*Channa species*), Puntius (Puntius species), Darkina (Esomus dentricus), Cuchia (Monopterus cuchia), Tilapia (*Oreochromis mosambicus*), Small carps (*Labeo species*), Moka (*Amblypharyngodon mola),* prawns (*Macro brachium species*), Gourami *(Colisa fasciatus*), smoked major Carps, smoked minor Carps (*Labeo species*), Koi (*Anabus testudineus*) etc.

Infrastructural facilities along the dry fish value chain

The dry fish value chain infrastructure includes, facilities at landing centre, processing facilities, storage, ice, salt and other input supply, packaging facilities, transport facilities, market shed, market storage facilities, electricity, communication facility and most important financial services etc. Two processing centers in coastal states were surveyed and it was observed that in the Veraval, Gujarat dry fish processing centre, infrastructural facilities in terms of processing sheds, drying space, storage space, tanks for cleaning and other processing facilities, road, transport facilities, banking services were available and are better in comparison to Dhiga dry fish processing centre. This was also noticed that in Veraval Dry Fish Processing Centre, processing of fish and drying were undertaken at larger scale and in more commercial way. In Digha, West Bengal, dry fish processing centre, processing were done at smaller scale, by poor people and facilities like processing shed, drying plate form, processing tanks, storage space, poor transportation facilities, road, etc. The most important observation from Dhiga was lack of banking facilities as well as credit facilities.

The infrastructural facility at Asia's biggest dry fish market at Jagiroad, Assam was poor in terms of small parking space, no common storage facilities. This whole market is operated by private business operators. In case of other wholesale and retail markets studied, the market infrastructure was found poor like some of the markets were operating in open space on street without enough parking spaces, storage facilities, with poor hygiene and sanitation and financial services etc. On an average about 7 to 10 per cent of total quantity of traded products, deteriorate in quality and some time it turns into waste, due to poor storage and handling of the product during wholesaling and retailing. In the dry fish markets banking facilities are available; however, financial services as per need of dry fish business concerns were not available. The businessmen in wholesaler and retailer markets required short terms to medium terms credit with easy and minimum processing time.

Economic analysis of dry fish value chain

Cost and margins of dry fish processors

Drying is core process and most important economic activities of dry fish value chain. The drying involves series of activities like purchase of wet fishes, transportation, sorting, washing, cleaning, salting, drying, packaging and sale etc. The variable costs incurred by dry fish processors were, purchase of raw fish, salt, ice, packaging material, cost of transportation etc. Fixed cost of processors included construction of bamboo platform for sun drying of fish, depreciation/annual rental of fixed assets like Rickshaw, Crates, processing tanks and the salary of monthly paid labours etc. The cost and margin of the processors of Digha and Veraval were computed and it is presented in Table 6.4. The average cost of processing of dry fish was Rs. 84,179.65/t in Digha, West Bengal and it was Rs. 84,034.06/t in Veraval, Gujarat. About 87 to 88 percent of total cost of processing of dry was contributed by cost of raw material (wet fish) alone. Other variable costs such as transportation cost and costs of salt, ice, rubber, medicines etc. contributed about 3-4% of the total cost. In fixed costs, wages of monthly labour was found to be highest and it accounted for about 5.3% in Digha and 6.05% in case of Veraval. The net incomes of processors' were estimated Rs. 5760.35/t in Digha and Rs. 29009.70/t in Veraval. Net margin of processor was 6% and 25% of total revenue in Digha and Veraval, respectively. The higher returns at Veraval may be because of scale of operation of dry fish processing in Veraval was higher and it was more commercialized as compared to the Digha. At Veraval mostly the fishes were salted before drying and moisture content is less. The species, quality and overall appearance of dry fish products of Veraval were well recognized by the buyers in the value chain. Whereas at Digha, most of the fishes were dried without salt and moisture content is slightly more. Therefore, the quality of dry fish of Veraval was better which fetched higher prices to the processors.

Table 6.4: Cost and Margin of Processors of Dry Fish

Particular	Cost and return (Rs./t)			
	Digha	%	Veraval	%
A. Variable cost				
Cost of raw fish	73328.74	87.11	74020.00	88.08
Transportation	2400.00	2.85	1600.00	1.90
Misc (salt, ice, rubber, medicines etc.)	921.68	1.09	849.64	1.01
Total variable cost	76650.42	91.06	76469.64	91.00
B. Fixed Cost				
Bamboo plate form	1328.02	1.58	512.89	0.61
Machineries	56.57	0.07	691.53	0.82
Misc (tank, bags, baskets etc)	664.42	0.79	399.11	0.47
Rent	530.34	0.63	345.77	0.41
Shed and other assets	486.58	0.58	528.64	0.63
Permanent labour	4463.30	5.30	5086.48	6.05
Total fixed cost	7529.23	8.94	7564.42	9.00
Total cost(A+B)	84179.65	100	84034.06	100
Total revenue (TR)	89940.00		113043.76	
Net income	5760.35	6.40	29009.70	25.66

Cost and margins of processors in Matka shidal

The matka shidal is prepared by secondary processing. The dried fishes particularly of Puntius species which is a fresh water fish used for preparation of puthi shidal. The dried fishes are mainly imported from West Bengal and Uttar Pradesh and also from Bangladesh. The processors of shidal purchase dry fishes in bulk quantity during season and they store it and utilize for preparation of shidal in whole season (November to April). The processing period of shidal varies from 3 months to 1 year depending on desired quality of shidal. It was reported by the processors that more time period of processing leads to better the quality and price of matka shidal. The cost and margin in processing of matka shidal has been calculated for the processors of Tripura and Manipur and it is presented in Table 6.5, Table 6.6 and Table 6.7. Since shidal is widely demanded fish product in the North East Region, Therefore, processors, process it in bulk quantity to meet out the market demand. In Tripura two varieties of shidal namely puthi shidal and bashpati shidal are processed and it is available in the markets. The total cost processing of puthi shidal was estimated to be Rs. 213647.5/t. Out of it, 89.9% was accounted by raw material i.e. dry fish utilized for shidal preparation and other items like earthen pots, cost of mustard oil and transportation were account for about 2% each of the total cost of processing of puthi shidal in Tripura. The processors earned on an average net margin of Rs. 52593.8/t in processing of puthi shidal, which was about 20% of the total revenue (Table 6.5). Similarly cost and margin of processing of bashpati shidal was estimated and it was found that processors were spending Rs. 171728.27 on processing of one tone bashpati shidal (Table 6.6). However, the cost of processing varies from Rs. 156757.9 per tonne in Teliamura market to Rs. 197221.1/t in Udaipur market of Tripura state. These variations in costs may be attributed to quality of dry fish that utilized for preparation of shidal and other inputs cost. On an average a bashpati shidal processors earn net return of Rs. 59616.17/t which is about 25.77% of total revenue.

The total cost of processing of matka shidal in Manipur was found to Rs. 2, 85,129.99/t (Table 6.7). It is observed from the table that the cost of processing of shidal is higher in Manipur as compared to Tripura. It was mainly because of cost of raw material in Manipur was relatively higher. The net return in processing of shidal in Manipur was Rs. 77065.31/t which constituted about 21.28% of total revenue. Above results shows that, shidal processing is highly profitable business in NE region and it provide employment to rural people specially women. It was further observed that the shidals from Tripura was traded to other states like Manipur, Meghalaya and Assam.

Table 6.5: Cost and Margin of Processors of Putti Shidal in Tripura

Head of expenditure and return	Cost/return(Rs./₹)			
	Gol Bazar	Udaipur	Teliamura	Overall
Variable cost items				
Cost of raw material	257692. (87.30)	259722.22(91.77)	154529(89.94)	192053.4(89.89)
Earthen pot	5000(1.69)	5000(1.77)	5000(2.91)	5000(2.34)
Mustard oil	4500(1.52)	4500(1.59)	4500(2.62)	4500(2.11)
Transportation cost	9846.15(3.34)	4222.22(1.49)	5311.59(3.09)	5767.98(2.70)
Loading unloading cost	205.77(0.07)	211.11(0.07)	354.26(0.21)	301.97(0.14)
Labour cost	17930.776.07)	9373.61(3.31)	2127.81(1.24)	6024.13(2.82)
Total variable cost	295175(100)	283029.16(100)	171822.66(100)	213647.5(100)
Average quantity of sale (kg/month)	4333.33	2571.43	3450	3315.38
Total revenue	360769.23	318333.33	226992.75	266241.3
Net return	65594.23(18.18)	35304.13(11.09)	55170.05(24.3)	52593.8(19.75)

(Figures in parentheses indicate percentage of total)

Table 6.6: Cost and margin of processors of Bashpati Shidal in Tripura

Head of expenditure and return	Cost/Return(Rs./t)				
	Golbazar	Udaipur	Teliamura	Kumarghat	Overall
Variable cost items					
Cost of raw material	150000 (80.5)	175465.12(88.97)	139132.88(88.76)	165652.17(92.21)	152676.06(88.91)
Earthen pot	5000 (2.68)	5004.65(2.54)	5009.01(3.20)	5000(2.78)	5006.40(2.92)
Mustard oil	4500 (2.42)	4504.19(2.28)	4508.11(2.88)	4500(2.50)	4505.76(2.62)
Transportation cost	8000(4.29)	4186.05(2.12)	5054.05(3.22)	3000(1.67)	4686.3(2.73)
Loading unloading cost	125(0.07)	213.37(0.11)	408.45(0.26)	250(0.14)	325.19(0.19)
Labour cost	18700(10.04)	7847.67(3.98)	2645.38(1.69)	1239.13(0.69)	4528.55(2.64)
Total variable cost	186325(100)	197221.05(100)	156757.88(100)	179641.3(100)	171728.27(100)
Average quantity of sale (kg/month)	3000	3071.43	2775	3066.67	2892.59
Total revenue	275000	229534.88	231081.08	222608.7	231344.43
Net return	88675(32.25)	32313.83(14.08)	74323.2(32.16)	42967.4(19.3)	59616.17(25.77)

(Figures in parentheses indicate percentage of total)

Table 6.7: Cost and margin in preparation of Puthi shidal in Manipur.

Details	Cost and return (Rs/t)	%
Variable cost items		
Cost of raw material	264375.00	92.72
Earthen pot	4001.18	1.40
Mustard oil	4284.64	1.50
Transportation cost	7749.95	0.40
Labour cost	3579.16	2.72
Misc(loading/unloading cost)	1140.06	1.26
Total variable cost	285129.99	100
Average quantity of sale (kg/month)	10835	
Total revenue	3,62,195.30	
Net return	77065.31	21.28

Cost and Margins of Processer of Smoked Fish in Manipur

The smoking is traditional practice used in Manipur to develop value added fish products. However, this technology is used in many part of the world, particularly in African Countries, for producing smoked fish at large scale and exports it to the US, UK and other European countries. The smoked fish is highly preferred fish products in Manipur. The small scale processors of Manipur use to smoke locally produced, small size freshwater fishes. The processing of smoked fish is mainly undertaken by women fishers both in group and at individual level. This is a seasonal activity performed for about five months in a year mainly in winter and summer seasons. On an average in a month about 391 kg of wet fishes are processed by a fishermen family (Table 6.8) and out of it they produce 156.58 kg of smoked fishes. Total cost and gross return on processing were Rs 1,57.265.3/- and Rs. 1,95,734.1 per tonne, respectively. The costs of processing include cost of raw material (wet fishes), firewood, transportation of raw fish and processed products and imputed value of family labour. Their net return after excluding the family labour was Rs. 70,400.7/t, constituting almost 36% of total revenue. A fishermen family earn net income of Rs. 11,023.34 per month. This finding clearly indicates that smoking is a viable and employment generating avenue for the fishermen family. It not only provides employment to the women of fishermen families but also other unemployed women who can buy the fishes and smoke it and sell it at better prices in the market. Further, with small amount of spending on regular basis in the business of smoking fish, women can earn sufficient income for their family. Though under existing practice adopted in smoking, only small quantity of fishes can be smoked by a women in a day. However, with proper technical and financial support and promotion of group/community. The processing capacity and income can be increased.

Table 6.8: Cost and margin of preparation of smoked fish in Manipur

Head of expenditure and return	Manipur	
	Cost/return (Rs. /t)	%
Cost of smoking		
Cost of raw material	80000.07	50.87
Firewood	33897.6	21.55
Marketing cost (transportation/fare)	11112.3	7.07
Imputed value of family labour	31931.91	20.30
Chulha	323.41	0.21
Total cost	157265.3	100
Return		
Quantity of raw fish (kg/month)	391	
Quantity of smoked fish (kg/month)	156.58	
Value of output (PXQ)	195734.1	100
Net return	38468.79	19.65
Net return excluding imputed value of family labour	70400.7	35.97
Net return excluding imputed value of family labour (Rs. /month)	11023.34	

Cost and Margins of Wholesalers in Dry Fish Value Chain

The cost and margins of wholesalers of all the selected markets of Assam, Manipur and Tripura were estimated is presented in Table 6.9. It is reflected from the table that on an average a wholesaler of dry fish in NE region was selling 8653.33 kg dry fish per month. This quantity however, varies from market to market and state to state. It was the lowest in Manipur with an average transaction of a dry fish wholesaler was 5158.9 kg whereas it was 11762.28 kg in Tripura. These figures are an indicative of quantum of trade and demand for dry fish in NE region of the country. At this stage of value wholesalers were incurring marketing cost of Rs. 230995.24/t out of which 94.48% were paid to previous agency (Traders) on purchase of dry fish. Remaining 5.52% of total cost, they spend on transportation, loading unloading, market fee, electricity charges and labourers etc. Total revenue of wholesalers was Rs. 252811.7/t. The wholesalers were earning Rs 21816.46/t as net margin which is about 9 % of total revenue. It was higher in case of Assam (15.78%) as compared to Tripura (7.7%) and Manipur (6.97%). Higher margin in Assam may be due to lower transaction cost such as transportation.

Cost and margins of retailer in dry fish value chain

After the wholesaler the next stage of the value chain is retailing. At this stage, retailers are core actor responsible for transaction of produce to ultimate consumers and for the relevant feed back of the consumers' choice to the processors. Further, based on retailers' feedback, processors of dry fishes configure their processing strategy and product mix of dry fishes. The cost and

Table 6.9: Cost and margin of wholesalers in different states of NE region

Head of expenditure/ return		Cost and return (Rs./t)							
		Assam		Manipur		Tripura		Overall	
		Cost	%	Cost	%	Cost	%	Cost	(%)
Cost items									
Variable cost	Value of purchased fish	111131	95.76	340543	94.33	203091	94.61	218255.1	94.48
	Packaging cost	250.9	0.22	0	0	0	0	250.9	0.11
	Transportation	1543.38	1.33	18735.6	5.19	6397.81	2.98	8892.27	3.85
	Loading unloading cost	1834.73	1.58	168.86	0.05	1990.38	0.93	1331.32	0.58
	Storing	242.74	0.21	0	0	312.64	0.15	277.69	0.12
	Misc/others	250	0.22	0	0	730.37	0.34	490.18	0.21
	Total variable cost	115253	99.31	359448	99.57	212522	99.00	229497.46	99.35
Fixed cost	Market fee	87.49	0.08	1037.71	0.29	143.12	0.07	422.78	0.18
	Electricity	98.89	0.09	249.04	0.07	177.66	0.08	175.19	0.08
	Permanent labour	617.48	0.53	265.06	0.07	1816.88	0.85	899.81	0.39
	Total fixed cost	803.86	0.69	1551.81	0.43	2137.66	1.00	1497.78	0.65
Total cost (TC)		116057	100	361000	100	214659	100	230995.24	100
Average quantity of sale (kg/month)		9038.81		5158.9		11762.28		8653.33	
Total revenue (TR)		137799.19		388057.92		232577.92		252811.7	
Net margin		21742.34		27058.18		17918.5		21816.46	
% margin of total revenue		15.78		6.97		7.70		8.63	

earning of retailers were calculated separately for all the selected markets and also it was calculated for all the three selected states of NE region by pooling of data for all the wholesalers selected in the state (Table 6.10). It is observed from the table that the retailers of dry fish in NE region sold 943.47 kg per month. However, quantity of sale was lowest in Manipur (366.06 kg/month) and highest in Assam (1279.09 Kg/month), followed by Tripura (1185.27 kg). The quantity of sale depends on several factors like type and location of market, concentration of retailers in the market, consumers demand etc. The smaller quantity of sale of retailers in Manipur may be due to many of the retailers of dry fish were used to sale dry fishes with other grocery items and therefore, their size of sale of dry fish was less.

The marketing cost of retailer was Rs. 295543.25/t out of which 94.36 per cent was accounted for purchased value of fish or it was paid to previous agency (wholesalers). Retailers of Manipur incurred the highest marketing cost (Rs. 444523/t) followed by Tripura (Rs. 264700/t) and it was the lowest in case of Assam (Rs. 168108/t). High cost in case of Manipur might be due to high transportation cost and the quality of fishes traded by business operator. The variable costs incurred by retailers included purchased value of fish from previous agency, transportation cost, loading and unloading, misc cost and collectively it constitutes 98.69% of total cost incurred by the retailers of dry fish. Fixed cost incurred by retailers were market fee, electricity charges, market fee/rent etc. which accounts 1.31 percent. On an average a retailer of dry fishes earned Rs. 42377.20/t as net margin. However, it varies from Rs. 49034.21/t in Manipur, Rs. 41305.77/t in Assam to Rs. 46090.11/t in case of Tripura. Percentage margin of the retailers ranges between 10 to 20 percent of the total revenue generated on the sale of dry fish retailing.

Financial inclusion of dry fish value chain actors

With this study it was found that 56%, 73% and 87% of the retailers in Manipur, Assam and Tripura, respectively possessed bank saving accounts. Similarly 92%, 82% and 64% of wholesalers of Assam, Manipur and Tripura respectively possessed savings account in the banks. In case of processors 50% in Tripura 72% in Assam and 100% in Manipur were reported to have saving bank accounts. These results revealed that about 75% of the chain actors possessed bank account or they were linked with financial institution.

Financing in Dry fish value chain

It was found that only 20% of processers and wholesalers in Assam and only 9.09% of the wholesalers in Manipur and 2.33% of processers, 7.55% of the retailers at Tripura borrowed credit from different agencies (Figure 6.5). Further,

Table 6.10: Cost and margin of retailers in different states of NE region

Head of expenditure/ return		Cost and return (Rs./t)							
		Assam		Manipur		Tripura		Overall	
		Cost	%	Cost	%	Cost	%	Cost	(%)
Cost items									
Variable cost	Value of purchased fish	159580	94.93	435061	97.87	241945	91.40	278862.43	94.36
	Transportation	2623.22	1.56	5334.11	1.20	4837.24	1.83	4264.86	1.44
	Loading unloading cost	2481.39	1.48	428.64	0.1	533.67	0.20	1148.07	0.39
	Misc/others	1220.55	0.73	0	0	778.47	0.29	999.51	0.34
	Total variable cost	165906	98.69	440824	99.17	248095	93.73	285275	96.53
Fixed cost	Market fee	710.52	0.42	2601.13	0.59	475.59	0.18	1262.41	0.43
	Electricity	557.83	0.33	1097.46	0.25	467.21	0.18	707.5	0.24
	Permanent labour	934.11	0.56	0	0	15662.8	5.92	8298.47	2.81
	Total fixed cost	2202.46	1.31	3698.59	0.83	16605.6	6.27	10268.38	3.47
Total cost (TC)		168108	100	444523	100	264700	100	295543.25	100
Average quantity of sale (kg/month)		1279.09		366.06		1185.27		943.47	
Total revenue (TR)		209413.79		493556.98		310790.58		337920.45	
Net margin		41305.77		49034.21		46090.11		42377.20	
% margin of total revenue		19.72		9.93		14.83		12.54	

it is observed that out of total chain actors borrowed loan, (about 53%) was taken from Government Banks, remaining 47% were taken loan from other sources like Private Banks (20%), money lenders (13%), MFIs (7%) and SHGs (7%). Further, exploratory analysis of borrowings of the chain actors revealed that in case of 75% borrowings, amount of loan was <= 5 Lakh. These borrowings were predominated by short term and medium term credit and period ranges between 1-5 years. 96% loans were taken for the business purposes. The interest rate was observed to be higher (24-30% per annum) in case of non institutional credit than the institutional credit where interest rate was reported to be 10-12% per annum. It is evident from these results that despite of 75% chain actors possessed banks' saving accounts, credit borrowers are quite low. Further, out of total borrowing, 47% loan was from non-institutional sources of credits where they paid high interest rate. This clearly indicates poor financial supports in dry fish value chain. Therefore, the strengthening of institutional financial services along the dry fish value chain may be helpful to the business operators.

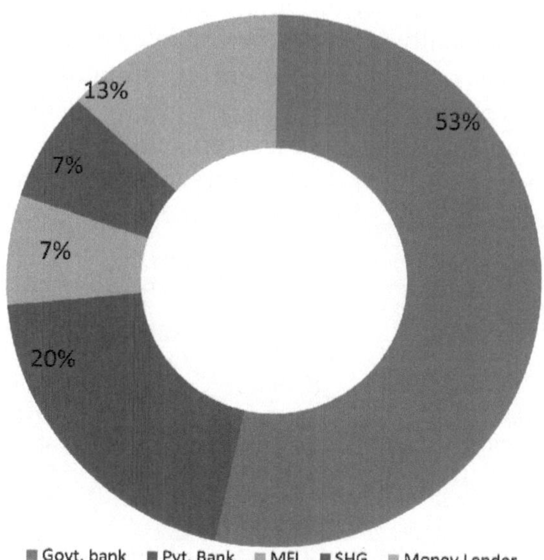

Fig. 6.5: Sources o Credit for Value Chain Actors of NE Region

Constraints faced by the Value chain actors

Constraints analysis of all categories of chain actors of dry fish value chain has been performed and it is presented in the following section.

A Processer

The major constraints faced by processors were analyze and is depicted through radar diagrams in Figure 6.6. It is observed from the figure that, low capital as a constraints reported by 60% processers of Digha, whereas fluctuation in demand of processed product as constraint, were reported by 50% of processors. About 30% processors of Digha reported non availability of wet fish of desired quality and quantity and price fluctuation at landing centres as major problem they faced in processing business. Fluctuation in prices at landing centre was due to competition for purchase of fishes among three categories of buyers at landing centre such as direct fish seller (wholesalers), raw fish processors and dry fish processors. The raw fish processors purchased high value and better quality fishes to meet out export demand. Due to this sometimes dry fish processors fail to compete with other buyers particularly in the time of shortages of fishes at landing centre. It affects their processing activities and income. In fish processing, processors has to procure wet fishes and other inputs almost on daily basis for which they need sufficient funds. One cycle of processing of dry fish starting from purchase of fish to final disposal of fish takes at least 10 days and to run parallel cycles, processer required money on regular basis. The processors indicated that they had the potential and ability expand their business of processing but due to paucity of fund sometimes they fails to utilize their existing resources optimally. That is why suitable credit supports and other financial services at the processing centre are crucial. Further, it was interesting to note in the prices of the processed products offered by the traders were not only depends on market demand but also on current prices of wet fishes at landing centre. Due to this the processors are exploited by the traders in the name of low market demand of processed products or low landing price of

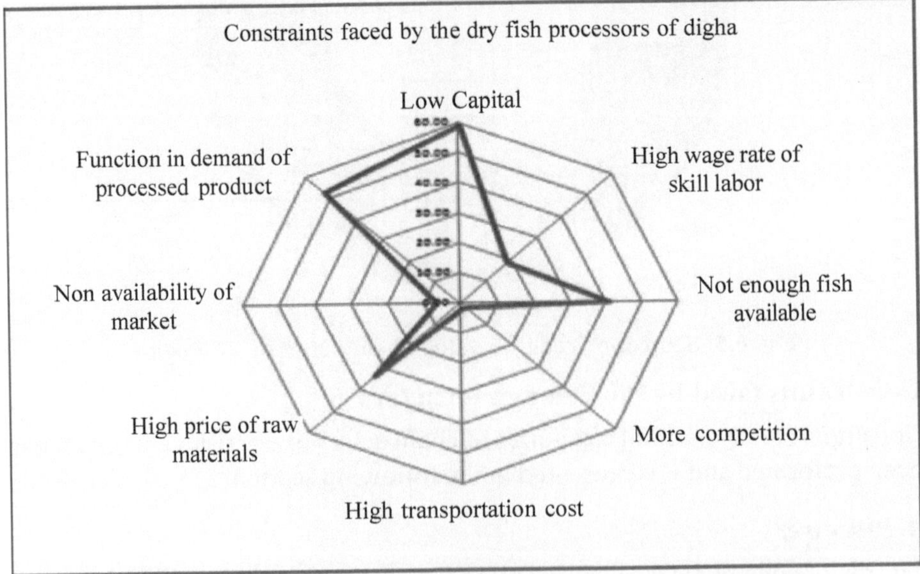

Fig. 6.6: Constraints Faced by the Dry Fish Processors of Digha

fishes as per their convenience. This strategy of traders crushes the profitability of the processors and sometimes it lead to loss of dry fish processors particularly at the time when the purchase of wet fishes at higher prices and during the processing when landing prices goes down. Therefore, through appropriate price mechanism and updated price information we can help to the dry fish processors in general and small scale processors in particular.

b. Wholesalers

The wholesalers in the dry fish value chain play a vital role in distribution of dry fish products in the North East region. They regulate supply of dry fish products. The constraints faced by this category of chain actors are presented in Figure 6.7. Most of the wholesalers reported that there was a shortage of supply of desired quality and quantity of dry fishes in comparison to demand. In addition to this, they also reported that due to lack of storage facilities and financial crisis during the peak periods there were several constraints in wholesale business. However, a significant proportion of chain actors also reported that fluctuation in prices of processed products at processing centres lead to instability of dry fish business and the problem of transport were vital in NE region. The financial services and storage facilities both are crucial at wholesale markets. Due to lack of proper storage facilities at wholesale and retail markets about 10-15% of total traded dry fish in NE region get spoiled during the trade.

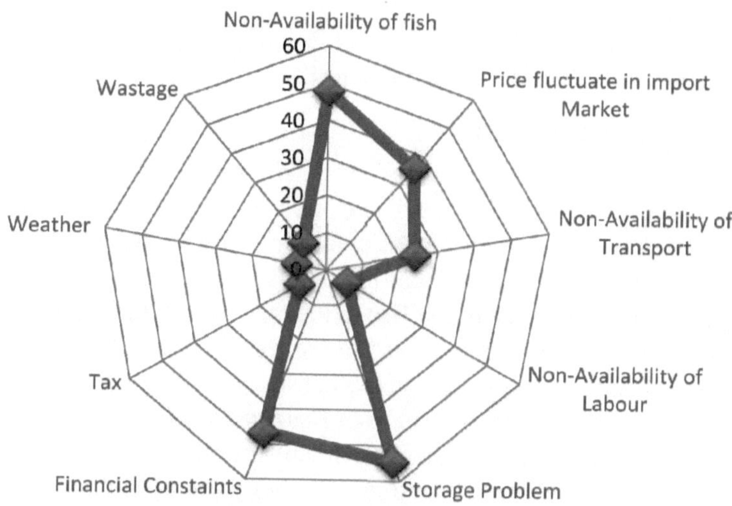

Fig. 6.7: Major Constraints Faced by Wholesalers

c. Retailers

The constraints faced by this category of chain actors are presented in Figure 6.8. It is observed from the figure that like wholesalers, retailers were also facing the problems of storage facilities, lack of fund to manage the business, price fluctuation and shortage of supply of desired quality and quantity of dry fishes. The shortage of supply of desired quality of fishes in desired quantity was reported by these two categories of chain actors indicating that demand for dry fish and other fish based processed products in the NE region is quite high. Hence, dry fish value chain can be expanded in the NE region with the support of financing and development of market infrastructural facilities.

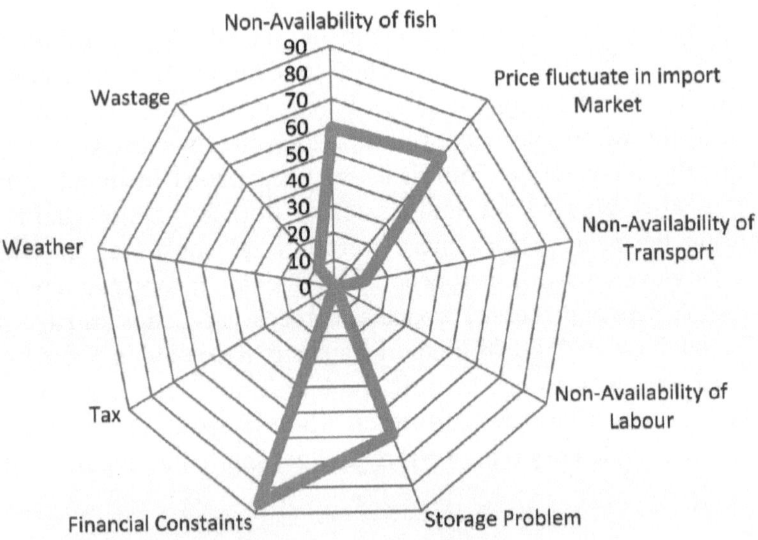

Fig. 6.8: Major Constraints Faced by Retailers

Risks in dry fish value chain

Fishes, being an extremely perishable food item, requires preservation for future uses. Fish drying is one of the old and most widely used methods of fish preservation. Sun drying is the most commonly use method. Fisher women around the sea used to dry small quantities of fish on the beach or on sands. But this traditional method of drying of fish in the open sun is hazardous as they are host to a lot of impurities. Sometimes the process takes nearly five days to dry and if during this period it rains whole lots get spoiled and it is a major risk in fish processing and is very common in case of East Coast. Further, during the rainy season, maggots /small bacteria develop in the partially dried fish which results in loss of most of the dry fishes as traders either do not purchases such fishes or they pay very low prices. It was also noticed that the fish processors in Digha were using pesticides/insecticides to control the fleas/maggots /small bacteria. Due to application of pesticides during dry fish processing in West Bengal and Orissa, dry fishes of these places were less preferred and had demand in the markets of NE region. Moreover, unprotected handling while drying affects the health of fisherwomen. The solar dryer is one of the promising technology through which many of these risk can be minimized and quality of dry fish products can be improved.

Conclusion

The dry fish production in India and its domestic as well as export demand is growing. To meet out growing demand for dry fish and other processed products, due emphasis on its production and trading is needed. In North East region of the country due to shortage of fresh fishes and food habit the demand for dry fish is quite high. This demand is met from dry fish imported from other states. In this study, value chain of dry fish was analysed with special emphasis on financial aspect of dry fish value chain. It was found that the network of dry fish value chain in the country is distributed from a Coastal states and other Northern states to remote corners of NE region. It is a well established value chain in which a series of core actors such dry fish processors, traders, wholesalers in assembling market and wholesalers in distributing market and retailers and number of supporting actors including input suppliers, labourers, financial institutions transport agencies etc. were functioning and supporting to whole value chain. More than 39 fish species in dried form and several fermented and smoked products are traded through dry fish value chain which contributes to price variation. The core chain actors involved in dry fish value chain are earning sufficient margins. The added value were found to be more in later stages of value chains rather than initial stages of the value chain whereas reverse to this more efforts in terms of value addition and movement of the products were made by the actors involved in initial stages of the value chain. Poor financial services along the value chain leads to higher dependency of chain actors on credit transaction which increases inefficiencies in the value chain. Based on these findings following measures are suggested for up gradation of the dry fish value chain. Based on above mentioned findings, specific recommendations of this study are as follows-

1. Technological and financial support for large scale solar driers to dry fish processors may be helpful in increasing the capacity, quality production of dry fishes and minimizing risks.

2. Micro-financing and the banking services need to be strengthened at dry fish processing centers of the coastal area.

3. The small scale enterprises such as shidal processing, smoking of fishes, drying of fishes, packaging of fishes and other value added activities needs technological and financial supports.

4. The technology related to Sidhal need to be refined according to local needs.

5. Large scale scientific storage facilities at major markets of dry fish in NE region can help in avoiding losses; deterioration in the quality of fishes during the trade and better distribution of fishes.
6. The market intelligence is needed to be strengthened the dry fish value chain which can help dry fish processors.
7. Financial services for the dry fish value chain actors which are presently week or missing at several nodes of dry fish value chain are required to be strengthened to avoid overdependence of the chain actors on credit transactions.
8. A comprehensive quality regulations and mechanism for the quality control is required to be developed and implemented.
9. Appropriate pricing mechanism at dry fish processing centre is to be developed in order to avoid exploitation of fish processer by the traders.
10. Whole dry fish value chain is governed and managed by private agencies where market charges, price fixation, quality regulation are arbitrary that causes imperfection in dry fish marketing system.
11. Group processing and trading can be encouraged though establishment of self-help groups.

References

Asian Development Bank (2013). Agricultural value chain for development. Learning Lesson, Manila.

Casuga, M.S., F.L. Paguia, K.A. Garabiag, M.T.J. Santos, C.S. Atienza, A.R. Garay, R.A. Fernandez and G.M. Guce. (2008). Financial access and inclusion in the agricultural value chain. APRACA FinFlower Publication: 2008/1. Asia Pacific Rural and Agricultural Credit Association (APRACA), Thailand.

FAO (2010). The state of world fisheries and aquaculture; p. 3.

FAO (2007). The state of world aquaculture and fisheries 2006.

Govt. of India (2014). Handbook on fisheries statistics 2014. Published by Dept of Animal Husbandry, Dairying and Fisheries, Ministry of Agriculture, Government of India, New Delhi. Pp 166.

GTZ Value Links (2008). Value Links Manual — The Methodology of Value Chain Promotion. GTZ, Germany.

Gopakumar, K. (1997). Tropical Fishery Products, Oxford and IBH Publishing Co, Pvt. Ltd. New Delhi, India. 1-10p.

Jallow, A. M. (1995). Contribution of improved chokoroven to artisanal fish smoking in the Gambia. Workshop on Seeking Improvements in Fish Technology in West

Kaplinsky, R. (2000). Spreading the gains from globalisation: What can be learned from value chain analysis? *Journal of Development Studies*, 37(2): 117-146.

Karthikeyan, M., Bahni Dhar and Kakati, Bipul Kumar (2012). Quality evaluation of smoked fish products from the Markets of Manipurj Inland Fish. Soc. India, 44 (1) : 37-46.

Lilabati, H. and Vishwanath, W. (2001). Biochemical and microbiological changes during storage of smoked Puntius sophoreobtained from market. *J. Food Sci. Technol.*38 (3) : 281-282.

Porter, M. E., The Competitive Advantage: Creating and Sustaining Superior Performance. NY: Free Press, (1985). (Republished with a new introduction, 1998.).

Ranjit Kumara, Khurshid Alam; Vijesh V. Krishnaand K. Srinivas (2012). Value Chain Analysis of Maize Seed Delivery System in Public and Private Sectors in Bihar. Agricultural Economics Research Review, 25 (Conference Number 2012): 387-398.

Singh, M. B., Sarojnalini, C. and Vishwanath, W. (1990). Nutritive value of sun dried Esomus danricus and smoked Lipidocephalus guntea. *Food Chem.* 36: 86-89.

Upadhyay, A.D.; D.K. Pandey; Bahni Dhar (2017).Value chain Analysis of dry fish in North East Region of India, in edited book Financing Agri Value Chains in India-Challenges and Opportunities (eds G. Mani, P.K. Joshi, and M.V. Ashok), Springer Nature Singapore Pvt. Lt. Singapore. pp 143-162.

7

Strategic Approaches for Holistic Development of Ornamental Fisheries Sector in India

B. Nightingale Devi, M. Krishnan and Nilesh Pawar

Introduction

Aquaculture has been recognised as an important livelihood creating employment opportunities in the remote rural villages and urban areas. Further, it makes crucial contribution to global economic security and growth and generates income both directly and indirectly through support services. Over the past six decades, the prospects of ornamental fish have been emerging as a lucrative commercial aqua-venture. The global ornamental fish trade is the fastest growing sector and a multimillion dollar business which provides avenues for overseas earnings. The developing countries play a crucial role in production of ornamental fishes as more that 60% of the production comes from the households of the developing countries (FAO, 2006). The estimated worth of international ornamental fish trade at retail price is US$ 8 billion and the entire sector including aquarium tanks, accessories, medicines, feeds, plants, etc is worth of US$ 20 billion. India's share in global export market is negligible as it is mainly dominated by the wild caught species mainly from the northeastern states of India and Western Ghats and estimated at Rs. 5.5. crores in 2009-10 (MPEDA, 2011).

Besides exports market, India's overall domestic trade is estimated to be nearly Rs. 15 Crore with an annual domestic demand of Rs. 50 Crore (Kurup, 2012; Felix, 2012) and the economic activities that drive the ornamental fisheries sector is mainly concentrated in the states of West Bengal, Maharashtra, Tamil Nadu and Kerala. The sector is recognised as low investment steady return activity suitable for adoption by economically weaker sections of the population (Ambilikumar and Anna Mercy, 2012). The ornamental fish business has attracted farmers, policy makers, corporate sector, researchers, and private investors and is gaining attention all over the country. Indeed, the sector faces an increasing presence of global players in distribution and commercialization channels associated with increasing consumer demand for eco-friendly and certified quality fishes. Moreover the sector has been changing from family-owned small business or homestead production to large firms engaged in production and distribution chains and hence, the sector has been becoming more technological and managerial intensive.

Despite its high potential and significance in both the domestic and international markets, the sector remains unorganised and the lack of information in respect of this sector's progress is astounding in its absence which may be attributed to the scattered and unorganised nature of the sector and the lack of awareness among the stakeholders for an organised business approach. Therefore, it is essential to study the multidimensional performance and role of the stakeholders engaged in this sector. The present study is aimed to study and analyse the ornamental fisheries sector in our country and address the major issues which the sector has to face in the near future and build a strategic planning framework for ornamental fisheries enterprise by considering all the roles and linkages of the stakeholders in general.

Study Area

The study was conducted in the three states of India *Viz.*West Bengal, Maharashtra and Tamil Nadu which is predominant in ornamental fish production and trade in the country (Devi, 2014). Primary information was gathered using structured questionnaire and by conducting focus group discussions with producers, entrepreneurs, investors as well as traders. A range of information and statistics from secondary sources such as fisheries departments, ICAR fisheries research institutes, Cooperative societies, MPEDA and NGOs helped to understand weakness and the opportunities which were not available at the primary level. The data were also collected from journals, annual reports, newspapers and Internet. The research tools applied in the study are SWOT analysis and TOWS matrix as it helps in assessing the technological, financial, marketing and sociological environment which help in developing the strategic framework for the sector (Narula *et al.*, 2005).

Overview of Ornamental Fisheries

India is known as a hotspot for indigenous ornamental fish resources and its congenial climatic conditions is ideal for breeding of variety of exotic species. The country has been identified as a thrust region as it offers vast scope for starting commercial endeavours in breeding, culture and trade of ornamental fishes. A different study reveals the potential of this sector in generation of employment and earning income as well as a reliable alternative for diversification of fisheries in the country (Devi *et al.*, 2013). The export market is dominated by the wild caught species which constitutes more that 90 percent mainly from the north-eastern states of India and Western Ghats. More than 30 freshwater species dominate the market such as guppies, platies, swordtail, mollies, neon tetra, angel fishes, gold fish, zebra danio, discus and bards (Silas *et al.*, 2011).

The domestic market consists mainly of locally breed exotic species and is worth around ' 50 crore and demand is increasing at 20 % annually (Felix *et al.*, 2013). The domestic ornamental fish trade is unorganised and highly complex and hence, it is essential to study the role of different stakeholders involved in the sector to explore the key stakeholders and strategic dimensions in the sector. The different stakeholders involved in the domestic ornamental fish production and trade is given in Figure 7.1.

The government of India plays a central role in the whole system by acting as a catalyst for fostering research through various research institutes, dissemination of knowledge, providing funding and developing marketing regulations. It is significant to make sure that all the stages of value chain are included in a model that is developed based on the requirements of specific location wherein all provision have been made starting from inputs to markets information by creating backward and forward linkages. As the consumer tastes and preferences are very important for the success of any endeavour, the recent development of concept of green certification system adds value to the product and enhances its consumer appeal and again, such guidelines ensures the adoption of sustainable and environmental sound practices which ensures a value chain system linking all segments of the sector from collector/producer to the consumer. This approach can also help in launching a revolution of a certified branded, organised item of trade made attractive like in the case of high end niche markets for agricultural products such as Food World, Godrej and Mother Dairy (Narula *et al.*, 2005).

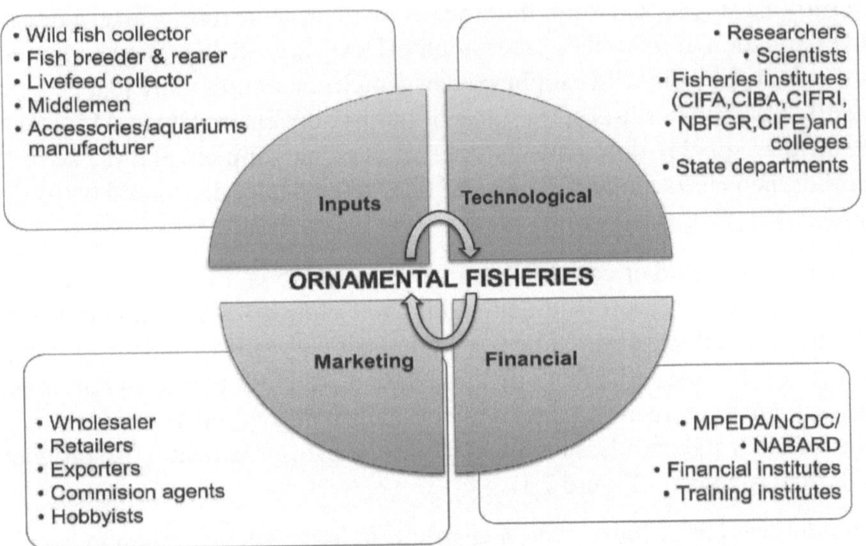

Fig. 7.1: Different stakeholders in ornamental fisheries sector in India

SWOT analysis

It is very important that the ornamental fisheries sector has to be developed considering the full gamut of these sectors strengths, weaknesses, opportunities and threats. It is very important that the concerned government institution as well as the private sector in both culture and trade work in tandem to ensure that it attains maximum efficiency in terms of output and optimal returns to investment that will ensure the sustainability and longevity of the sector. The results have been summarised in Table 7.1 which reveals that the ornamental fisheries is marked with a number of internal strengths and weakness and external opportunities and threats.

Strengths

India is known for its rich biodiversity in aquatic fauna that can be practically exploited for ornamental fisheries sector. Owing to availability of low cost labour in the country it is possible for the sector to keep the procurement cost at the minimum. These enable the trader to have substantial margins over cost price when it deals with wild collection. Though systematic knowledge base exists, the database needs regular updating since nature throw ups several new species. Number of research institution and organisations can take up ornamental fisheries seriously and both from the point of view of research as well as exploiting the opportunity of creating a new livelihoods at low cost. The scope for ornamental

Table 7.1: SWOT Analysis - Ornamental Fisheries in India

Strengths	Weakness	Opportunities	Threats
• Biodiversity hotspot • Increase in number of breeder or producers • Varieties of species with different agro climatic condition • Sellers market:	• Availability of resources • Various research institutes & organisations • Keen interest of the funding agencies • Dependent on natural collection • Underdeveloped captive breeding technologies • Poor awareness • Limited capability & infrastructure to produce quality fish on mass scale • Poor extension and training • Locational disadvantages • Weak association & coordination in value chain	• Favourable government vision • Scope for developing breeding technology • Initiatives of Green Certification • Private sector keen to join ornamental fish marketing chain. • Unlimited scope for expanding markets • Increased awareness among consumers on variety and quality	• Unorganized sector • Destructive exploitation • Threats from extinction of indigenous species • Unexpected natural calamities

fisheries has also been realised by the financial institutions. Agencies like NABARD and NFDB have come out with several schemes that support the sector. Institutionalization of ornamental fisheries under special programmes like establishment of aqua-estate or SEPZ (Special Export Promotion Zone) has ensured regular supply of popular species to the market throughout the year. This has also led to increase in number of breeders and rearers. Looking at its tremendous scope, the investors do not shy away in investing and have great expectation to cover up both the domestic as well as the international markets.

Weakness

The sector is besieged by people with inadequate awareness of the dangers of wild collection of ornamental fishes. Under these circumstances these people do not realise the importance of preservation of natural resources and do not impose on themselves any restriction to capture even the red listed species. Ornamental fish production requires specialised handling and keen involvement of technicians. The sector suffers from the lack of qualified and experienced technicians which leads to less than full realisation of the potential of the sectors. Post harvest mortality, loss of germplasm and inability to produce on a mass scale are major issues. Again, the inability of the sector to meet the specific demands of the export market is only because of the way this sector has developed in an unorganised and non-institutionalised manner. The state Department of Fisheries of the State Governments who are responsible for training and extension are themselves severely constrained with the lack of staff and technical personnel. Locational disadvantages, lack of investment power, poor infrastructural facilities, lack of an organisational structure due to weak association and coordination between stakeholders and poor quality consciousness are some of the other weaknesses of this sector that needs to be surmounted.

Opportunities

The Government of India has now realised the importance of ornamental fisheries and hence, specific capacity building programmes, technology development and extension and refinement are being taken up seriously by various government institutions engaged in research as well as financing the sector. A very strict protocol of green certification for freshwater fishes has been implemented for ornamental fisheries sector by the Government of India initiated by Marine Product Export Development Authority (MPEDA). Important thrusts are also made for diversification of species to meet the demand of ornamental fishes in the world market. Changing consumer taste and preferences are being taken care of by ensuring quality fish seed production. Promotional institution such as

MPEDA has taken the initiatives to introduce the concept of producer companies in areas of production concentration. Such initiatives have enabled the producers to meet the demand in both the domestic and export markets. Breeding of new varieties are based on the recommendation of exporters who specifically identify the species that have a growing demand in the international market. Different governmental schemes also has been come out to part finance or fully subsidised investment on ornamental fish breeding and culture through institutions like MPEDA, NFDB and NABARD.

Threats

The ornamental fisheries sector is greatly threatened. There is massive exploitation of natural resources which happen because of the unorganised nature of the sector. Threats of overexploitation have only increased over time because of competition from other countries but, unfortunately such threats could lead to practical extinction of the indigenous species. Facilitating functions and also lack of infrastructure development resulting in high electricity charges, irregular supply of good potable water and unexpected natural calamities has increased the threats to the sub sector.

TOWS matrix

The TOWS matrix is an important decision making tool that helps in developing four types of strategies by matching internal strengths and weaknesses with the environmental opportunities and threats : SO (Strength Opportunities), ST (Strength-Threats), OT (Opportunities- Threats). These four sets of strategies have been worked out for ornamental fisheries and key strategic dimensions have been identified.

The four set of strategies have been worked out in Table 7.2 for ornamental fish production and trade.

***Strength-Opportunities Strategies*:** Strength is internal while opportunities are external Therefore, it is important to match the best of strengths with the best of opportunities and deliver a set of strategies that will enable the growth of the ornamental fisheries sector. The SO strategies to develop the ornamental fisheries include strategies to tackle the domestic market, off season production and supply of ornamental fish, eco-friendly production by following the green certification guidelines to utilize the resources in sustainable way. Research in production, marketing and post harvest management can enhance the quality to meet the demands for quality produce. Expansion of ornamental fish production locations with proper institutional arrangement by adapting a suitable business model based on the location, site specific characteristics and nature of investment will ensure reliable returns to investment.

Table 7.2: The Tows Matrix – Ornamental Fisheries in India

	STRENGTHS	WEAKNESSES
	• Biodiversity hotspot • Availability of resources • Various research institutes & organisations • Keen interest of the funding agencies • Increase in number of breeder or producers • Varieties of species with different agro climatic condition • Expectation and willingness of the producer to achieve and cover up markets	• Natural dependency • Few captive breeding technologies • Poor awareness • Limited capability & infrastructure to produce quality fish on mass scale • Poor extension and training • Locational disadvantages • Weak association & coordination between sellers
OPPORTUNITIES • Favourable government vision • Scope for developing breeding technologies for specific species • Initiatives of Green Certification • Private sector keen to join ornamental fish marketing chain. • Demand in both the markets • Increased awareness among consumers on variety and quality	SO STRATEGIES • Off season supply to meet the demand • An umbrella brand like green certification to fetch better price and to create niche market • Resources and knowledge base be utilized for research in production, marketing and post harvest management • Expansion of production locations with proper institutional arrangement	WO STRATEGIES • Strengthening the existing linkages with different stakeholders. • Building market information and intelligence system • Governmental support to strengthen the infrastructure • Encouragement of private investment to reduce the locational disadvantages.
THREATS • Unorganized sector • Destructive exploitation • Threats from extinction of indigenous species • Unexpected natural calamities	ST STRATEGIES • Standardised breeding technology for certain important species. • Quality and consumer oriented production with proper intervention of the institutions • Environmental friendly and sustainably production of ornamental fish • Compliance with standard record keepings	WT STRATEGIES • Proper institutional arrangement for ornamental fish marketing. • Used of genetically & biotechnological tools for conservation. • Enforcement of specific regulations and legislation.

Note: SO- Strength Opportunity WO- Weakness Opportunity ST-Strength threat WT-Weakness threat

Weaknesses- Opportunities Strategies: It is necessary to weigh the best of opportunities against the worst weaknesses that impedes the growth of the ornamental fisheries sector. The opportunity of improving technology in breeding and rearing of select ornamental fishes must counter act the key weakness in the system that is the post harvest mortality as well as absence of low cost technology that will enable mass production of ornamental fishes. Use of knowledge base of various research institutes and private sector for extension support and linkages with the small farmers is required. Other strategies to iron out weakness in this sector includes building market information and intelligence systems through provision of specific market intelligence, strengthening basic and specialised infrastructure with the help of central government schemes and support and creation of hub for input supply for resource extension support, Marketing cooperatives, SHGs and Producer companies like in case of various agricultural crops should be promoted to reduce producer risks and enhance profits. Effective enforcement of green certification guidelines can help in meeting the quality demand both for domestic and export market. The locational disadvantage that hinders the marketing segment can be sort out with the help of private investors by arranging contract farming which will help to reach the produce at final destination within the shortest marketing channel.

Strength- Threats Strategies: In this scenario the best of strengths which is an internal environment will be used to surmount the primary threats. Introduction of policy led growth which has led to creation of infrastructure for ornamental fisheries would help overcome the add on cost like high electricity charges and non-availability of potable water. Development and dissemination of standardised breeding protocols through various research institutes will help to meet the consumer preferences. Proper maintenance of records and documents is important and maintenance of sector at par with the agricultural sector is required.

Weaknesses- Threats Strategies: Weakness is an internal environment characteristic while threats are external. It is necessary to reduce the internal weakness in order to avoid the external threats. An internal weakness such as having no effective regulation is a threat in the face of possible rejection of consignment in the export market as well as competition with other countries. Green certification guidelines by MPEDA would help overcome these issues by helping the producer to follow a strict protocol which is internationally acceptable and that which will enhance the production of quality ornamental fishes. This would completely eliminate the external threats of rejection of consignments and can provide a means for maintaining world class quality throughout supply chain as well as proper institutional arrangement for ornamental fish marketing.

Consequently the key strategic dimensions have been identified in figure 7.2 showing the role and linkages of different stakeholders. The key strategic dimensions identified have to be addressed by the researchers, policy makers, scientists working in field of ornamental fisheries in the country. All the key stakeholders have to play an important role in the whole system along with the government occupying a central role and joining hands with other institutes and private sector to develop public private model like producer companies and contract farming like in case of agricultural and other allied activities where one has to work individually as well in collaboration with different stakeholders to make the system efficient, quality conscious and cost competitive.

Figure 7.2 gives the strategic framework for the ornamental fisheries development in India. The ornamental fish production and marketing system in India is a function of local demand as well as in house technology. The ornamental fisheries sector is ideal for promotion as a livelihood enterprise for enhancement of employment and income opportunities especially in regions of availability of essential resources for ornamental fish production like potable water and highly bio-diverse regions. The opportunities for scaling up this enterprise are directly related to the strength of the organisational structure and the institutional arrangement that enables this up scaling. It can be seen from the figure that research, innovative production and marketing arrangement with the right doses of government regulations will enable this fisheries subsector to attain formidable heights.

Organisational structure like individual enterprises, cooperatives, self-help groups and the recently evolved concept of Producer Company are different types of aqua-business models. The concept of Producer Company will enable concerted development of the various individual units of ornamental fisheries to develop in tandem because of its inherent advantages.

Based on above two results, a strategic framework for holistic development of ornamental fisheries sector in India has been developed showing the role and linkages of different stakeholders (Figure 7.2).

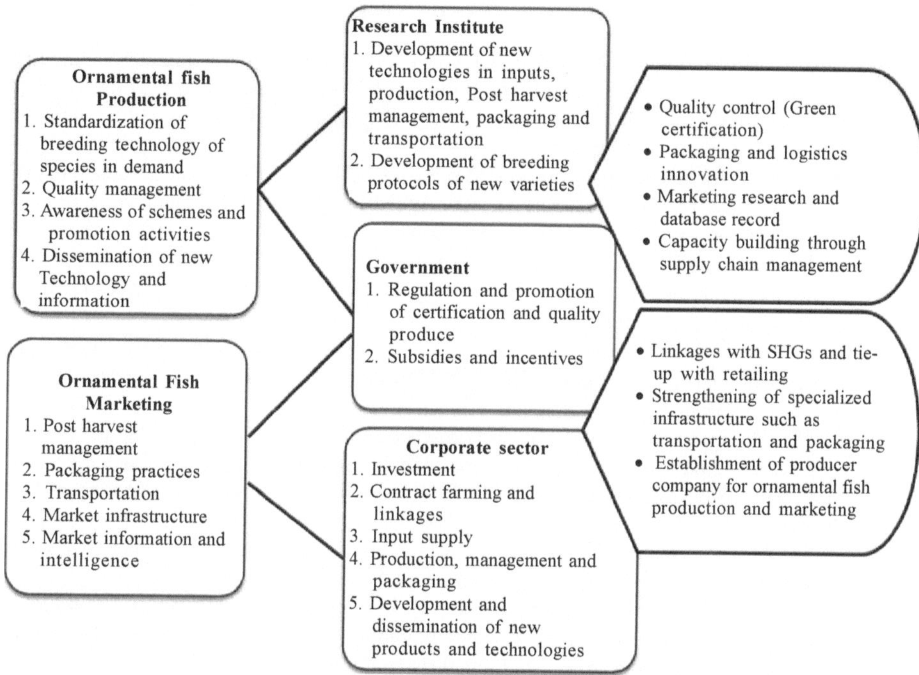

Fig. 7.2: Strategic Framework for Holistic Development Ornamental Fisheries

Conclusion

The key strategic dimensions identified have to be addressed by researchers, scientists and other policy makers for development of the ornamental fisheries sector in India. Numbers of stakeholders are involved in the sector both directly and indirectly in providing support services. All the key stakeholders have to play an important role in the whole system centered on government schemes and policy and joining hands with the other research institutes which can endow the producers and private or corporate sector with new technologies to develop public private partnership model where each one has to work individually as well as in collaboration of the other stakeholders either by arranging contract farming system or by establishing backward and forward linkages to make the system efficient, quality conscious and cost competitive. Strengthening of infrastructure, knowledge-base, technological up gradation and linkages between markets that is environmentally sound should be the priority areas where the efforts must be given by all the concerned agencies in order to make a prominent place in global market.

References

Ambilikumar, V. and Mercy A. T. V. (2012). Rural employment through ornamental fish culture – A developmental Model. In: Souvenir- Ornamentals Kerala 2012. International seminar on ornamental fish breeding, farming and trade. Department of Fisheries, Government of Kerala, India. pp: 36-38.

Devi, B.N, Krishnan, M., Venugopalan, R. and Mahapatra, B.K. (2013). Artificial Neural network for synergy analysis of input market in ornamental fish trade in Mumbai. *Agricultural Economics Research Review*. 26(01). pp: 83-90.

Devi B.N. (2014). Organisational structure, market dynamics and livelihood in ornamental fish production and trade. Unpublished Ph.D. thesis, Division of Fisheries Economic extension and statistics, CIFE, Mumbai.

FAO, (2006). Fishstat Plus: Universal software for fishery statistical time series. Fisheries Department, Fishery Information, Data and Statistics Unit. Rome (www.fao.org/fi/statist/FISOFT/FISHPLUS.asp).

Felix, S. (2012). Mass production technology for ornamental fishes using advanced culture systems. In Souvenir- Ornamentals Kerala 2012. International seminar on ornamental fish breeding, farming and trade. Department of Fisheries, Government of Kerala, India. pp: 17.

Felix, S., Anna Mercy, T.V. and Swain, S.K. (2013). Ornamental Aquaculture Technology and Trade in India. In: Daya Publishing House New Delhi -110002. pp: 3-38.

Kurup, B. M., Harikrishnan, M. and Renjithkumar, C. R. (2012). Breeding farming and trade of ornamental fishes in India- prospects and challenges. In: Souvenir- Ornamentals Kerala 2012. International seminar on ornamental fish breeding, farming and trade. Department of Fisheries, Government of Kerala, India.

MPEDA (2011). Prospect of ornamental fisheries in West Bengal. MPEDA Newsletters, Kochi. pp: 1-4.

Narula, S.A. Sikka, B.K. and Singh, S., (2005). Strategic Framework for Horti-Business in Uttaranchal: the tows matrix approach. *Indian Journal of Agricultural Marketing* (Conf. spl.) 19(2). pp: 222-225.

Silas, E.G., Gopalakrishnan, A., Ramachandran, A., Anna Mercy, T.V., Kripan Sarkar, Pushpangadan, K.R., Anil Kumar, P., Ram Mohan, M.K. & Anikuttan, K.K. (2011). Guidelines for Green Certification of Freshwater Ornamental Fish. The Marine Products Export Development Authority, Kochi, India. pp: 106.

8

Assessment of Socio-Economic Improvement of Women Engaged in Ornamental Fish Trade at Hatibagan and Dasnagar, West Bengal, India

Ambalika Ghosh, B.K. Mahapatra and A.K. Roy

Introduction

Ornamental fish is often used as a generic term to describe aquatic animals kept in the aquarium hobby, including fishes, invertebrates such as corals, crustaceans (e.g., crabs, hermit crabs, shrimps), molluscs (e.g., snails, clams, scallops), and also live rock. Live rock is a general term for any type of rock encrusted with, and containing within its orifices, a wide variety of marine organisms including algae and colorful sessile invertebrates. Live rock serves as the principal biological and chemical filter in many marine-type tanks, and the encrusted organisms usually provide much of the background coloration in the tank. Ornamental fish culture is fast emerging as a major branch of aquaculture globally. Aquarium keeping is the second largest hobby in the world next to photography and the ornamental fish and aquatic plant industry is fast gaining importance due to its tremendous economic opportunities and prospects. Ornamental fish keeping is a popular hobby in developed countries and is gaining popularity in many developing countries. The growing interest in aquarium fishes has resulted in steady increase in aquarium fish trade, globally. According to Food and Agriculture Organization (FAO) (2006), export earnings from ornamental fish trade is US $ 251 million and more than 60% of the production comes from households of developing countries. The wholesale value of the global ornamental fish trade is estimated to be US $ 14 Billion. More than 2500 species of ornamental fishes are traded and some of 30-35 species of fresh water dominate the market. The trade with an annual growth rate of 8 percent offers a lot of scope for development. India's share in global ornamental fish trade is negligible and at present the ornamental fish export from India is dominated by the wild caught species. The top exporting country is Singapore, followed by Hong Kong, Malaysia, Thailand, Philippines, Sri Lanka, Taiwan, Indonesia and India. The largest importer of ornamental fish is USA. Europe and Japan, China and South Africa are the emerging markets of ornamental fish trade. The aesthetic, attractive colour and appearance of the fish appeal to children and aged alike. As the day passed, ornamental fish keeping become an interesting activity for many, in the process, generating income for the unemployed youth and farmers. The concept of entrepreneurship development, through ornamental fish farming is gaining popularity day-by-day. Many numbers of people are entering into this business of culturing, breeding and farming. In India, out of total export of ornamental fish, 95% is based on wild collection. Majority of the indigenous ornamental fish trade is from the North Eastern states and Southern states in India. Fish species grown for their importance, can be successfully bred in India. The ornamental fish trade depends upon supply and demand, which is possible only by mass breeding. Ornamental fish market linkage exists between buyers and sellers as well as domestic and international market. In order to strengthen Indian position in international ornamental fish trade, MPEDA has implemented several strategies to adopt in terms of technology, infrastructure in order of to develop export demand based production for major importers in EU, USA and Japan. The distribution and marketing channel for the ornamental fish and plants in India is developing. The exporter (or) wholesaler plays on important role in promoting breeders and consumers. Wholesaler usually sells the fishes to local retailers and in turn, retailers directly sales to local customers, hobbyists etc. India has a rich resource of rivers, lakes, reservoirs and other water bodies suitable for rearing ornamental fishes. Indian waters possess a rich diversity of ornamental fishes, with over 300 varieties of indigenous species. However, the popularity and trade of ornamental fish in India is dominated by the exotic species that are bred in captivity. The importance of international markets as a source of live, ornamental fish supply is growing due to market demand. The wild-caught and cultured ornamental species industry represents an important natural resource-based industry in some pockets of West Bengal. Assessing the determinants of import, supply and export demand of individual species price, would help growers of ornamental species better determine which new species or subspecies to introduce into the market. Providing such an analysis on a species and country basis would be invaluable. Therefore, a thorough literature scanning from different journals and Aquatic Science & Fisheries Abstract (ASFA) database reveals the status of present national and international scenario of ornamental fish culture and marketing is presented in the following section.

Global Scenario: Reports are available on the trade of aquatic organisms for home and public aquariums and water gardens, along with associated equipment and accessories, has become a multi-billion dollar industry known as the Aquatic-Ornamental Industry (AOI) (Larkin, 2003; Wabnitz et al., 2003; Pelicice and Agostinho, 2005; Prang, 2007). The recent estimated value of the AOI was in the vicinity of 15 billion US dollars (Prang, 2007; Wittington and Chong, 2007). Within the AOI, among a myriad of aquatic organisms traded, freshwater and marine fish species are the most popular and dominant groups (Lecchini et al., 2006; Moreau and Coomes, 2006). And on a unit weight basis, ornamental fish form the most valuable fisheries commodity in the world (Hardy, 2003).

The Indian Scenario: Researchers report that the Western Ghats and the North-Eastern Region of India are considered to be two of the 34 'hotspot' areas of the world for biodiversity conservation with a variety of vegetation types, climatic zones and remarkable endemism. These regions are also endowed with a variety of brilliantly coloured ornamental fishes. With an abundance of these living jewels, India has a distinctive edge over other countries in trade potential. However, the Indian export of ornamental fish during 2009-10 was only US$ 1.17 million (Rs.55 million) (source MPEDA). Collection from the wild, handling, transport, holding, breeding and culture facilities, conditioning for export, infrastructure and maintenance of records to conform to a value chain system for delivering healthy ornamental fishes to the trade and the hobbyist have been taken into consideration and reported (Sailas et al.2011).

West Bengal Scenario: Reports are available on the total population is involved in ornamental fisheries and related activity in West Bengal, foreign exchange earnings from export of ornamental fishes; Ornamental Fish trade; the ornamental fish industry; Concentration of ornamental fish breeders; in five districts viz. Uttar Dinajpur, Howrah, South 24 Pargans, Nodia and Murshidabad; the retailers in Baranagar and Hatibagan regionof Kolkatal Co-opratives dealng with ornamental fish in Howrah and Nadia; value additions by stakeholders in the form of garding quarantine and packing; species wise sorting by size, grading, proper packing and quarantine; retailers attractive display, advertising and attractive packing (Annual Report 2009-10; Ayyappan et al., 2006; Basu et al; DAHDF 1999; Dutta et al., 2015; Goswami et al., 2012; MPEDA, 2012; Handbook on Fisheries Statistics, 2014; Bhaskar et al., 2000; Jonathan et al., 2010; Kumar 2010; Kumar et al., 2004; Madhu et al., 2009; Kukherjee et al., 2002; Kukherjee M. et al., 2000; NABARD; Panigrahi et al., 2009; Ponniah and Sood, 2002; Ponniah et al., 2002; Pramod et al., 2009; Sarkar et al., 2002; Singh and Prusty, 2008; Nagesh, et al., 2007; Usha et al., 2013; and Welcome, 1988).

The present study aims to assess the involvement of women in various activities of culture, breeding, rearing, and marketing of ornamental fish as these were not earlier covered by the researchers mentioned above. Another emerging business of aquarium accessories manufacture and marketing are also covered that were not reported earlier in details. Earlier research work has focused on breeding, rearing and maintenance of aquarium business service where successful ornamental fish aquarium business enterprises employ women to support their business activity. Nowadays aquarium shop owners are initiating with new and innovative ideas to manufacture aquarium accessories aiming for import substitution from country and abroad to improve the sales volume on aquarium business service with higher returns. It is aimed at identifying manufacture and sales of various aquarium related products to satisfy the aquarium users based on consumer preference particularly involvement of women in this trade in the selected pockets of Kolkata and Howrah, West Bengal.

Materials and Methods

Methodological aspect of the study is concerned it relates to the methods adopted in selection of the study area, drawing the sample respondents, schedule-cum-questionnaire used in collecting the desired information both at macro and micro levels to arrive at analytical conclusion. The methodological details containing the research design, methods and tools and the analytical procedures, etc. have been outlined in sub-sections as below.

Data collection is based on two unique methodology prescribed for the study. The primary data have been collected through survey among women retailers at Hatibagan market, Kolkata dealing with aquarium and accessories associated with ornamental fish keeping and trade. The other group of women engaged as wage earners in various manufacturing small scale industries located in a cluster at Daspara, Howrah, and West Bengal. Third group consists of women working as support in live ornamental fish trade. In order to collect data, field survey was conducted administering pre tested questionnaire during a period from January 2014 to December 2015.

Sampling and Sample size: The purposive sampling methods have been adopted to select sample respondents. The sample size is 50 individual women respondents dealing with ornamental fish and aquarium keeping accessories items from two clusters of trade activities like Hatibagan market, Kolkata and Dasnagar, Howrah, West Bengal.

Data analysis, interpretation and presentation: The chapter is based on primary data collected from women engaged in various activities related to ornamental fish trade, marketing, manufacturing of accessories at Hatibagan market, Kolkata and Dasnagar located at Howrah, West Bengal on various socio economic aspects. The raw data is compiled, analyzed, interpreted and presented in the form of tables, graphs, and charts for easy visualization and comprehension. Various statistical methods like, descriptive statistics, correlation, regression, test statistics, trend line fitting-test, F-test etc were performed using SPSS (16.0) and MS Excel, Trend lines with time series data on resources, production, import etc are also fitted for model building and forecasting.

Results and Discussions

Before proceeding to the main analytical report, it was thought imperative to furnish survey report on various activities that women are involved through visuals to have an idea about how women are actively participating in various trades, marketing, manufacturing units of accessories required for ornamental fish keeping. Colourful and fancy fishes known as ornamental fishes, aquarium fishes, or live jewels is one of the oldest and most popular hobbies in the world. The growing interest in aquarium fishes has resulted in steady increase in aquarium fish trade globally. The ornamental fish trade with a huge turnover offers lot of scope for development. There is very good domestic market too, which is mainly based on domestically bred exotic species. The earning potential of this industry and the relatively simple techniques involved in the growing of these fish has helped the aquarium industry to evolve and provide rural employment and subsequent economic upliftment in many countries including India. The potential for the development of ornamental fish trade in India is immense, though it is still in a nascent stage. The Government of India has identified this sector as one of the thrust areas for development to augment exports. The turn of the century has seen a spurt in the collection, culture and trade in freshwater ornamental fishes. This type of study is first of its kind covering details of activities where women can get gainful employment for their livelihood in rural areas of West Bengal. In the following sections we will present how women are engaged in various activities followed by socio-economic conditions of women with respect to various parameters.

Survey of Hatibagan Market Exclusively Dealing with Ornamental Fish Keeping Accessories

A large number of very beautiful ornamental fish species are still neglected which are easily available in the natural waters of coastal Bay of Bengal. Due to congenial climatic condition, Kolkata and its surrounding districts have emerged as promising breeding centers for ornamental fish where a considerable number of small fish farmers and amateurs are engaged in this trade (Dutta *et al.*, 2013).

Fig. 8.1: Hatibagan Market Exclusively Dealing with Live Ornamental Fish and Related Accessories

The domestic market for ornamental fishes in India is estimated at Rs 20 crores and the domestic trade is at growing annual rate of 20%. Availability of considerable number of indigenous ornamental fish of high value in the country has contributed greatly for the development of ornamental fish industry in India. However, there is a great demand for exotic fishes due to its variety of color, shape, appearance, etc.

Women Engaged at Work at Hatibagan Market and Dasnagar Works

It is wide believed that rural women have a major role in sustainable fisheries and aquaculture, but they often have unequal access to resources and services that they need to be successful and often suffer greater poverty and hunger than men as a consequence. Therefore, it is aimed at gathering evidence and information on how poor women access and use resources related to ornamental fish and accessory business and how they are benefitting from the activities. The social norms that constrain poor women's and men's opportunities need to be better understood minimizing the gender gap.

Fig. 8.2: Women Engaged at Work at Hatibagan Market and Dasnagar Works

Office Maintained by Women Entrepreneurs at Dasnagar, Howrah

A lucrative export market and high domestic demand has made ornamental fish industry in West Bengal a potential source for income generation. This is an example how women are coming forward in various ornamental fish and accessory business to have independent livelihood option. It has been recognized ornamental fish keeping hobby is growing at a faster pace within the country and abroad resulting in the growth accessory trade and business as it stimulates growth of a number of subsidiary industries facilitating scope for employment particularly women .The domestic market for ornamental fishes in India is estimated at Rs 20 crores and the domestic trade is at growing annual rate of 20%. Availability of considerable number of indigenous ornamental fish of high value in the country has contributed greatly for the development of ornamental fish industry in India. However, there is a great demand for exotic fishes due to its variety of color, shape, appearance, etc. It has been estimated that more than 300 species of exotic variety are already present in the ornamental fish trade in India and still there is great market demand for exotic fishes in international market

Fig. 8.3: Office of Women Entrepreneurs at Dasnagar, Howrah

Live Feed Sale for Aquarium Fishes at Hatibagan Market

Feeding ornamental fishes in captivity is a very delicate and important issue. Live food is the most preferred feed. At Hatibagan market many people are engaged in collection and selling of preferred live feed. The trade in live aquatic ornamental animals for the aquarium trade is a global multi-million dollar industry, which can provide economic incentives for habitat conservation. However, little is known about the scale of the international trade in many species, and there are concerns that trade in some species might not be sustainable, given factors such as their biology, distribution, conservation status and ability to survive in captivity. It is widely believed that the trade in wild-caught freshwater and marine aquarium species, if managed sustainably, can present a valuable opportunity for income generation and support to livelihoods, while at the same time providing an alternative to environmentally destructive activities (Junk, 1984; Chao and Prang, 1997; Ng and Tan, 1997; Brummet, 2005; Calado, 2006; Moreau and Coomes, 2007).

Fig. 8.4: Live Feed Sale for Aquarium Fishes at Hatibagan Market, Kolkata

There are mainly two categories of sources available to feed the ornamental fish species at different stages of their life.

*1. **Natural feed***

*2. **Artificial/Supplementary feed***

Preferred live feed is tubifex a natural feed besides phytoplanktons and zooplanktons which are abundantly available in natural ecosystem of the fish and on which fishes thrive and sustain themselves. However, in the present context tubifex are cultured or collected from natural water bodies and sold to the market. Artificial feed is on the other hand composed of calculated nutrients as per the requirement of the fish or its spawn in proportionate amounts using various feed ingredients of both plant and animal origin. However, for feeding aquarium fishes a variety of packaged feed are noticed to be sold at Hatibagan market.

Rearing Facilities of Aquariums Fishes for Sale at Dasnagar, Howrah

Most of the ornamental fishes available are wild catching and some economically important species are bread and reared indoor at a number of places at Dasnagar, Howrah for supply to different places. It is reported that the facilities for the production of aquarium fish are often small compared to major food-fish production operations (Chapman 2000). Fishes are typically raised in small vats and outdoor-ponds, usually in conjunction with indoor facilities that house many small tanks and aquaria. There is a growing interest in the cultivation of ornamental fish in indoor facilities using water-recirculation systems.

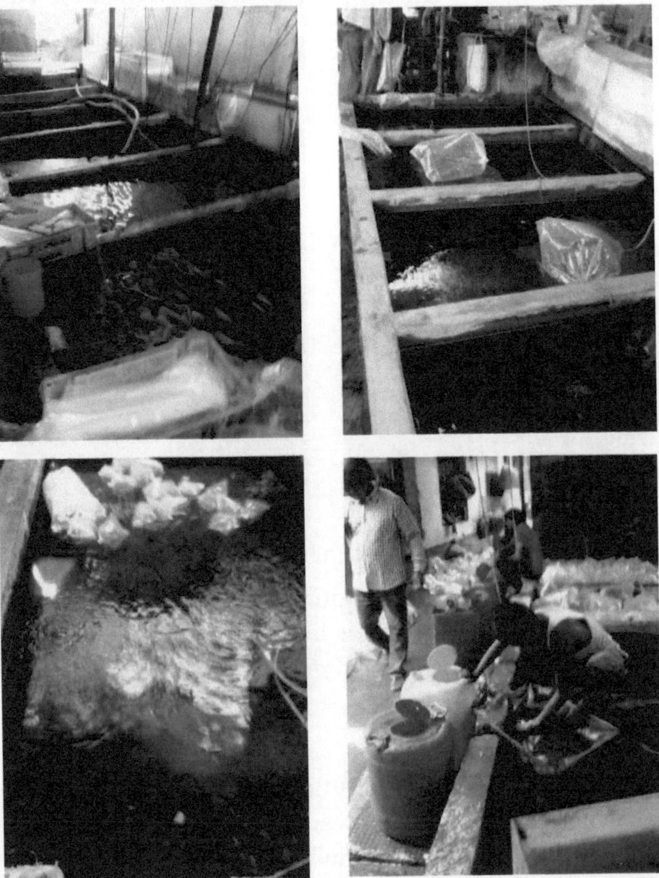

Fig. 8.5: Rearing Facilities of Ornamental Fishes at Dasnagar, Howrah, W.B.

Developments in husbandry methods and water filtration technologies: Farming ornamental fish in indoor facilities using water-recirculation systems can further minimize any potential impact to the environment by minimizing the escape of farmed fish and reducing water use. The small-scale nature of an ornamental fish farming system, both outdoors or indoors, minimizes any adverse impact to the environment, and seeks to optimize the use of land, labor, capital, and operational costs. In the wild, most freshwater aquarium more recently the culture of ornamental fish has shifted to regions near consumer centers. For example for the European market, many aquarium fish are now cultivated in countries such as the Czech Republic, Spain, Israel, Belgium, and Holland. Producing fish close to consumer centers is becoming more profitable because transport costs are greatly reduced.

Similar practices of Development of Captive Breeding Techniques for Marine Ornamental Fish are reported by Jonathan A. Moorhead and Chaoshu Zeng, 2010.Ornamental Fish Farming—A Lucrative Business for Rural Womenfolk of Some Villages of Howrah, West Bengal is reported Ghosh and Debnath. Ornamental fish farming- successful small scale aqua business in India is opined by Ghosh et al., 2003.

At Dasnagar, Howrah all sorts of infrastructure from rearing, oxygen packaging for export to different places is noticed. A number of enterprises have come up in this area who have standardize the ornamental fish export along with feed and accessories manufactures locally claimed to have made import substitution.

Packaging of Live Ornamental Fish for Export at Dasnagar

This is a very important activity observed in an indoor facility of an entrepreneur. The live fishes are packed in double plastic bags, filled with water to one third of their capacity are be tightly sealed and packed inside a cardboard box lined with Styrofoam. Each bag contains only one species. Care is taken that the bags are sufficiently transparent to enable proper inspection and identification of the fishes and free from any extraneous matter, unapproved plant material, pests or unauthorized species. It is reported that Ornamental Fish Farming is a Lucrative Business for Rural Womenfolk of Some Villages of Howrah, West Bengal (Ghosh and Debnath, 1998).The global trade of ornamental fishes including accessories and fish feed is estimated to be worth more than USD 15 billion with an annual growth of 8%. Around 500 million fishes are traded annually by 145 countries, of which 80-85% is tropical species. Domestic market for ornamental fish in India is much promising. At present, the demand for quality tropical fish far exceeds the supply (Handbook on Fisheries Statistics 2014). Singapore serves as the largest worldwide center and clearinghouse for the import, and export of both fresh and marine ornamental fish. Los Angeles,

Miami, and Tampa are the major centers for the ornamental fish trade in the US. From these ports of entry, large importers, wholesalers, or trans-shippers may distribute fish to smaller franchises that sell to retailers. That deal directly with home aquarists. In the exporting, importing, and larger warehouses, fishes are acclimated and water changes are provided for the shipping containers. To reduce transport costs, the trans-shipper primarily consolidates orders, receives, and redistributes the consignments generally without acclimatization of the fish. However, depending upon length of time between shipments and the number of fish sent to customers, some trans-shippers will also redistribute fish, re-oxygenate bags and do water changes for the transporting container.Welcomme, 1988, reported on "International Introductions of Inland Ornamental Fishes of Coastal West Bengal, India—Prospects of Conservation and Involvement of Local Fishermen Aquatic Species. Madhu et al., 2009 outlined present scenario of marine ornamental fish trade in India, Captive breeding, culture, and trade and management strategies. Mandal et al., 2007 studied Agribusiness Opportunities of Ornamental Fisheries in North-Eastern Region of India. Mukherjee, et al. 2007 observed that Ornamental fish farming – A new hope in microenterprise development in peri-urban area of Kolkata.

Fig. 8.6: Packaging of Live Ornamental Fish for Export from Dasnagar, Howrah, W.B.

Aquariums at Display for sale at Hatibagan Market

Being an aquarist and fish hobbyist is enjoyable, relaxing, and an educational and rewarding experience. Aquarium keeping is amongst the most popular of hobbies with millions of enthusiasts worldwide. The vast majority of ornamental fishes in the aquarium trade is of freshwater origin and farm-raised.

Fig. 8.7: Aquariums at Display for Sale at Hatibagan Market

Most people choose fish, corals, and other invertebrates from tropical warm waters because they tend to be more colorful an aquarium enthusiast can easily become overwhelmed by the endless variety of fish, invertebrates including corals, anemones, mollusks, plants, and live rock available, and ultimately forget to consider their source and method of collection. Although many species in the hobby have been domesticated and are produced on farms, it is important to remember that many species are also collected from the wild and are not in limitless supply. To help promote resource sustainability, care is necessary.

Display Accessories in Hatibagan Market

Most of the 'live rock', corals, and invertebrates are also collected from the wild. The use of 'live rock' has increased drastically due to the rise in popularity of reef tanks. Collecting or mining of 'live rock' and many types of coral is conducted by snorkeling or wading, often using a hammer and chisel to remove pieces from the reef. Problems with wild-collection methods include generation of rubble and habitat damage. Fiji and Indonesia are currently the world's largest suppliers of 'live rock' and coral. The United States is the major importer of 'live rock' and corals (Cato and Brown, 2003). Culture of live rock and corals has increased with culture operations currently in the Indo-Pacific (Fiji, the Solomon Islands, Indonesia, and the Philippines) and the United States (Florida and Hawaii). Propagation of 'live rock' and many types of corals is conducted through a variety of methods. The most popular method utilizes pieces of limestone or other rock (or substrate), simply placed in selected areas on the sea floor, and allowed to be 'seeded' by a myriad of naturally occurring organisms, forming the 'live rock'.

Fig. 8.8: Display Accessories at Hatibagan Market, Kolkata

The 'live rock' can then be harvested after the organisms have achieved marketable size characteristics. More complex and labor intensive methods involve attaching coral fragments to selected pieces of rock, which are placed back into the wild or cultured in land-based operations. The clippings from the new growth are used in the reef tanks. The original parent stock may be left at sea to continue to grow and may also be used to seed previously damaged natural reefs.

High numbers of herbivorous animals, such as hermit crabs, snails, and shrimps, are harvested annually for the aquarium trade. Observations and management of a mini reef tank in our laboratory highlights the essential role of these invertebrates in minimizing growth of algae and other organisms. Without grazers, algae may quickly dominate a reef. Since animals occupying different tropic levels have valuable roles in the wild, overfishing of key species may have important ecological consequences. Fortunately, there is an increased interest and technology emerging to culture these valuable and highly sought invertebrates.

Findings of Women Engaged in Various Activities at Hatibagan Market, Kolkata and Dasnagar, Howrah

***West Bengal Scenario*:** Since the nearest international airport to the collection sites is located at Kolkata, the trades of this species have thrived in West Bengal. Few traders and exporters of Kolkata regulate the market of this industry. At present, there is no virtually technological, infrastructural and institutional support from the state for promoting this activity; as a result the industry is far from being organized. Despite these bottlenecks, it is encouraging to note that enlarged the number of entrepreneurs in West Bengal have started breeding and rearing this ornamental species. These are more than 500 units, which have come up in the districts of Howrah, Hooghly and 24 Parganas (North and South) Mukharjee et al.2007.

The fisheries dept. of govt. of West Bengal has already taken some steps to encourage the upliftment of the marginal farmers both breeders and collectors in technical as well as financial matters. These are: i) set up at least 400 women operated co-operative society regarding ornamental fish farming for the proper upliftment of the women of these areas. ii) Decided to establish two big aquarium fish market one at South 24-parganas and another at Howrah and in future they have plans to establish a large collection rearing center at North Bengal to proper rearing of the wild catch. The Marine Product Export Development Authority (MPEDA) and BFDA, Govt. of W.B. started some financial schemes for marginal farmers for construction of ornamental fish farms and maintenance. In a recent study Goswami and Zade, 2015 made in depth Analysis of International Trade and Economical & Commercial Scope of Ornamental Fishes.

Species diversity of ornamental Fishes at Hatibagan Market

Hatibagan market is well known for assemblance of variety of ornamental fish species that are brought from various remote locations of West Bengal, North East States and Orissa for marketing. Brilliant colours and unique features attract domestic purchasers as well as exporters. There is a mix of fresh water, marine and the brackish water attractive species which are dearer to the hobbyists. India exports over three hundred varieties of fresh water fishes today. The fresh water catfish, angel, molly, arowana, gold fish, tetras and gouramis showed comparatively higher breeders' share in consumers' rupee. Wholesalers were earning comparatively higher annual profit than the other stakeholders due to moderate initial investment and also due to the comparatively lower risk involved. The tropical ornamental fishes from north eastern and southern provinces of India are in great demand in the hobbyist's market. Export of marine ornamental fish is yet to take off from India. Prominent among the fresh water Indian Ornamentals them are Loaches, Eels, barbs, catfish, Goby. India also exports tank raised varieties of fishes such as gold fish, Mollies, Guppies, Platties, Sword tails, Tetra, Angel, Gourami, African Cichlids and fighters. Different colours with various patterns of fishes are the hall mark of these varieties. Species of ornamental fishes that are identified at the market is furnished below with *Scientific names are as follows*

Acauthocobitis botia;Acanthophthalmus kuhlii; Amblyceps mangois; Amblypharyngodon mola; Anabas teuudinens; Anguilia bengalensis; Aorichthys aor; Aplocheilus panchax; Aplocheilus blocki; Aplocheilus lineatus; Badis badis; Barbus schberti; Barilius bakeri; Barilius harna; Barilius canarensis; Barilius gatensis; Bardius shacra; Batasio travancoria; Botia acanthocobitis; Botia lohachata; Botia rostrata; Botia striata; Botio dario; Chanda nama; Chanda ranga; Chanda wolfi; Chichlasoma meeki; Clarius batrachus; Colisa chuna; Colisa fasciata; Colisa labios; Colisa lalius; Colisa sota; Ctenops nobilis; Danio devario; Danio malabaricus; Danio neaglierriensis; Datniodes quadrifasciatus; Oxyaleotris marmoratus; Etroplus maculates; Etroplus suratensis; Esomus danricus; Gagata cenia; Golius; Goblus sandaundio; Onoproctoptems curmuca; Gyptothorax cavia; Hara horai; Horabagrus nigricollaris; Harabagarus brachysoma ;Labeo boga; Labeo boggut; Labeo calbasu; Labeo dyocheilus; Labeo rohita; Laguvia shawl; Laubuca dadiburjori; Laubuca laubuca; Lepidocephalus guntea; Lycodontis tile Macrognathus aculeatus; Macropodus cupuanus dayi; Mastacembelus aculeata; Mastacembelus armatus; Mastacembelus pancalus; Monoterus cuchia; Monotetrus travancorieus; Mystus gulio; Mystus montanus; Mystus tengara; Mystus vittatus; Nandus nandus; Nangra itchkeea; Noemacheilus botia;

Noemacheilus corica; Ncomacheilus savona; Neomacheilus beavani; Neoma cheilusscaturigina; Notopterus notopterus; Puntius sophore; Puntius stigma; Puntius tambraparnel; Puntius vittatus; Rasbora daiorus; Rasbora maculate; Rhinomugil corsula; Rita rita; Rohtee alfrediana; Schismatorhynchus nukta ; Scisor rhapdophorus; Sctaophagus argus; Sctophagus rubrifrons; Sillaginopsis panijus; Stigmatogobius sadanundio; Tetradon cutcutia; Tetradon steindachneri; Tor khudree; Tor mussullah; Toxotes aculator; Trichogaster leeri; Wallago attu and Xenentodon cancilla.

Ornamental fish is often used as a generic term to describe aquatic animals kept in the aquarium hobby, including fishes, invertebrates such as corals, crustaceans (e.g., crabs, hermit crabs, shrimps), molluses (e.g., snails, clams, scallops), and also live rock. Live rock is a general term for any type of rock encrusted with, and containing within its orifices, a wide variety of marine organisms including algae and colorful sessile invertebrates. Live rock serves as the principal biological and chemical filter in any marine-type tanks, and the encrusted organisms usually provide much of the background coloration in the tank. It is reported that Indian ornamental fisheries sector is small but it constitutes one of the vibrant sub-sectors of fisheries and aquaculture. One hundred and eighty-seven species are traded from India, of which 85% are wild fishes. The major share of export of ornamental fishes is from the wild, particularly from the eastern and north-eastern states of our country. West Bengal is the hub for ornamental fish trade. Mumbai mainly attributes angel and tetra breeding centre, whereas Chennai farmers devote much of their time to mollies and gold fish breeding and culture. Due to the proximity to sea, collection of sea angels and other allied species have become popular in Chennai. Loaches, gouramis, barbs, Dario, eels, catfishes, air breathing fishes and glass fishes are the common indigenous fishes which are being traded significantly. India is actively involved in trading cultured ornamental fish varieties like molly, guppy, platy, sword tail, barbs, cichlids, angels, Siamese fighter, tetras, gold fish, manila carps and sharks (Ghosh *et al.*, 2010). The major breeding centers are located across Kolkata, Mumbai and Chennai as these metros have become major export points. High numbers of herbivorous animals, such as hermit rabs, snails, and shrimps, are harvested annually for the aquarium trade. Observations and management of a mini reef tank in our laboratory highlights the essential role of these invertebrates in minimizing growth of algae and other organisms. Without grazers, algae may quickly dominate a reef. Since animals occupying different tropic levels have valuable roles in the wild, overfishing of key species may have important ecological consequences.

Demographic Profile of Women Traders

***Classification of Women Traders According to Age*:** This is an important aspect, as several other studies have also indicated the greater presence of women, widows and women heads of household in different type of trading / working activities. The relationship between gender, age and economic activities in the Ornamental fish trade is an aspect that needs to be better understood. The results of the present study are presented in the Table 8.1 below.

Table 8.1: Classification of Women Traders According to Age

Age Class (Years)	Frequency	Cumulative Frequency	Percentage (%)
10 to 20	7	7	14
21-30	20	27	40
31-40	17	44	34
41-50	3	47	6
51—60	2	49	4
61-70	1	50	2
Total	50		

Observation: Maximum of the women traders in Ornamental fish fall in the age category of 21-30 years, representing 40% of the total women traders surveyed followed by 31-41 age group representing 34%. Only 2% is observed in the age group up to 61-70 years, which is the minimum. It is interesting to note that 74% of the women traders fall under the age group of 21-40 years. This is indicative of the fact that younger women like the business of Ornamental fish.

Classification of Women Traders According to Caste: Frequency distribution table is arranged to understand the caste of women involved in trading of Ornamental fish (Table 8.2).

Table 8.2: Classification of Women traders according to Caste

Caste	Scheduled Caste	Scheduled Tribe	Other Backward Class	Others
Frequency	7	1	6	36
Percentage (%)	14	2	12	72

Observation: It is evident that maximum numbers of women (72%) from General caste are involved in the Ornamental fish trade. 14% of scheduled caste and 2% of scheduled tribe category and other backward class represents 12%. This is justified with the reality that this type of business requires skill as well as investment that is absent in scheduled caste category.

Classification of Women Traders According to Religion: Frequency distribution table is arranged according to religion to understand the religion of women involved in Ornamental fish trading (Table 8.3).

Table 8.3: Classification of Women traders according to Religion

Religion	Hindu	Muslim	Christian	Others
Frequency	50	-	-	-
Percentage (%)	100	0	0	0

Observation: As far as religion is concerned, it is found that only Hindu community (100%) represents this business.

Classification of Women Traders According to Education: Education is very much important for any activity. Therefore, frequency distribution table is arranged according to education in order to understand the educational status of women involved in Ornamental fish trading (Table 8.4).

Table 8.4: Classification of Women Traders According to Education

Education →	Primary	Secondary	Graduate	Post Graduate	Others
Frequency →	14	33	3	0	0
Percentage →	28	66	6	—	—

Observation: It terms of literacy, education up to primary level being done by 28%, secondary level (66%) and only 6% is found in graduate level, carrying on the trading of Ornamental fish. This revealed women, with a standard of general education, like this business for independent earning due to less opportunity in employment.

Marital Status of Women traders: Marital status indicates that 48% are married and 52% are unmarried. Thus it is revealed that involvement in other work and or becoming a housewife does not prevent to carry on this business as this can be done part time in addition to household work.

Bank / Post Office Accounts Maintained by Women traders: All women traders possess either bank or post office account for saving their small saving. It is definitely a sign of self dependence and reliance of women engaged in this trade.

Particulars of Family Members of Women Traders: Particular numbers of family members is tabulated as well as a figure showing the ratio of male and female is sketched out to get an understanding of family structure (Table 8.5).

Table 8.5: Particulars of family members of Women traders

No. of family members	Male	Percentage (%)	Female	Percentage (%)
1	15	30	13	26
2	29	58	34	68
3	6	12	2	4
4	0	-	1	2
Total family	50		50	

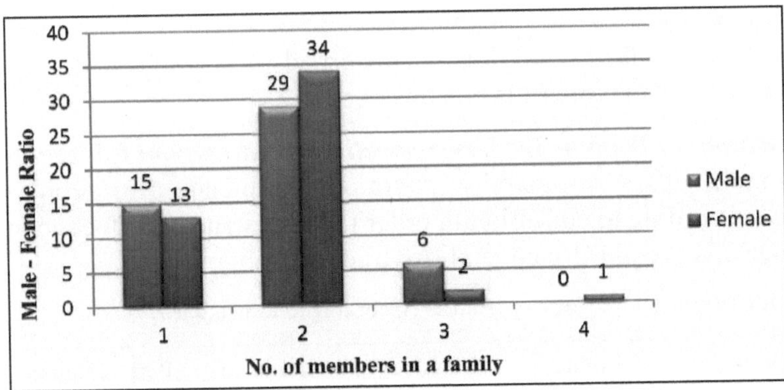

Fig. 8.9: Male-Female Ratio within the Family Members

Observation: From Table 8.5 and Figure 8.9 it is seen that maximum numbers of families have only two members followed by one and next followed by three members. Families with above three members are very less. This gives a clear picture that size of the family members is in control. Male –female ratio is shown in Figure 8.9.

Occupation of Family Members of Women Traders: Apart from self, how other family members (both male and female) are engaged, are also compiled in the following Table 8.6 to understand occupation pattern of women family members.

Table 8.6: Occupation of family members of Women Traders

Occupation	Male	% Male	Female	% Female	Total (Male + Female)
Ornamental Fish trade	10	10.98	50	54.95	60
Business	3	3.30	0	0	3
Labour	31	34.07	1	1.10	32
Service	5	5.49	0	0	5
Factory worker	3	3.30	0	0	3
Household / Housewife / Aged	7	7.69	33	36.26	40
Others	2	2.20	0	0	2
Minor / Child / Student	30	32.97	7	7.69	37
Total	91		91		182

The Figure 8.10 and Figure 8.11 on the occupation level of the male and female members of the women traders reveals that 11% of the male members is engaged in ornamental fish trading; 34% is engaged as labour and 33% are minor / child / student; 8% is in household activities or aged; 6% is in service; 3% is engaged either as Factory worker or in business; only 2% is engaged in other work. Whereas in case of female members 55% is engaged in ornamental fish related activities, 36% is in household activities; 8% is minor / child / student and only 1% are working as labour. Few facts, emerged from the study are that i) Majority in female members is in ornamental fish related activities but in contrast majority in male members are engaged in labour work; ii) Un-employment is more in case of female compared to male and iii) Male members are found involved in business or in service or factory worker or others but these types of occupation are absent in female members.

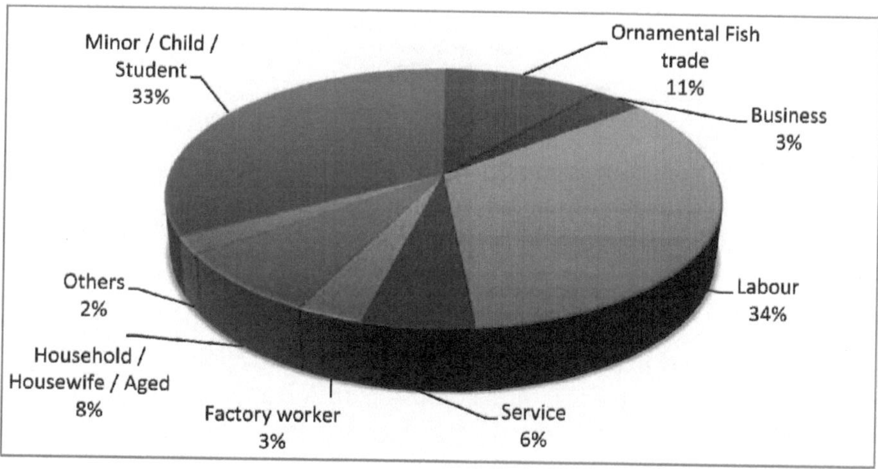

Fig. 8.10: Occupation of Male family members

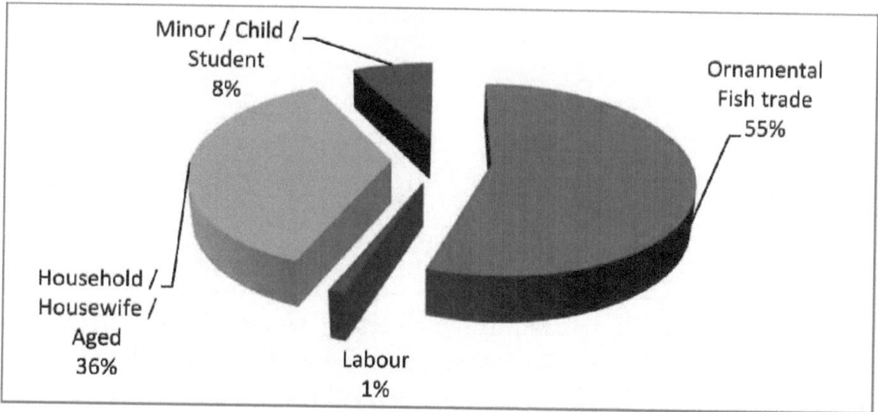

Fig. 8.11: Occupation of Female family members

Socio-Economic Profile / Livelihood Profile Women Traders

Types of Household of Women traders: Women engaged in ornamental fish retailing activities are coming mostly from urban and suburbs of city. Surveys conducted to understand the types of household they live in are displayed in Figure 8.12 to have a grasp of their living condition.

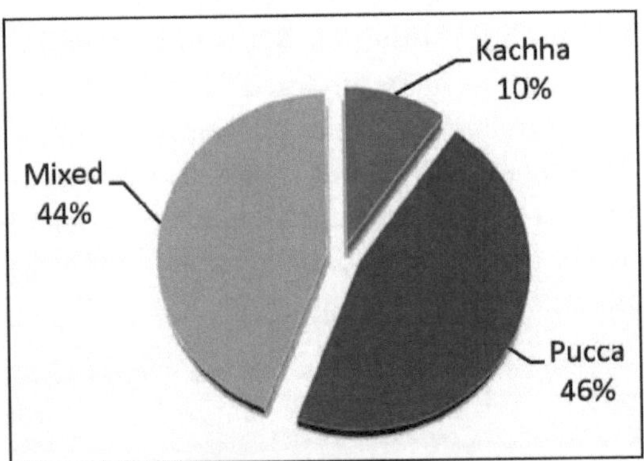

Fig. 8.12: Types of Household of Women in Ornamental fish trading

The scenario of types of household, where the women traders live, suggest that maximum (46%) live in pucca house, followed by 44% live in mixed house built as kachha and pucca, and only 10% live in kachha houses. This is indicative of the fact that women engaged in ornamental fish trade are better off compared to those engaged in other fishery activities.

Possession of Assets in the family of Women traders: Details of assets owned by family of women traders are compiled in a Table 8.7 form for easy comprehension.

Table 8.7: Possession of Assets in the family of Women traders

Assets	Frequency	Percentage (%)
Bicycle	38	76
Motorcycle	10	20
Car / Jeep / Auto	1	2
Television	43	86
Radio	14	28
Sewing Machine	8	16
Fishing Net	4	8
Fishing Boat / Handi	2	4

Possession of Assets by the family members of women in ornamental fish activities reveals that 36% families have television, 32% have bicycles, 11% possess radio, 8% motorcycles, 7% sewing machine, 3% fishing net and only 1% have car in possession. This indicates bicycle / motorcycle is mostly used for transport / conveyance, television / radio for recreation in leisure time and sewing machine to meet household requirement in need.

Family members of Women in Ornamental fish activities, associated with this trade: This is a very important piece of information to unearth how many family members of the women in ornamental fish activities are associated with this trade. Survey data is presented below clarifies the scenario (Figure 8.13).

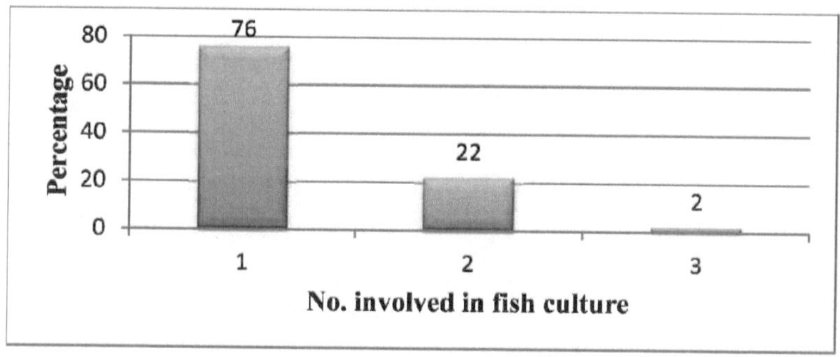

Fig. 8.13: Distribution of No. Family members involved in Fish Culture

Number of family members involved in ornamental fish activities indicates that model class being one family member with 38 families, two with 11 families, and only one family has three working members.

Distribution of Dwelling Land Holding of Women Traders: Women traders hardly have any land except their dwelling houses. The following frequency Table 8.8 gives a complete picture about the land holding structure.

Table 8.8: Distribution of Dwelling Land holding by Women Traders

Land holding (Sq. feet)	Frequency	Percentage (%)	Remarks
100 - 300	8	16.67	Sl. No. 6 holds 10 kactha of land (the highest of the survey samples) and sl. no. 8 has no land; she lives in a rented house.
300- 500	20	41.66	
500 - 700	11	22.92	
700 - 900	6	12.50	
900 - 1100	2	4.17	
1100 - 1300	0	0	
1300 -1500	1	2.08	
Total	48		

One women trader is found possessing 10 (ten) kactha of land (the highest of the survey samples) and one has no land owned by her. She lives in a rented house. It is interesting to note that modal class of land holding being 300 to 500 sq. ft. with highest frequency of 20, 500 to 700 sq. ft. with frequency 11, 100 to 300 sq. ft. with frequency 8, and 700 to 900 sq. ft. with frequency 6, 900 to 1100 with frequency 2 and only one holds 1300 to 1500 sq. ft. It is clear that about 65% of women traders are living in houses having a total area ranging from 300 to 700 sq. ft.

Possession of Land Holding by Women Traders: Women in ornamental fish activities were asked to ascertain whether they live in their own or leased houses. The following Table 8.9 provides answer.

Table 8.9: Possession of Land holding by Women traders

Land	Frequency	Percentage (%)
Owned	49	98
Leased	0	-

Out of fifty women interviewed only one was found living in a rented house. All others are staying in their puce or mixed (puce and kasha) small houses.

Numbers of Earning Members in the Family of Women Traders: The following frequency table gives a complete picture of numbers of earning members in the family of the women traders.

Table 8.10: No. of Earning Members in the Family of Women Traders

Earning members	Frequency	Percentage (%)
1	2	4
2	42	84
3	5	10
4	1	2

An attempt was made to ascertain the number of earning members in the family of women in ornamental fish activities. The tabulated data in form of frequency table reveals that 84% of the women possessing two earning members, 10% of the families have three earning members, 4% with one earning member and only 2% possessing four earning members for the family. However, model class is 84% with two earning members is indicative of the fact that in a family 2-3 earning members tried to meet their ends meet (Table 8.10).

Annual Income Distribution of the Family of Women Traders: Collected data from individual women traders are classified into several groups along with frequency to easily understand annual income which is a very important indicator of socio-economic condition.

Out of 50 (fifty) sample survey average annual Income of 4 (four) families are more than Rs.1, 00, 000/-. Annual income of others is given in Table 8.11.

Table 8.11: Annual Income distribution in the family of Women traders

Average annual Income (Rs.)	Frequency	Cumulative frequency	Percentage (%)
Up to Rs. 50, 000/-	33	33	66%
Rs. 50, 000/- to Rs. 75, 000/-	9	42	18%
Rs. 75, 000/- to Rs. 1, 00, 000/-	4	46	8%
Rs. 1, 00, 000/ and above	4	50	8%

To unearth the annual income distribution of the women in ornamental fish activities, raw data compiled in the form of frequency table which identifies that model income class is up to Rs. 50,000 with highest frequency of thirty-three and income class with lowest frequency as Rs. 75,000 to 1,00,000 and Rs. 1, 00, 000/ and above each being four. It is also apparent that annual income of 66% families of women traders is less than Rs. 50,000 even though income from the business is reported to be lucrative (Table 8.11).

Family Expenditure Pattern of the Women Traders: Item-wise expenditure data were collected from the women to understand the pattern of expenditure incurred by the women traders to meet various needs of the family. Basic statistics like mean, standard deviation, range and coefficient of variation were computed for each item of expenditure to understand extent of variation in expenditure among items as well as among women traders (Table 8.12).

Table 8.12: Aggregate Annual Family Expenditure of the Women Traders for Different Items

Items of Expenditure	Average annual spent (Rs)				
a) Recurring Expenses	Mean (Rs.)	S.D. (Rs.)	Range (Rs.)		C.V. (%)
			MAX	MIN	
Purchase of food	12,230.00	20,754.45	150,000.00	3,000.00	169.70
School fees	2,905.13	2,677.47	12,000.00	0.00	92.16
Medical expenses	4,994.00	3,571.05	20,000.00	200.00	71.50
Clothing	7,490.00	6,459.02	45,000.00	1,200.00	86.23
Household saving	9,559.09	20,955.78	125,000.00	300.00	219.22
Religious function	5,306.25	7,706.29	50,000.00	0.00	145.23
Fish culture operation	50,611.11	65,632.11	400,000.00	0.00	129.68
Any others	4,266.67	4,408.23	25,000.00	0.00	103.32
b) Non – Recurring Expenses					
House Building	11,500.00	31.498.24	200,000.00	0.00	273.90
Vehicle	3,783.33	7.982.73	50,000.00	300.00	211.00
Land development for fish pond / fencing etc.	11,500.00	10,403.97	60,000.00	0.00	90.47
Other house hold goods	4,397.92	4,059.39	25,000.00	0.00	92.30
Fish culture implements	6,372.00	9,063.49	50,000.00	0.00	142.24
Total (a + b)	74,900.00	149.615.87	1,000,000.00	9,000.00	199.75

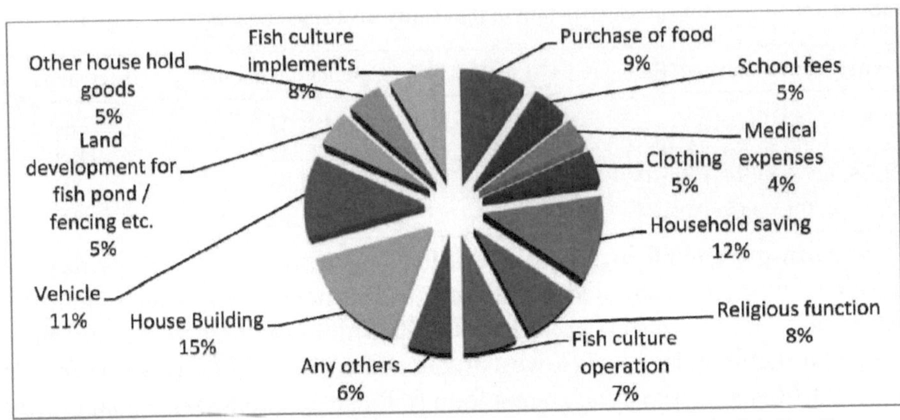

Fig. 8.14: Proportion of Expenditure of Women in Ornamental Fish Trading (Recurring and Non-Recurring)

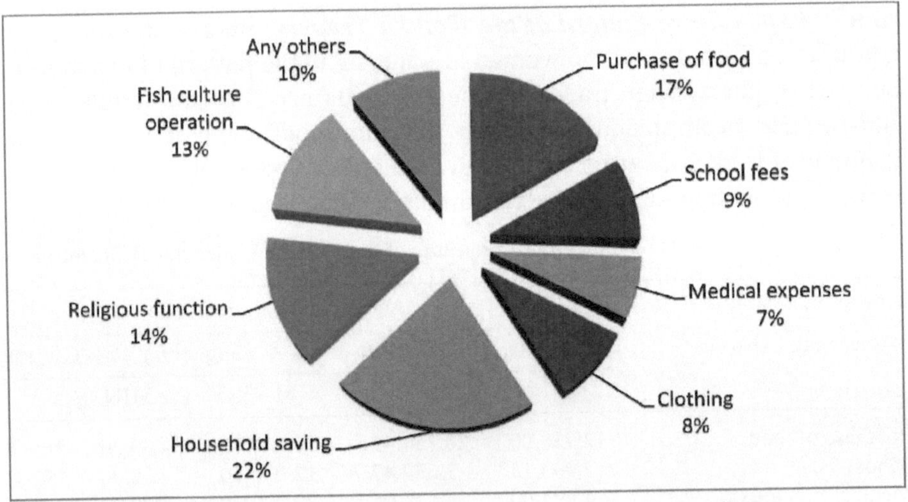

Fig. 8.15: Average Annual Item-wise Recurring Expenditure

Recurring Expenditure: Within this group it is evident that women traders make savings up to 22%. They spend 17% on purchasing food items followed by religious functions (14%), fish culture operations (13%), school fees (9%), clothing (8%), medical expenses (7%) and others (10%) (Figure 8.15).

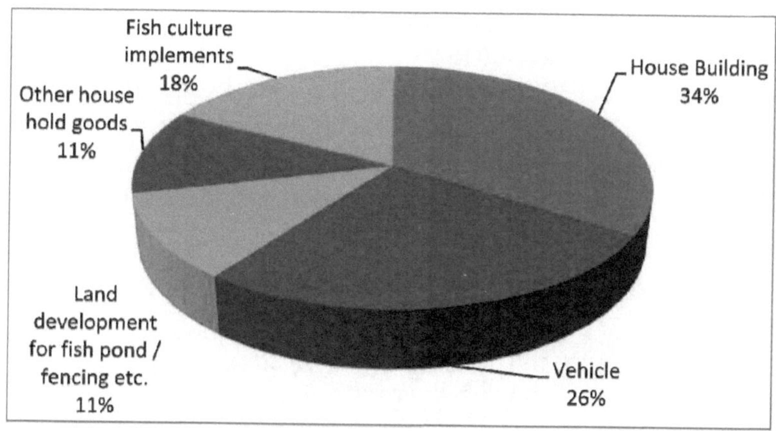

Fig. 8.16: Average Annual Item-wise Non-Recurring Expenditure

Non-recurring Expenditure: In this group, the largest expenses of the women traders are on housing (34%) followed by vehicle (26%), fish culture implements (18%), household goods (11%) and land development etc. (11%) (Figure 8.16).

Recurring and Non-recurring Expenditure Taking into consideration both the recurring and non-recurring expenses together, it is observed that maximum expensed are incurred on house building (15%) followed by saving (12%), vehicle (11%). The next large expenses are made for fish culture implements (8%), purchasing of food (9%), religious function (8%), fish culture implements (8%), land development etc. (5%), household goods (5%), school fees (5%), clothing (8%), medical (4%), and others (6%).

Source of Credit Taken by Women Traders: Some women traders have taken loan to run business. The following Fig.8.17 tells sources of credit taken. Out of 50 respondents 42 heads have taken no credit from any source and 8 heads have taken credit either from money lenders (62%) or from NGO (38%). No credit is taken from Relatives, Friends, Banks, Traders, or credit society (Figure 8.17).

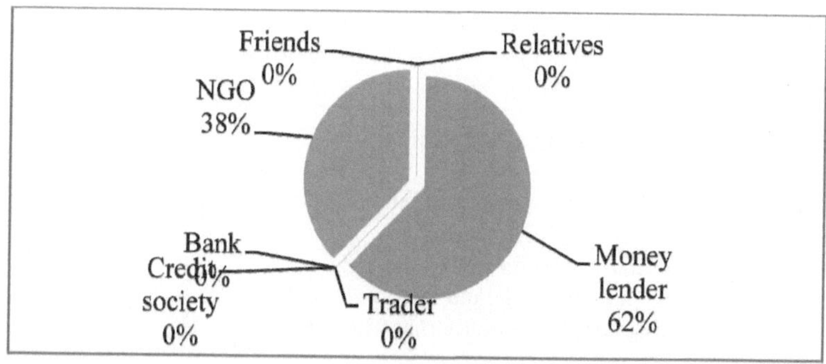

Fig. 8.17: Sources of Credit taken by Women in Ornamental Fish Trading

Average Amount Borrowed by Women Traders: Average amount borrowed is Rs. 14,437 range being Rs 500 to Rs. 30,000 with a C.V of about 91%.Period of loan ranged between 12 to 84 months. Interest rate varied between 14 to 15%9 (Table 8.13). High CV value of 92.52% for amount borrowed reflects inconsistency and similar is the case period of credit. It is quite usual women have taken according to their need for a chosen period keeping in mind high interest rate and repayment liability.

Table 8.13: Average Amount Borrowed by Women Traders

Descriptive Statistics		Amount borrowed (Rs)	Period of credit (Months)	Interest rate (%)
Mean		14,437.50	39.00	14.80
SD		13,069.70	37.26	0.45
RANGE	MAX	30,000.00	84.00	15.00
	MIN	500.00	12.00	14.00
CV% [SD/Mean x 100]		90.52	95.54	3.04

Support from other Family Members to Women Traders: Types of support received by women traders is available in the Table 8.14 below.

Table 8.14: Support from other Family Members to Women Traders

Type of Support		Frequency	Percentage (%)
Psychological / moral support		5	10
Direct assistance	Financial	5	10
	Physical – viz., look after / feeding / marketing etc.	9	18
No support		31	62

Activity Profile

Activity of women fish traders: Activity of women fish traders is summarized in Table 8.15 below.

Table 8.15: Activity of Women Fish Traders

Type of Activity	Frequency	Percentage (%)
Ornamental fish culture / farming / rearing / trading	11	22
Ornamental fish purchase and retail	1	2
Ornamental fish capturing from local river and sale to exporter	1	2
Making accessories for Ornamental fish rearing	6	12
Making accessories decorative items for aquarium	4	8
Packaging of accessories / feed for O. F. and supply to market	3	6
Contract worker for making accessories for Ornamental fish	24	48
Total	50	

Statistics of Average Initial Investment and Average Monthly Income of Women Traders: Mean, standard deviation and Coefficient of variation of Average Initial investment and Average monthly income of Women Retailers is presented in Table 8.16 below.

Descriptive statistics provides the central values of parameters and also relative measure of variation with respect to initial investment and monthly income table 20; high initial investment does not reflect in proportionate income.

Reasons for Adoption of Fish Trading as the Activity as Primary Income: Summarized answers are in Table 8.17 below.

The respondents were asked to justify the reasons for adoption of fish trading as the primary income earning activity with ranking the answers are summarized in the following table. 58% ranked one and have expressed it as gender friendly and 12% have cited it as ease of work and in convenient timing and 6% have declared it as regular and growing income potential being the reason for adoption of fish retailing as primary activity. Only 2% has stated it as family enterprise.

Sources of Motivation for Fish Trading Activity: Women fish traders were asked about sources of motivation for fish trading activity. Answers were summarized in table 22 below. Respondents (80%) were motivated in this trade for ease of operations, 72% are motivated good local resources, 62% are motivated realizing market potential and 20% are motivated by influence of neighbours / participating members or department of fisheries to take up as a profession.

Sources of Information for Fish Trading Activity: Responses of various sources of information along with frequency is presented in table 23. It is distinct that regarding fish trading activities information was mostly (78%) available from Department of fisheries and (72%) available from NGO's (Table 8.18)

Table 8.16: Descriptive Statistics of Average Initial Investment and Average Monthly income of Women Traders

Average Initial investment (Rs.)					Average Monthly Income (Rs.)				
Mean	Range		SD	CV (%)	Mean	Range		SD	CV (%)
	Max	Min				Max	Min		
67,000.00	300,000.00	10,000.00	114,463.97	1.71	3,465.00	41,600.00	500.00	5,662.97	163.43

Table 8.17: Reasons for Adoption of Fish Trading as the Activity as Primary Income (rank in order of reason)

Reasons for Adoption as Primary earning activity		RANKS							Total
		1	2	3	4	5	6	7	
Ease of work and convenient timing	Frequency	6	26	18	0	0	0	0	50
	Percentage (%)	12	52	36	-	-	-	-	
Regular and growing income potential	Frequency	3	9	18	13	4	3	0	50
	Percentage (%)	6	18	36	26	8	6	-	
Family enterprise	Frequency	1	0	1	5	9	33	1	50
	Percentage (%)	2	-	2	10	18	66	2	
Social and family relationships strengthened	Frequency	0	0	1	4	31	11	3	50
	Percentage (%)	-	-	2	8	62	22	6	
Improvement in social status	Frequency	0	6	8	26	5	1	4	50
	Percentage (%)	-	12	16	52	10	2	8	
Incentives provided by the Govt.	Frequency	0	2	1	6	2	4	35	50
	Percentage (%)	-	4	2	12	4	8	70	
Gender friendly	Frequency	29	10	8	3	0	0	0	50
	Percentage (%)	58	20	16	6	-	-	-	

Table 8.18: Sources of Information for Fish Trading Activity

Sources of Information	Frequency	Percentage (%)
Department of Fisheries	39	78
Institutional agencies like MPEDA/NFDB/KVK	8	16
Universities and research institutes	1	2
Media support (Print/Visual/Audio)	2	4
NGO involvement	36	72
Friends and relatives	1	2

Women Fish Traders having Debt: An attempt was made to study the debt scenario of the women traders while conducting survey. Details of debt structure are presented in table 19 below for comprehension. It is found that only 8 traders have taken debt out of 50 respondents. Model debt class belongs within Rs. 5000 to Rs. 10000 constituting 50% of the total who taken loan for their activities (Table 8.19 & Figure 8.18).

Table 8.19: Women Fish Traders having Debt

Average range of debt.	Frequency	Percentage (%)
Up to Rs. 1, 000/-	1	12.5
Rs. 1, 000/- to Rs. 5, 000/-	0	-
Rs. 5, 000/- to Rs. 10, 000/-	4	50
Rs. 10, 000/- to Rs. 15, 000/-	0	-
Rs. 15, 000/- to Rs. 20, 000/-	0	-
Rs. 20, 000/- to Rs. 25, 000/-	0	-
Rs. 25, 000/- to Rs. 30, 000/-	3	37.5
Above Rs. 30, 000/	Nil	—
Total	8	

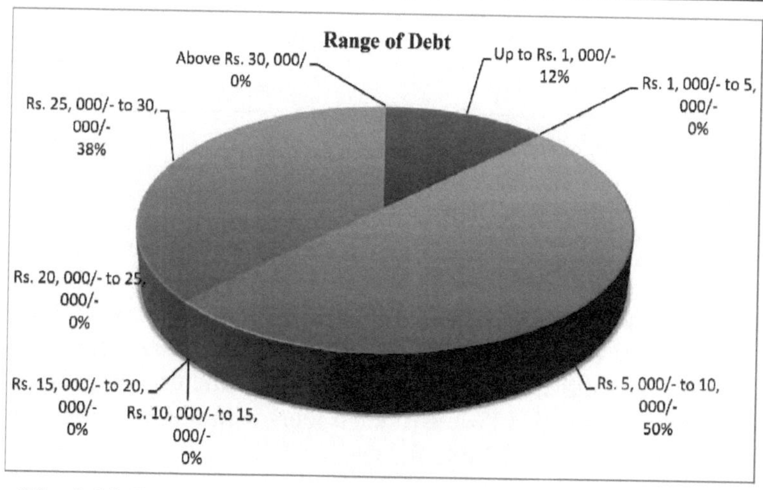

Fig. 8.18: Range of Debt taken by Women in Ornamental fish trading

Exposure to Technical Knowhow: As women are involved in trading activities for a long time with some of them doing so for generations, it is expected a lot technicality will be known to them. Answers from them revel that 78% are aware of production technology, 72% are in marketing and 16% have exposed their trading as earning and income generation.

Awareness of Operational Knowhow: While talking about the types of 'Awareness of Operational Knowhow' varied answers erupted from target respondents that are tabulated in the Table 8.20.

Table 8.20: Awareness of Operational Knowhow

Types of Awareness of Operational Knowhow	Frequency	Percentage (%)
Sources of resources	38/50	76
Training support	13/50	26
Market availability	27/50	54
Financial assistance	15/50	30
Total	93	

It is quite interesting to note that 76% knows about the 'sources of resources', 54% become aware of 'Market availability' while 30% regarding 'financial assistance' and 26% about 'training support'. This is obvious because of the fact that due to keen involvement of fish trading the women are aware of sources of resources of fish, necessary feed & accessories and availability at markets spread over different locations.

Constraints and Problems in Fish Trading: Like all other business, ornamental fish trading is also having a lot of constraints as experienced by women traders that came out during interview as follows.

Table 8.21: Constraints and Problems in Trading Activity

Constraints and problems in activity	Frequency	Percentage (%)
Infrastructural constraints	9	18
Technical constraints	7	14
Economic constraints	40	80
Marketing constraints	34	68
Socio psychological constraints	11	22

Most prominent constraint in ornamental fish trading as faced by the women traders is 'Economic constraints' (80%), followed by 'Marketing constraints' (68%), Socio psychological constraints (22%), Infrastructural constraints (18%) and 'Technical constraints' (14%). Women traders with small capital investment somehow run their business (Table 8.21).

Possible way out – Idea & Measures Taken by Women Traders: Most of the respondents (48%) prefer to have permanent job / extra work / increase of remuneration in making accessories / decorative items / packaging of fish feed etc. 30% prefer to remain silent or have no idea, 14% keenly demand for Govt. assistance (Table 8.22).

Table 8.22: Possible Way out – Idea & Measures Taken by Women Traders

Views / Opinion from women retailers	Frequency	Percentage (%)
Setting up infrastructure for manufacturing low cost accessories for aquarium / opening modern hatchery	2	4
Exploring all India market to compete Chennai market	2	4
Permanent job / Regular work / extra work / Increase of remuneration	24	48
Government assistance / saving money for self independence	7	14
Nil / No idea	15	30
Total	50	

Future Planning of Women Traders: Every individual has his/her own thought. Therefore, it was attempted to extract maximum information from the experienced women traders about their idea about future planning. Views / opinions are presented in the frequency Table 8.23 below.

Table 8.23: Future Planning of Women Traders

Views / Opinion from women retailers	Frequency	Percentage (%)
Setting up a small unit of aquarium business	3	6
To seek Bank loan for business	1	2
Setting up breeding unit / Making of Dice set	3	6
Participation at A I O F / Survey in India & abroad	2	4
Nil	41	82
Total	50	

Out of 50 only 9 responded to questions about the future planning. As many as 6% expressed that they want to set up a unit of ornamental fish aquarium business or manufacturing dice set for accessories or a breeding unit. 4% likes to participate at A I O F or survey work in India & abroad and 2% want to seek government loan for enhancing business.

Socio Economic Development / Improvement (Past Activity)

Socio Economic Development / Improvement of Women Traders: Women traders were asked a number of questions on socio economic improvement due to ornamental fish trading and three options were there to answer whether they agree, disagree or no change due to past activity. Responses are summarized in the Table 8.24 below.

Table 8.24: Socio-economic Development / Improvement of Women Traders

Question	Agree (frequency)	Percentage (%)	Disagree (frequency)	Percentage (%)	No change (frequency)	Percentage (%)
Increase family income	48	96	1	2	1	2
Improvement in standard of living	47	94	0	—	3	6
Saving capacity	38	76	8	16	4	8
Asset creation	13	26	32	64	5	10
Participation in decision making	16	32	19	38	15	30
Improved relationship with stakeholders	6	12	8	16	36	72
Increase the level of social interaction / participation within the community	33	66	3	6	14	28
Increase in knowledge level	45	90	4	8	1	2
Secured primary occupation	24	48	23	46	3	6
Potentiality for fulfilling future need	15	30	28	56	7	14
Development in infrastructure	1	2	8	16	41	82
Upliftment of status in the area / community	6	12	5	10	39	78
Total	292		139		169	

It is interesting to note that 96% respondents agreed to the fact that due to this trading, family income have increased followed by 94% agreed to that standard of living improved considerably. 90% women agreed of increasing knowledge level and 76% declared for enhancing saving capacity. 66% considered in increase the level of social interaction / participation within the community, 48% divulged the trading as a secured primary occupation. 32% agreed of improve participation in decision making, 30% expressed the trading as potential for fulfilling future need, 26% as asset creation. The remaining disagreed to the facts that there are any improvements in asset creation (64%), potentiality in fulfilling future need (56%), secured primary occupation (46%), participation in decision making (38%), relation with stakeholders / infrastructure / saving capacity (16% - in each case) etc. Likewise, 82% reported that there are no changes in development in infrastructure, 78% found no upliftment of status in the area / community, 72% considered no improvement in relationship with stakeholders, 30% stated no change towards participation in decision making, 28% found no increase in the level of social interaction / participation within the community 14% described the trade as no potentiality in fulfilling future need.

Fish Biodiversity and its Conservation

Fish Biodiversity and its Conservation: Majority (86%) opined that during the past 20 years' availability of fish has increased while 8% expressed decline and 6% preferred not to make any comment.

Indigenous Technology (ITK) used by Women Traders: Regarding use of indigenous technology, 50% of the women fish traders considered that locally made accessories for ornamental fish are capable to replace the products from China and Chennai. 18% reported that they used ITK for manufacturing accessories for aquarium, 14% liked not to make any comment. 6% divulged that earthen pots are used for rearing ornamental fish, clean glass bottle for rearing fighter fish and conventional packaging system are used for Oxygen / fish feed. 4% informed of using ITK traps are used for catching ornamental fish. 2% demanded for setting up of large scale infrastructure by Govt. for rearing & trading of fish using ITK.

Suggestion for Conservation of Fish Species in the Area: Details of suggestions, received from respondents through interview, are summarized and tabulated at Table 8.25 below for attention.

Table 8.25: Suggestion for Conservation of Fish Species in the Area

Suggestions	Frequency	Percentage (%)
Development of Farm in Co-Operative sector	8	16
Govt. initiative for construction of hatchery / rearing pond / export hub of Ornamental fish	1	2
Initiative / management from Government for infrastructure for large scale manufacturing of accessories for Ornamental fish	4	8
Govt. initiative / support for training / financial assistance for the trade for employment generation	6	12
Govt. assistance – Increase of remuneration – job security for worker of making accessories for O F	16	32
Pollution control in water bodies and prevention for wild catch of rare specimens	5	10
Nil / None / No comment	10	20
Total	50	

Observation: Out of 50 respondents, 32% demanded Govt. assistance – Increase of remuneration to worker making accessories for Ornamental fish along with job security. 20% could not say anything or remained silent. 16% wants development of farm in co-operative sector, 10% suggested for pollution control in water bodies and prevention for wild catch of rare specimens, 8% opined that initiative / management from Government for infrastructure for large scale manufacturing of accessories for Ornamental fish is very much required and 2% likes for Govt. initiative for construction of hatchery / rearing pond / export hub of Ornamental fish.

Observations on Species Diversity by Women Traders: It is interesting to note that only 29 women traders, out of total 50, responded the question on fish species observed by them during their survey but no answer is found against the issue on fish species generally comes to the market, but not observed during survey. Maximum stated that 20-25 species of ornamental fish *viz.*, Red tail Shark, Gold fish, Black Molly, Guppy, Gold fish etc. and 25-30 varieties of ornamental fish accessories observed in market.

Conclusion and Recommendations

The present field study revealed many interesting socio-economic parameters of women engaged in ornamental fish and aquarium accessory manufacturing and marketing at Hatibagan Market, Kolkata as well as Dasnagar, Howrah, West Bengal. Interesting to note that many accessories are manufacturing at Howrah and these are import substitution as these were imported traditionally from China to cater to the need of local hobbyists. Keeping colourful and fancy fishes known as ornamental fishes, aquarium fishes, or live jewels is one of the oldest and most popular hobbies in the world. The growing interest in aquarium fishes has resulted in steady increase in aquarium fish trade globally. The ornamental fish trade with a turnover of US $ 6 Billion and an annual growth rate of 8 percent offers lot of scope for development. Unfortunately, India's share in ornamental fish trade is estimated to be less than 1 % of the global trade. The major part of the export trade is based on wild collection. There is very good domestic market too, which is mainly based on domestically bred exotic species. The earning potential of this industry and the relatively simple techniques involved in the growing of these fish has helped the aquarium industry to evolve and provide rural employment and subsequent economic upliftment in many countries including India. Ornamental fish trade has much more to offer than what is exploited now as far as India is concerned. As the trade expands, issues such as quality, environmental concerns, habitat protection and sustainability of the resource come to the fore simultaneously. Green certification is the certification given to a product to ensure its environmental and socioeconomic sustainability. It ensures product quality, safety and traceability. The potential for the development of ornamental fish trade in India is immense, though it is still in a nascent stage. The Government of India has identified this sector as one of the thrust areas for development to augment exports. The turn of the century has seen a spurt in the collection, culture and trade in freshwater ornamental fishes. Aqua-shows have now become an annual feature in some states as in Kerala where Government support for such an activity is in vogue. For the trade to prosper, the three pre-requisites are quality, quantity and sustainability. The fish species diversity of the rivers and streams of the Western Ghats and North East India are well recognized with as many as 68 per cent of the 327-species listed from the former and over 50 per cent of the 350 or so listed from North East India being endemics. Of these, 40 to 50 per cent are ornamentals, some fetching very high prices in the international market. Ninety per cent of the freshwater ornamental fishes exported from India are wild caught indigenous species. The total marine products exported from India in 2009-2010 was about Rs.100485 million (US $ @ 2132 million) of which ornamental fish formed only a minuscule of hardly Rs.55 million (US $ @ 1.17 million). We are today in an unenviable position, so that we could consider policies on environmental and human management approaches to make the growing industry of freshwater ornamental fish trade sustainable, eco-friendly and at the same time monitor resilience of the resources. We have to create awareness among local communities and stakeholders to desist from unlawful and illegal practices of catching ornamental fishes from the wild. The trade should also encourage protection of the habitat for an eco-friendly approach. Presently exotic species dominate our domestic aquarium trade, and some are being bred for export. Unregulated/illegal introduction of exotic species and control or elimination of invasive species poses problems and these issues have to be addressed.

Rural women have a major role in sustainable fisheries and aquaculture, but they often have unequal access to resources and services that they need to be successful and often suffer greater poverty and hunger than men as a consequence. Closing the gender gap in access to important resources can improve productivity and increase incomes and food security. The social norms that constrain poor women's and men's opportunities need to be better understood. World Fish gender equity research is aimed at generating evidence and information on how poor men and women access and use resources, who has power and makes decisions, whose priorities are being addressed and who is impacted by, or benefiting from, different development alternatives.

India needs to improve their export contribution and earn foreign exchange, to improve substantially in global trade contribution. In spite of having enormous potential, India is not up to the mark in this respect. But proper utilization of resources pursuing gender equity will lead India to a higher position. The findings of the socio-economic condition of women engaged in ornamental fish rearing, trade and marketing may be useful for policy decision to improve their condition leading to sustainable development of this sector.

References

Annual Report (2009-10). Government of India, Ministry of Agriculture Department of Animal Husbandry, Dairying & Fisheries, New Guidelines for import of ornamental fish, Page no. 1-2.

Anonymous, (2008). Department of Agriculture, Fisheries and Forestry - Operational Procedures

Ayyappan S., Jena J. K., Gopalakrishnan A. and Pandey A. K. (2006). *Handbook of fisheries and aquaculture.* Directorate of Information and Publications of Agriculture, Indian Council of Agricultural Research, New Delhi, 22: p. 354.

Basu A., Dutta D. and Banerjee S., (2012). Indigenous ornamental fishes of west Bengal, Aquaculture Research Unit, Department of Zoology, University of Calcutta, West Bengal, India, *Recent Research in Science and Technology*, 4(11), 12-21.

Benjamin, R. P. (2012). A highly profitable hobby. *The Hindu,* Vishakapatnam, India, p.12. http://www.thehindu.com/news/cities/visakhapatnam/a-highly-profitable-hobby/article 3399986.ece

Burka J. F., Hammell K. L., Horsberg T. E., Johnsons G. R., Rainnie, D. J. and Spears, D. J., Canberra, AQUAVETPLAN Enterprise Manual, Version 2.0, 2015.ACT. http://www.daff.gov.au/aquavetplan.

Bhaskar, S., P. S. R. Reddy, B. E. Barithy, R. Subramanian and R. S. Lazarus (1989). Exotic fresh water aquarium fishes and other role in aquarium fish in India. Proceedings Special Publication 1, Indian Branch, Asian Fisheries Society, Mangalore, 1989, pp. 35-39.

DAHDF (1999). Guidelines for the Import of Ornamental Fishes in to India. *Report of Department of Animal Husbandry, Dairying and Fisheries (DAHDF),* India, p. 1-2. http://d a h d. n i c. i n / d a h d / W r i t e R e a d D a t a / N e w %20 Guidelines%20for%2 0import %20of% 20fundamental %20fish.pdf

Department of Fisheries, Government of Kerala, Thiruvananthapuram, India, (2006), 95–102,

Dutta Abir Lal, Debargha Chakraborty, Sajal Kumar Dey, Ashim Kumar Manna, Pankaj Kumar Manna (2013). Ornamental Fishes of Coastal West Bengal, India — Prospects of Conservation and Involvement of Local Fishermen *Natural Resources*, 2013, 4, 155-162 http://dx.doi.org/10.4236/nr.2013.42020 Published Online June 2013 (http://www.scirp.org/journal/nr) 155

Ghosh A., B. K. Mahapatra and N. C. Dutta (2000). Studies on native ornamental fish of West Bengal with a note on their conservation," *Environment and Ecology*, 20(4):787-793.

Ghosh Abalika, Mahapatra B. K. and Datta N. C., (2003). Ornamental fish farming- successful small scale aqua business in India, *Aquacult. Asia*, 8(3): 14-16.

Ghosh Subir (2010). Freshwater ornamental fishes- A viable economic activity in India. *CSG*, http://www.csgroupinfo.com

Ghosh, A. and S. Debnath (1998). Ornamental fish farming—a lucrative business for rural womenfolk of some vil- lages of Howrah, West Bengal, *National Workshop on Aquaculture Economics*, 6-8 October 1998, Bhubaneswar.

Ghosh, S.K. and Sureshbabu, P.C> 2006. Techno-economic viability of freshwater ornamental fisheries shcemes in West Bengal, In: Ornamentals Kerala, 2006, Souvenier Department of Fisheries, Government of Kerala, Thiruvananthapuram: 95-112

Ghosh, S.K. and Sureshbabu, P.C. (2006) Techno economic viability of freshwater

Goswami Chandasudha, and V. S. Zade (2015). Analysis of international trade and economical & commercial scope of ornamental fishes. International Journal of Engineering and Applied Sciences (IJEAS) 2(5):

Goswami U.C., Basistha S.K., Bora D., Shyamkumar K., Saikia B. and Changsan K. (2012). Fish diversity of North East India, inclusive of the Himalayan and Indo Burma biodiversity hotspots zones: A Checklist on their taxonomic status, economic importance, geographical distribution, present status and prevailing threats, *International Journal of Biodiversity and Conservation, 4(15):* 592-613.

Guidelines for Green Certification of Freshwater Ornamental Fish. The Marine Products Export Development Authority, Cochin

Handbook on Fisheries Statistics (2014). August 2014. Dept of Animal Husbandry, Dairying & Fisheries, Ministry of Agriculture, Govt. of India, New Delhi

Jameson, J. D. and P. Sonthanam (1994). Ornamental fish culture. The national symposium for aquaculture for 2000 (AD), Madurai Kamraj University, India Press, Madurai, pp. 72-87.

Jonathan A. Moorhead and Chaoshu Zeng. (2010). Development of captive breeding techniques for marine ornamental fish: A Review. *Reviews in Fisheries Science*, 18(4):315–343,

Kumar Anjani (2010). Exports of livestock products from India: Performance, Competitiveness and Determinants. *Agrl.Econ. Res. Rev.*, 23: 57-67.

Kumar N.R, Rai M. (2007). Performance, competitiveness and determinants of tomato export from India. *Agricultural Economics Research Review*; 20:551-562.

Madhu K, Madhu R, Gopakumar G. (2009). Present scenario of marine ornamental fish trade in India, Captive breeding, culture, and trade and management strategies. *Fishing chimes;* 28:10-11.

Mahapatra, B. K., (1999) "Ornamental Fish Culture," Fisheries Research Station, Directorate of Fisheries, Government of West Bengal, Kalyani, Publication No. 10, 1999.

Manda, 1. S., Mahapatra B. K., Tripathi A. K., Verma M.R., Datta K.K., Ngachan S.V. (2007). Agribusiness opportunities of ornamental fisheries in North-Eastern Region of India, Agricultural Economics Research Review, 20 (Conference Issue): 471-488 Govt. of India,

Mandal Banani, Arunava Mukherjee, Subrata Sarkar and Samir Banerjee (2012). Study on the ornamental fin fish of Indian Sundarbans with special reference to few floral sources for carotenoid pigmentation. *World Journal of Fish and Marine Sciences*, 4 (6): 566-576.

MPEDA (2014) , Ornamental Fish Assistance Schemes 1-2. http://www.mpeda.com/subsidy/brofd.pdf.

Mukherjee M. (2000). Problem and prospects of aquarium fish trade in W.B., Fishing Chimes, April-2000, page 90-93.

Mukherjee M. Data A, Sen S, and Banerjee R. (2002). Ornamental fish farming – A new hope in microenterprise development in peri-urban area of Kolkata, HRD in Fisheries and Aquaculture for Eastern and North eastern India, CIFE, Kolkata Centre, 14 – 15 March, 2002, pp. 48 –59.

Mukherjee M., Datta S. and Datta A. (2002). The present status of ornamental fish industry in West Bengal – Its natural resource and marketing, Office of the Deputy Director of fisheries (Mand P), Govt. of West Bengal, Captain Bhery, E. M. Bypass, Kolkata – 700 039, India, VII (2), 8-11.

Mukherjee Madhumita, Sayantani Datta and Arindam Datta (2002). The Present Status of Ornamental Fish Industry in West Bengal – Its Natural Resource and Marketing. Office of the Deputy Director of fisheries (M&P), Govt. of West Bengal, Captain Bhery, E. M. Bypass, Kolkata – 700 039, India

Mukherjee Madhumita. (1999). West Bengal "Ugly Darlings" – trash or ornaments? Aquaculture Asia, April – June 1999.Page. 51 – 52.

Nagesh, T. S., J. Barman and D. Jana, (2004). Karyomor-phological Study of three Freshwater Ornamental Perches of West Bengal," *Journal of Inland Fish Society India*, 36(2):45-48.

NABARD (2007). Guidelines: Model bankable projects: Fisheries. *National Bank for Agriculture and Rural development (NABARD),* India. http://www.nabard.org modelbankprojects/fish_ornamental_fish.asp.

Ornamental fisheries schemes in West Bengal, In: International Conference Sustainable Ornamental Fisheries, March 23-25 Kochi, Kerala, India.

Panigrahi A. K., Dutta S. and Ghosh I. (2009). Selective study on the availability of Indigenous fish with ornamental value in some district of West Bengal, Sustainable Aquaculture Asia, XIV (4), 13-15.

Ponniah, A. G & Sood, N. (2002). Aquatic exotics and quarantine guidelines. NBFGR Special Publication No. 4, xii + 97p. National Bureau of Fish Genetic Resources, Lucknow, U.P., India.

Ponniah, A. G Unnithan, V. K. & Sood, N. (2002). National strategic plan for aquatic exotics and quarantine. NBFGR Special Publication No. 3, xiii + 119p. National Bureau of Fish Genetic Resources, Lucknow, U. P., India.'

Pramod, P.K., Sajeevan, T. P., Ramachandran, A., Sunesh Thampy S. &Somnath Pai (2010). Effects of two anesthetics on the water quality parameters during simulated transport of a tropical ornamental fish Puntius filamentosus(Valenciennes). *North American Journal of Aquaculture*, 72, pp. 290-297

Prathvi Rani, Sheela Immanuel, Nalini Ranjan Kumar (2014). Ornamental fish exports from India: performance, competitiveness and determinants. *International Journal of Fisheries and Aquatic Studies*, 1(4): 85-92.

Prathvi Rani, Sheela Immanuel, P. S. Ananthan, S. N. Ojha, N. R. Kumar and M. Krishnan (2013). Export performance of Indian ornamental fish - an analysis of growth, destination and diversity. *Indian J. Fish.*, 60(3): 81-86.

Pushpangadan K.R., Anil Kumar P., Ram Mohan, M.K. & Anikuttan, K.K. (2011). Ramachandran, A. (Ed.) (2002). Breeding, farming and management of ornamental fishes. School of Industrial Fisheries, Cochin University of Science and Technology (CUSAT), Kochi 682 016, Kerala, India, 203 p.

Rajan, M. R. and N. Karpagam (2004). Effect of different concentrations of vitamin c on feed utilization and breeding of certain live, breading ornamental fishes. *Environment and Ecology*, 22(1):31-37.

Sarma, S., B. K. Bhattacharyya, S. G. S. Zaildi and A. Landge, (2004). "Indigenous Ornamental Fish Biodiversity of Central Brahmaputra Valley Zone Assam," *Journal of Inland Fisheries Society of India*, 36(1):29-35.

Saha Manab Kumar and Bidhan C. Patra, (2013).Customers Preference for Aquarium Keeping: Market survey, Special Emphasis on Indigenous Ornamental Fishes in four District of West Bengal, India.

Salim, S. S. and Biradar, R. S. (2009). Indian shrimp trade: Reflections and prospects in the Post-WTO era. *Asian Fish Sci.*, (22): 805-821.

Sarkar U.K. and Lakra W.S. (2010). Small indigenous freshwater fish species of India: Significance, conservation and utilization, *National Bureau of Fish Genetic Resources,* Aquaculture Asia Magazine, XV (3), 34-35.

Silas, E.G., Gopalakrishnan, A., Ramachandran, A., Anna Mercy, T.V., Kripan Sarkar, (2011). Guidelines for green certificate for ornamental fish, MPEDA, Kochi

Singh, S. N. and Prusty, A. K. (2008).Ornamental fish trade: The Indian scenario in retrospect, its status and prospectus *vis-à-vis global demand.* Central Inland Fisheries Research Institute, Gujarat, India, p. 112-117.

Tissera K. (2010). Global trade in ornamental fishes- 1998 to 2007, International Aqua show 12-14 February, Cochin, Kerala, India (2010)

Usha Anandhi D. and Sharath Y.G., Ornamental Fish Fauna of Adda Hole: Kabbinale Forest Range, Southern Western Ghats, Karnataka, India, (2013). International Research Journal of Biologicla Sciences, Vol. 2(11), 60-64.

Welcomme, R. L. (1988). international introductions of inland ornamental fishes of coastal West Bengal, India—Prospects of Conservation and Involvement of Local Fishermen Aquatic Species," FAO Fish Technical Paper, pp. 294-318.

9

Estimation of Food Deprivation Undernourished and Elasticity of Income and Expenditure Behavior of Fish Farmers of Tripura

A.D. Upadhyay, A.B. Patel and Late M.C. Nandisha

Introduction

Access to adequate food, which is one of the fore-most basic needs of life, should be birthright of every single human being on this earth but in our country one fourth population is still underfed. The food insecurity situation is more rampant in the North-eastern region of the country. The proportion of population below poverty line (BPL) was reported at 31.4 per cent in Arunachal Pradesh, 34.0 per cent in Assam, 37.9 per cent in Manipur, 16.10 per cent in Meghalaya, 15.4 per cent in Mizoram, 8.8 per cent in Nagaland and 40.0 per cent in Tripura, in the year 2015 (RBI, 2015). At national lavel it was reported to be 37.20 per cent. By comparing BPL population in NE Region states and national level poverty it was found that several states of this region like Nagaland, Mizoram and Meghalaya were far better as in these states BPL percentage were very less than the national average, whereas in the states like Tripura, Manipur, Assam and Arunachal Pradesh BPL percentages were close to National level (37.20%). It has now been well established fact that the availability of food grains may not be the real problem of food security; it is prevailing poverty that comes in the way of achieving household's food security. Hence, household food security is a function not only availability of food but also of the purchasing power of household. Any effort to achieve sustainable development demands concerted efforts to reduce poverty, including finding the solutions to hunger and malnutrition. In this study, an attempt has been made to analyze the status of food deprivation among fish farmers of West Tripura district of Tripura and to examine consumption expenditure behavior fish farming families of the study area.

Conceptual Framework

The 1996 World Food Summit (WFS) adopted the definition of food security "Food security exists when all people at all times have physical and economic access to sufficient, safe and nutritious food to meet their dietary needs and food preferences for an active and healthy life." This definition has been identified with the four dimensions of food security: availability, access, stability and utilisation. The access of food is one of the important dimensions of food security and it is highly dependent on level of income of the people. The poverty level is also defined in terms of per capita energy requirements; in India earlier poverty was defined by the Planning Commission based on per capita incomes that enable a person to achieve daily energy intake of 2400 Kcal in rural areas and 2100 Kcal in urban areas. While proportion of people living below the poverty 'line' can be adjudged against these standards, we need to have a weighted average of energy requirement for the entire population so that nation can plan for production and availability of at least that much energy giving foodstuffs. FAO has fixed India's minimum requirement as 1890 Kcal per capita per day. This minimum calorie requirement calculation was based on the age, sex, consumption, the lowest acceptable weight for the typical height of the group in a country and the light activity norms (FAO, 2000). The ICMR Expert Group (1990) has concluded that on the basis of the present recommended dietary allowance (RDA), it ought to be 2200 Kcal per capita per day. On the other hand the National Nutrition Monitoring Bureau (NNMB) of the National Institute of Nutrition has used per capita consumption unit requirement of 2400 Kcal of energy and 60 grams of protein. The Working Group in Agriculture has used the weighted average per capita requirement of 2200 Kcal energy & 50 grams of proteins. In this study, 2200 Kcal per capita per day energy requirement was used as the cut-off point for determining the prevalence of food insecurity in fish farming community.

Methodology

Sampling and Data collection

West Tripura district of Tripura was purposively selected because this district encompasses more area under aquaculture and better representation of Tribals (about 40 per cent) in the fish farming community. There are 16 rural development blocks in this district; all blocks were included for the sampling of fish farmers. Now from each block, four villages were selected on the basis of aquaculture area and number of tribal fish farmers, at last all together 1242 respondents were selected using simple random sampling method. The sampled fish farmer comprised 58 % non tribal and 42% tribal fish farmers. For the detailed diagnostic study on food deprivation and household's expenditure behaviors, the primary data on demographic feature, monthly family consumption of food items both purchased as well as farm produced, annual income from different sources, monthly expenditures on food and non food items were collected through personal interview and study pertained to the period 2006. The pretested schedule including open ended type of questions have been used to canvass the primary data from the respondents.

Estimation of food deprivation

The estimate of the proportion of the population below minimum level of dietary energy consumption has been defined within a probability distribution framework (Loganaden Naiken, 1998)

$$P(U) = P(X<r_L) = + \int_{x<r_L} f(X)\, dX = F_X(r_L)$$

Where,

P (U)- the proportion of undernourished in total population

x -refer to the dietary energy consumption

r_{L}- is a cut-off point reflecting the minimum energy requirement (2200 K.Cal per capita per day)

f(x)- is the density function of dietary energy consumption.

Fx- is the cumulative distribution function

The density function f(x) is assumed to be log normal so that the parameters μ and σ^2 can be estimated on the basis of the mean $\overline{x}$ and the coefficient of variation, CV(x). The mean of dietary energy consumption ($\overline{x}$) was estimated by aggregating the food component of all commodities after conversion into energy values. The coefficient of variation, CV(x) of the household per caput dietary energy consumption is formulated as follows:

$$CV(X) = \sqrt{CV^2(x/v) + CV^2(x/r)}$$

Where, CV(X) is the total CV of the household per caput dietary energy consumption, CV (x|v) is the component due to household per caput income (v) and CV (x|r) is the component due to energy requirement (r). CV (x|r) is considered to be a fixed component and is estimated to correspond to about 0.20. CV (x|v) is however, estimated on the basis of household survey data.

Estimation of the Proportion of Undernourished

The density function of dietary energy consumption, f(x), is assumed to be log normal with parameters μ and σ^2. These parameters were estimated on the basis of the x and the coefficient of variation, CV(x) as follows-

$\sigma = [\log e\ (CV^2\ (x) + 1)]^{0.5}$ and

$\mu = \log e\ x - \sigma^2 / 2$

The proportion of population below r_L was evaluated as-

$\Phi\ [(\log e\ r_L - \mu) / \sigma)]$

Where Φ = standard normal cumulative distribution

Estimation of elasticity

In order to see the expenditure behavior of fish farming households, they have been classified into nine sub classes based on per capita income. Further, the log linear form of equation given below was used to measure the Engel's elasticity (income elasticity) and hosehold size elasticity (Pradhan and Panda, 2000) as it was observed that semi log and double log forms were more successful than other functional forms in case of cross section data.

$\text{Log}\ f\ (Yi) = \alpha i + \beta i \log(X) + \gamma i \log(n)$

Yi = *f* Consumption expenditure on i^{th} commodity

X = *f* Households income

n = *f* Family size

β = *f* Engel's Elasticity and

γ = Household size elasticity

Results and Discussion

Food Consumption Pattern

Food habit of tribal differs from the non tribal fish farmers. Rice is staple food of both tribal and non tribal fish farmers and it constitutes about 84 and 68 per cent in their daily food intake, respectively (Table 9.1). The consumption of dry fish was higher (0.45 g/capita/day) in tribal families than non tribal farmers (0.29 g/capita/day). However, non tribal fish farmers were found in a better position in terms of food consumption i.e. pulse, vegetables, milk, fat (oil), fresh fish, meat and eggs etc. It is also reflected from table that the fish farming community of study area were quite below than the ICMR norms in terms of consumption of pulses, milk, meat, eggs and oil. However, the protein requirement was compensated by consumption of fish.

Table 9.1: Food Consumption Pattern (per capita/day) of Different Fish Farming Community in West Tripura

Food items	Consumption of different food items (g/capita/day)			Per capita requirement (ICMR Norms)
	Non-tribal	Tribal	Overall	
Rice	469.85	590.55	519.93	420
Pulses	22.68	10.72	17.72	40
Vegetable	257.75	234.29	248.01	125
Milk	108.61	27.12	74.8	150
Fish	61.31	30	48.32	25
Dry fish	0.29	0.45	0.36	-
Meat	23.4	21.09	22.44	25
Egg (in no.)	0.37	0.15	0.28	45
Oil (in ml)	20.1	11.34	16.47	22

Nutritional Intake and Food Deprivation

The mean level of per capita calorie intake of fish farming community was estimated to be 2266 Kcal /day and it was slightly lower than the reported state average (2293 Kcal /day) by NSSO, 2000, for the rural areas (Table 9.2). However, calorie intake of fish farmers was at par with reported national average of calorie intake (2149 Kcal/day). The calorie intake in tribal family was relatively higher than of non tribal families. The same is also reflected from the Table 9.2 as per capita rice consumption of tribals is more that non tribals. However, protein intake was found much higher (67 g/day) in fish farming families than the state rural average i.e. 53.3 g/day and national average 59.1 g/days, (NSS, 2000). It is also interesting to note that among main source of protein, fish constitutes about 16 % in non tribal and 8.37% in case of tribal fish farmers. Whereas, as per NSSO report, 2000, protein from all animal sources shared only 10.8 per cent of total protein intake in rural areas of Tripura and at national level animal protein constitutes only 4.04 per cent. It is confirmed from above results that the protein intake particularly from animal sources in fish farming families is better. Hence, the problem of nutritional security in terms of protein of remote areas of the country can be addressed through promoting fish farming.

The segment of the population, whose calorie intake was below recommended dietary allowance (RDA), was considered as underfed. It was found that 54 per cent of non tribal families and 42 per cent of tribal families were below the cut off level (r_L) calorie intake (2200 Kcal/day). Overall proportion of fish farming families below recommended dietary allowance (RDA) was 48 per cent, which was significantly higher than state average 34.44 per cent, reported by NSSO, 2000. The tribal fish farming families are at par in both energy and protein intake, whereas, calories intake of non tribal were below to the state average and protein intake was at par. The Figure 9.1 and Figure 9.2 that majority of the fish farming families' calorie intake lies in range of 1800 to 2700 Kcal/day.

Table 9.2: Nutritional Intake (per capita/day) in Fish Farming Families of Tripura

Category	Nutritional intake (per capita/day)			Nutritional Intake reported by NSSO (1999-2000)	
	Non-tribal	Tribal	Overall	Rural Tripura	All India
Energy (Kcal/day)	2205	2332	2266	2293	2149
Fat (g)	30	18	25	49.5	36.1
Protein (g)	69	65	67	53.3	59.1
% of protein intake from fish	16.07	8.37	12.97	10.8*	4.04*
Proportion families below minimum energy requirement	54	42	48	34.44	-

% of total intake of proteinfrom egg, fish & meat

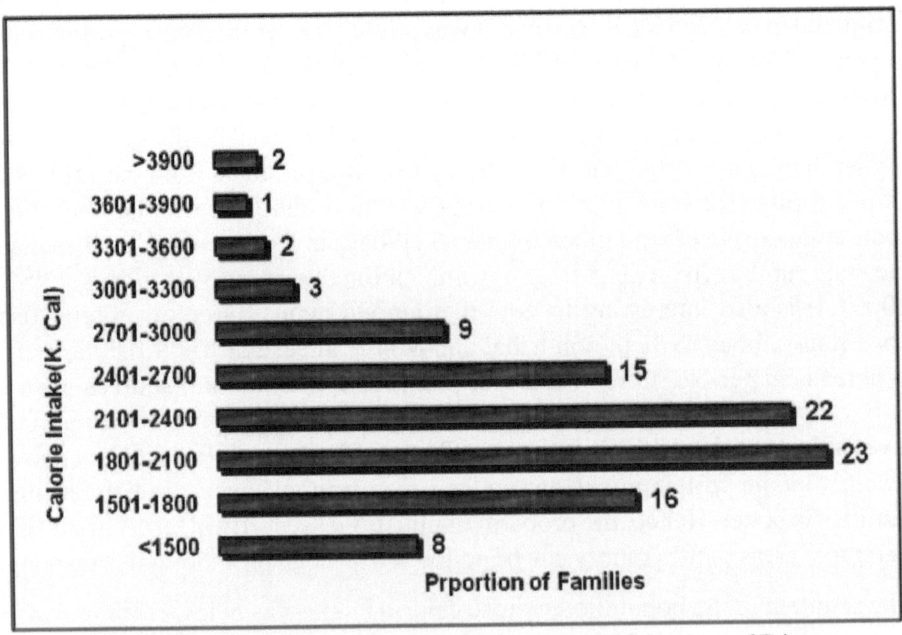

Fig. 9.1. Dietery Energy Consumption of Non Tribal Fish Farmers of Tripura

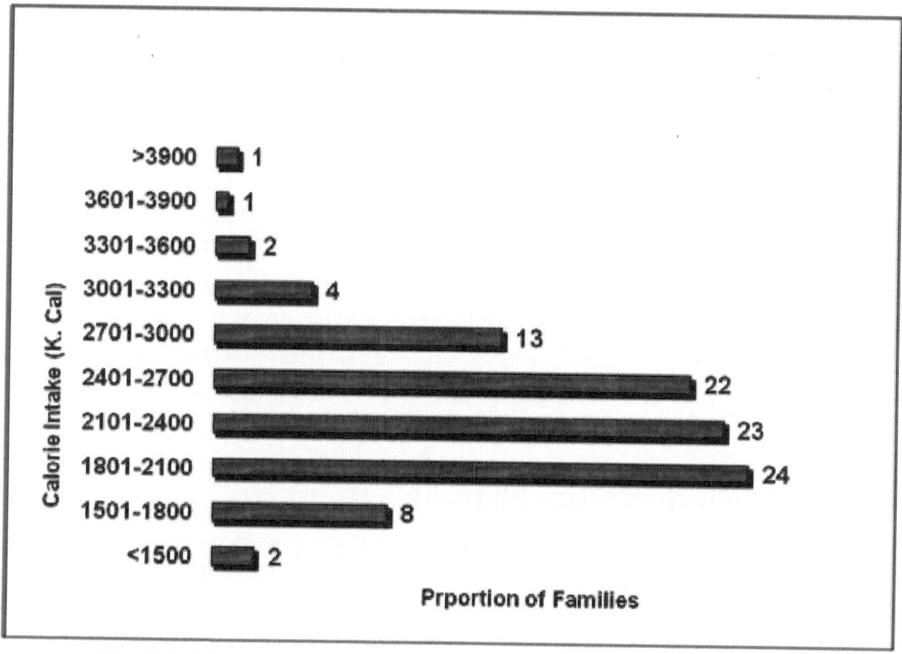

Fig. 9.2. Dietery Energy Consumption of Tribal Fish Farmers of Tripura

Factors Responsible for Low Energy Intake

In order to visualize the causes of low energy intake among fish farming families, they are further classified

Table 9.3: Factors Responsible for Low Energy Intake in Fish Farmer's Families

Energy consumption class (K.Cal)	Average family size (no)	Per capita land availability (ha)	Per capita water area availability (ha)
1500	7	0.3620	0.0117
1501-1800	6.23	0.3053	0.0139
1801-2100	5.74	0.3786	0.0187
2101-2400	5.67	0.4311	0.0168
2401-2700	5.54	0.5108	0.0180
2701-3000	5.98	0.5888	0.0198
3001-3300	5.51	0.6453	0.0200
3301-3600	5.27	0.6300	0.0251
3601-3900	4.25	0.7795	0.0293
3900	4.17	0.8932	0.0231

Income and Expenditure Pattern

In order to examine the income and expenditure pattern of both categories of fish farmers, they were further classified into different income class. The family incomes of both tribal and non tribal fish farmers are shown in Table 9.4. It is observed from the table that in the study area almost 80 per cent of households were having their monthly income below Rs. 7500. Family incomes of non tribal fish farmers were consistently higher in all income classes than the tribal fish farmers.

It is observed that the consumption expenditures of both groups increased with increase in households' income (Table 9.5 & 9.6). However, the proportionate increase in expenditure on education and health were faster than the food items. It is obvious that after meeting out the food requirements, education and health were in their priorities. Consumption expenditure of non tribal peoples on pulses, vegetables, milk and fish were significantly higher in all income classes than the tribal peoples. This may be because of their higher income.

Income and Households' size Elasticity

It has been postulated that the households' consumption expenditures for specific commodity is linear homogenous function of its income and family size. However, it is depicted from the Table 9.7 that elasticity of family size was higher than the income elasticity's for all commodities in both tribal and non tribal families. This reflects that the family consumption expenditure in study areas were more responsive to the family size than the family incomes. Engel's elasticity and households size elasticity tends to move in opposite directions. Engel's elasticity was low in case of rice whereas household size elasticity was very high for rice indicating that quantum of food consumption expenditure was more responsive to the family size than the income. Income elasticity for fish and milk in non tribal group was found higher than the tribal group on the other hand Income elasticity for meat was grater in case of tribal families. This indicates their tastes and preferences for these food items.

Table 9.4: Average Monthly Income (Rs) of Tribal and Non Tribal Families

Income class	Tribal fish farmers				Non Tribal fish farmers		
	Percapita Income	Number of families	Cumulative Proportion	Households' income	Number of families	Cumulative Proportion	Households' income
1	<300	73	15	1164	96	14	1299
2	301-600	110	38	2410	193	41	2840
3	601-900	89	57	4050	135	60	4394
4	901-1200	62	70	6219	94	74	6354
5	1201-1500	47	80	7259	75	84	7415
6	1501-1800	36	87	10708	32	89	8429
7	1801-2100	23	92	10051	30	93	10174
8	2101-2400	10	94	12747	14	95	10721
9	>2400	27	100	19376	34	100	35980
	Total	477			703		

Table 9.5: Monthly Consumption Expenditure (Rs) on Different Food and Non Food Items by Tribal Households

Income class	Expenditure on food items (Rs)								Expenditure on non food items (Rs)					Total expenditures (Rs)(A+B)
	Rice	Pulse	Vegetables	Milk	Fish	Meat	Misc.	Total expenditure on food(A)	Cloth	Health	Education	Misc.	Total expenditure on non food(B)	
1	1397	79	384	34	462	322	103	2781	309	166	289	478	1242	4023
2	1476	68	405	36	362	249	95	2691	251	123	399	323	1096	3787
3	1504	64	499	58	408	312	95	2940	271	263	403	364	1301	4241
4	1603	82	501	87	452	370	114	3209	331	200	583	506	1620	4829
5	1516	67	591	94	437	321	101	3127	456	134	1114	511	2215	5342
6	1798	95	713	144	509	483	135	3877	355	205	1066	704	2330	6207
7	1546	84	750	95	549	540	155	3719	250	472	959	731	2412	6131
8	1440	112	722	91	791	579	147	3882	360	339	910	1707	3316	7198
9	1344	123	650	153	555	502	134	3461	324	282	771	695	2072	5533

Table 9.6: Monthly Consumption Expenditure (Rs.) on Different Food and Non Food Items by Non Tribal Fish Farmers

Income class	Expenditure on food items (Rs)								Expenditure on non food items (Rs)					Total expenditures (Rs)(A+B)
	Rice	Pulse	Vegetables	Milk	Fish	Meat	Misc.	Total expenditure on food(A)	Cloth	Health	Education	Misc.	Total expenditure on non food(B)	
1	1193	162	481	194	545	383	178	3136	230	271	317	468	1286	4422
2	1323	167	498	220	588	486	181	3463	258	308	414	498	1478	4941
3	1255	183	561	246	659	425	200	3529	262	337	464	563	1626	5155
4	1285	180	599	326	749	493	205	3837	405	444	390	663	1902	5739
5	1213	172	655	357	730	485	220	3832	397	426	581	846	2250	6082
6	1088	183	644	340	744	445	229	3673	383	520	743	984	2630	6303
7	1128	178	746	340	660	564	236	3852	310	602	956	1048	2916	6768
8	970	161	554	387	916	403	219	3610	306	409	611	1423	2749	6359
9	870	158	595	358	849	576	220	3626	319	493	476	1384	2672	6298

Table 9.7: Engel's elasticity and family size elasticity for food and not food items

Commodity	Tribal		Non Tribal	
	β	Γ	β	γ
Rice	0.094736*** (2.729576)	0.791479*** (9.499392)	0.030261* (1.722657)	0.87536*** (24.45067)
Fish	0.054908 (1.102428)	0.615591*** (5.148568)	0.21163*** (7.14397)	0.548271*** (9.081187)
Milk	0.095019 (1.343551)	0.235935 (1.38968)	0.157041*** (4.740611)	0.414649*** (6.141711)
Vegetables	0.158371*** (3.833932)	0.685613*** (6.913928)	0.152797*** (4.892218)	0.493237*** (7.748752)
Meat	0.207904*** (4.038663)	0.543298*** (4.396339	0.178664*** (4.738454)	0.359302*** (4.675675)
Cloth	-0.04409 (-0.66598)	0.749113*** (4.713619)	0.249336*** (6.441334)	0.580793*** (7.362023)
Education	0.269271*** (3.117712)	-0.10868 (-0.52417)	0.271933*** (5.69963)	0.265669*** (2.732195)
Health	0.159419** (2.18586)	0.525623*** (3.00215)	0.25626*** (4.968002)	0.385696*** (3.668871)

***,**,* indicates significant at 1, 5 and 10 per cent probability level and figures in parentheses shows t values)

Conclusion

In the current study, food security status and income and expenditure patterns of selected fish farmers of West Tripura district were examined. It was found from the study that about 48 per cent fish farming families were below the minimum energy requirement 2200 K.Cal per capita per day, recommended by ICMR Expert Group (1990). This proportion was higher than the state average reported by the NSSO, 2000. It is concluded from the study that the large segment of fish farming community also suffers with dietary energy supplies. However, protein intake of fish farming families of Tripura, were more than ICMR recommendation. This nutritional security was mainly due to presence of aquaculture in their farming system. The nutritional security in terms of protein of remote areas of the country can be addressed through emphasizing or popularizing fish farming. From this study it is observed that non tribal fish farmers were better in terms of income. Due to this consumption expenditure on food items as well as non food items of non tribal families were more than that of tribal families.

References

Barah, B.C. (2004). Food security and public distribution system today: failures and successes, Ed.: Amalesh Banerjee. Kanishka publishing, New Delhi, 374 p.

Gillespie and Mason (1991). Indian experience on household food and nutrition security http://www.fao.org/docrep/x0172e/x0172e01.htm#P20_7375.

ICMR (1990). Report of Committee on Dietary Allowances, New Delhi,

Lorstad, M. (1974). On estimating incidence of under nutrition. FAO Nutrition Newsletter, 12(1): 1-11.

Mishra, S.K. (1999). Rural development in the North-Eastern Region of India: Constraints and prospects. MPRA Paper No. 1833.

Naiken, L. (1998). On certain statistical issues arising from the use of energy requirements in estimating the prevalence of energy inadequacy (under nutrition). *Journal of Indian Society of Agricultural Statistics*, L1(2.3): 113-128.

NSSO (2001). Nutritional intake in India, 1999-2000. Report No.471/55/1.0/9, National Sample Survey Organization, MSPI, GOI, pp. A-168.

Pradhan B.K., Roy P.K., Saluja, M.R. and Venkatraman, Shanta (1998). Income, expenditure and social sectors indicators of households in rural and urban India, Paper presented at the Micro Impacts of macro-economic and adjustment policies, III Annual Meeting, Nov.2-6, Nepal.

Pradhan, N.B. and Panda, B.K. (2000). Econometric studies of economic reforms in India ed. V.V. Somayajulu, Academic Foundation, Gaziabad, p 564.

Sukhatme, P.V. (1961). The world's hunger and future needs in food supplies. *Journal of the Royal Statistical Society*, Series A, 124: 463-585.

References

10

Assessment of Appropriateness of Indigenous Technical Knowledge (ITK) in Fisheries

B. Nightingale Devi, S.K. Mishra, M. Krishnan and Nilesh Pawar

Introduction

The better quality of life for the Indians who in majority depend on agricultural would be impossible by keeping the rich tradition of Indigenous Technical Knowledge (ITK) aside (Das, et al., 2002). According to Farrington and Martin (1988), ITK is based on knowledge, beliefs and customs which are internally consistent and logical to those holding them but at odds with the objectively deduced findings of formal science". Indigenous Technical Knowledge (ITK) is the knowledge that the indigenous and local community accumulates over generations of living in local environment. It encompasses all forms of knowledge, experience, technologies, know-how, skills, beliefs and practices. These are also termed as "indigenous knowledge", "people's knowledge" "traditional knowledge" or "local knowledge" and are limited within a community. It is unique to a given culture, location or society.

Fishermen all over the world use some kind of traditional knowledge. ITK provides valuable insight into sustainable aquaculture, because it passes through considerable adaptation, up gradation and modification over a period of time and carried on from one generation to another as a family technology. Naturally, ITK fits well to a particular environment rather than modern technological knowledge. This knowledge has been developed outside the formal education system and refers to the large body of knowledge and skills which includes variety of fish culture practices, when and where to catch, how to process fish with a cost effective methods, and how to maintain their environment in a state of equilibrium. ITK is collective in nature and is often considered the property of the entire community, and does not belong to any single individual within the community.

Scattered efforts have been made by ICAR, SAUs, other research institutes and NGOs for documenting ITKs in agriculture and allied disciplines like fisheries (Das et al., 2002 and 2003). But hardly any effort has been made to assess and validate the appropriateness of those ITKs. Inventories of ITKs alone will carry no meaning and solve any problem, unless they are properly used. Once the knowledge is systematically documented, they should be prioritized by experts in consultations with potential users and subjected to scientific validation to see their scientific rationality (Devi et al., 2009). It will also support to develop best alternative eco-friendly practices for sustainable aquaculture practices. Based on this background, the present study was undertaken in the state of Manipur to formulate a methodology for assessing appropriateness of Indigenous Technical Knowledge in Fisheries and to assess the appropriateness of some most widely used Indigenous Technical Knowledge in Fisheries of Manipur.

ITKs of Manipur

Manipur was purposively selected for the study as it is very rich in fisheries as well as natural resources and popular in its fisheries related activities among the northeastern states of India. Topographically, Manipur is constituted by two distinct geographic features, i.e., an elevated central plain forming a valley and rows of mountains on all sides, representing 10% and 90% of the total geographical area of the state, respectively. While the valley is very rich in lakes, ponds and wetlands, the mountainous regions are drained by three river systems, viz, the Barak river system, the Manipur river system and the Yu-river system. The study was restricted to the central valley of the state which comprises of four districts, namely, Imphal West, Imphal East, Bishnupur and Thoubal district. From each district, two fishing villages were selected. Looking into the nature of the study, the information was collected by personal interview, field observations and focus group discussions with about ten to twelve fish farmers & farm women selected randomly from each of the eight selected villages. A total of forty five numbers of ITKs were collected, documented and later categorized into ten groups/aspects, viz, i) Fishing methods, ii) Fishing traps, iii) Fish aggregating devices, iv) Fish processing and storage, v) Fish health management, vi) Construction of gear and craft, vii) Preservation of gear and craft materials, viii) Paddy-cum-Fish culture, ix) Rituals, Belief and Customs based on Fish and x) Others (Unclassified).

A. Assessment of Appropriateness

To make use of the ITKs, assessment of their appropriateness and scientific value is equally important besides collection and documentation. Appropriateness of ITK means identifying its potentially useful rational knowledge embedded in it and evaluating its effectiveness for making use of it in developmental process. In view of the above and in the light of the purpose of the present study, it was felt essential to assess the appropriateness of documented ITKs of fisheries to establish their rationale and validity. While assessing the appropriateness, characteristics of innovations (Rogers, 1969) and parameters of ITKs (Nirmale, 2002) were taken into consideration and accordingly, six parameters were identified, namely, (i) Cost-effectiveness, (ii) Materials availability, (iii) Social acceptability, (iv) Cultural appropriateness, (v) Environmental compatibility and (vi) Scientific value. The assessment of appropriateness of the identified ITKs was made based on responses of experts on these six parameters.

B. Parameters for assessing the Appropriateness

For assessing the appropriateness and scientific value of the ITKs in fisheries, selected parameters were operationalized as given below:

1) ***Cost Effectiveness***: It is the degree to which the fishermen find any ITK method or practice cheaper as compared to any other similar practice and affordable within their existing purchasing capacity. In other words, cheaper the practices or methods more will be the score (Table 10.1).

2) ***Materials Availability***: It is the degree to which the fishermen would find the materials required for any ITK method or practice are easily available in sufficient quantity in the local area of fishing villages. In other words, easy and sufficient availability of materials in the area, more will be the score.

3) ***Social Acceptability***: It is the degree to which fishermen would find any ITK practice or method suitable, comfortable and useful to their existing fishing conditions. In other words, more acceptability and usefulness by the society, more will be the score.

4) ***Cultural Appropriateness***: It is the degree to which any ITK practice or method is in consonance, compatible, matching and appropriate with the existing socio-cultural norms and values of the fishing communities. In other words, more appropriate to the culture, more will be the score.

5) ***Environmental Compatibility***: It is the degree to which any ITK practice or method is friendly and compatible to the environment i.e., without having any disturbance or negative effect on different components of the

environment, viz., soil, water, air, flora and fauna. In other words, more the environmental friendly, more will be the score.

6) ***Scientific Value*:** It is the degree to which any ITK practice or method is consistent, rational and agreement with the established and proven scientific theories, laws and principles. In other words, more the scientific values more will be the score.

Table 10.1: Scoring Pattern of the Parameters for Assessing the Appropriateness

S.No.	Parameters	Continuum Scale	Score Assigned
1	Cost effectiveness	Very High Cost	1
		High Cost	2
		Moderate Cost	3
		Cheap	4
		Very Cheap	5
2	Materials Availability	Very Less Availability	1
		Less Availability	2
		Somewhat Availability	3
		Easy Availability	4
		Very Easy Availability	5
3	Social Acceptability	Very Low Acceptability	1
		Low Acceptability	2
		Moderate Acceptability	3
		High Acceptability	4
		Very High Acceptability	5
4	Cultural Appropriateness	Very Less Appropriate	1
		Less Appropriate	2
		Moderate Appropriate	3
		High Appropriate	4
		Very High Appropriate	5
5	Environmental Compatibility	Very Low Compatibility	1
		Low Compatibility	2
		Moderate Compatibility	3
		High Compatibility	4
		Very High Compatibility	5
6	Scientific Value	Very Low Scientific Value	1
		Low Scientific Value	2
		Moderate Scientific Value	3
		High Scientific Value	4
		Very High Scientific Value	5

C. Selection of ITKs for assessment of appropriateness

Before assessment, the collected ITKs were prioritized based on the frequency commonly used by fishermen. Then, twelve commonly used ITKs under following four aspects were selected for subjecting to assessment of their appropriateness and scientific value.

1. ***Fish Aggregating Devices (FADs)***
   - a) Phoom-namba - Transplantation of Phoomdis (floating mass of aquatic weeds) for FAD in lake
   - b) Kao - branches, twigs in bamboo-made triangular-shaped structures as FAD in river
   - c) FAD for Grass carp - Putting of grasses in selected area before catching the fish
   - d) Use of "Ising kambong" (Hygoryza sp.) as FAD for air-breathing fishes in low-lying areas
2. ***Fishing Traps***
   - a) Operation of Box Traps (Taijeps) with identification marks of different fishermen community in Lakes as well as in Ponds.
   - b) Trap preservation - bamboo Splits are heavily smoked before making traps to avoid from infestation of insects and to increase its efficiency and longevity.
3. ***Fish Health Management***
   - a) Mixture of turmeric powder and lime are put in pond as prophylactic measures to prevent from EUS (Epizootic Ulcerative syndrome)
   - b) Banana stem are used in pond to improve water quality
4. ***Fish Processing and Preservation***
   - a) Hentak - (fermented fish paste) - Alocasia micrrohiza is used to accelerate fermentation process.
   - b) Ngari - (semi-fermented fish) - earthen pot is used as container for fermentation process.
   - c) Nganam - (steamed fish product) - turmeric leaves are used as preservatives
   - d) Fern (Microlepia strigosa) gives golden yellow color when its fume is passed through smoked fish.

D. Collection of Responses from Experts

The assessment of appropriateness and scientific value of these twelve identified ITKs were made based on responses of experts on six parameters, namely, (i) Cost Effectiveness, (ii) Materials Availability, (iii) Social Acceptability (iv) Cultural Appropriateness, (v) Environmental Compatibility and (vi) Scientific Value. The responses were collected by sending a "Response Sheet" containing the 5-point continuum scale through emails and by hand (for ICAR-CIFE, Mumbai) along with the brief description of the ITKs to 90 experts which included scientists, researchers, extension personnel, academician and subject matter specialists of different recognized institutes, namely, ICAR-Central Institute of Fisheries Education (ICAR-CIFE), Mumbai; ICAR-Central Marine Fisheries Research Institute (ICAR-CMFRI), Cochin; ICAR-Central Institute of Freshwater Aquaculture (ICAR-CIFA), Bhubaneswar; ICAR-Project Directorate of Cold Water Fisheries (ICAR-PDCWF), Dehradun; College of Fisheries, Tripura; Institute of Bio-resources and Sustainable Development (IBSD-DBT), Imphal and Department of Fisheries, Manipur. Response Sheets were received by return mails, post and by hand. Out of 90 experts, responses from 44 experts were received, analyzed and discussed.

E. Scoring Techniques for Assessment of Appropriateness

Experts were asked to give scores in a five point continuum scale, i.e., 1, 2, 3, 4 or 5 as per the weight age they give to each of the ITKs, based on their expertise and conscience, in an ascending order as given in the Table 10.1. Then, all the scores of a particular parameter of any ITK are added and average scores were calculated for all the parameters. To assess the overall appropriateness of these ITKs, average scores of all the six parameters are added. For easier discussion, the percentages were also calculated and tabulated.

Findings and Discussion

Finding in the Table 10.2 revealed the appropriateness of twelve selected ITKs as assessed by fisheries experts from different organizations. According to the experts, "Phoom namba" as a FAD very appropriately fits (4.20/5) to the cultural aspects of fishing community, whereas, they perceived, it is moderately compatible (3.00/5) with the environment. This might be due to the fact that "Phoom namba" being a transplantation of phoomdis (floating mass of aquatic weeds) may results in the depletion of water resources and may cause different water polluting factors and create environmental problem. They commented that considering the condition of the Loktak Lake of Manipur today, this method of FAD is environmentally and socially not sound and the Loktak Development Authority (LDA) is spending crores of Rupees annually for removing these phoomdis from the lake.

Table 10.2: Assessment of Appropriateness of ITKs in Fisheries
Distribution of Average Score of Experts on the Appropriateness of Selected ITKs in Fisheries

Sl.No.		Selected ITK Methods and Practices Average scores of the Parameters for the Assessment of Appropriateness (out of 5 scores)						Overall Appropriateness (out of 30 scores)	Percentage of Overall Appropr iateness
1	**Fish Aggregating Devices (FADs)**	Cost Effectiveness	Materials Availability	Social Acceptability	Cultural Appropri ateness	Environ-mental Soundness	Scientific Value		
a	Phoom-namba - Transplantation of Phoom (aquatic plants) for FAD in lake	3.53	4.07	3.93	4.20	3.00	3.80	22.53	75.10
b	Kao – branches & twigs in bamboo made triangular shaped structures in river	4.47	4.33	3.87	4.00	3.87	3.47	24.00	80.00
c	FAD for grass carp – Putting of grasses in selected area before catching the fish	4.07	4.20	4.07	3.93	3.93	4.33	24.53	81.77
d	Use of Ising kambong (Hygoryza sp.) as FAD for air breathing fishes in low lying areas	3.93	3.93	3.53	3.40	3.80	3.80	22.40	74.67
2. Fish Traps									
a	Operation of Box traps (Taijeps) with identification marks in ponds as well as in lakes	4.53	4.33	3.60	4.13	3.27	4.27	24.14	80.47

Contd.

b Splits of bamboo are heavily smoked before making traps	4.60	4.33	4.20	4.20	3.93	4.33	25.60	85.33
3. Fish Health Management								
a Turmeric powder and lime are put in ponds as prophylactic measures to prevent from EUS	4.33	4.13	3.93	3.73	4.13	4.87	25.13	83.77
b Banana stems are used in ponds to improve water quality	3.73	3.67	3.40	3.53	3.80	3.53	21.67	72.23
4. Fish Processing and Preservation								
a Hentak (fermented fish paste) – Alocasia microrhiza is used to accelerate fermentation	4.27	4.60	4.27	4.53	4.47	4.13	26.27	87.57
b Ngari (semi-fermented fish)- earthen pot is used as container for fermentation process	4.60	4.33	4.13	4.80	4.53	4.47	26.87	89.57
c Nganam (steamed fish product)- turmeric leaves are used as preservatives	4.33	4.47	4.60	4.53	4.40	4.20	26.53	88.43
d Fern (Microlepia strigosa) gives golden yellow colour when its fumes are passed through smoked fishes	4.20	4.07	4.20	4.53	4.27	4.07	25.33	84.43

As per the table, "Kao" as a FAD in riverine system was rated as a highly cost-effective (4.47/5) indigenous method. Materials for preparing Kao are also easily available (4.33/5). As per the experts, it possesses a moderate scientific value (3.47/5). As far as FAD for Grass carp i.e. "Bunch of Grass" is concerned, it was adjudged as having very high scientific value (4.33)/5 as it is taking the advantage of the feeding habit of the Grass carp and also the materials are very easily available (4.20/5).

The materials for Hygoryza sp. as FAD for air breathing fishes are also easily available (3.93/5) and cost-effective (3.93/5), but might not be culturally very appropriate (3.40/5) and it is limited to only some places of Nambol Thiyam pat under the Bishnupur district. The results in the table also reveal that 'Operation of Box traps (Taijep) in ponds and lakes' as fishing traps is very cost effective (4.53/5) and also possess a very good scientific value (4.27/5). But, the experts felt that these traps might not be so environmental friendly (3.27/5) and the social acceptability might also be moderate (3.60/5). The ITK practice of 'Heavy smoking' of bamboo splits" before making traps was assessed as very cost-effective (4.60/5) as well as having very high scientific value (4.33/5). However, the environmental compatibility got less score of 3.93/5. The expert gave a viewed that keeping the bamboo splits over traditional fireplace accelerates drying by reducing moisture content thereby enhancing longevity.

Two ITKs, relating to fish health management practices, were also assessed. The scores in the table indicate that 'application of turmeric powder and lime in ponds to prevent EUS disease of fish' was adjudged by experts as having very high scientific value (4.87/5) as well as a highly cost effective practice (4.33/5). The social acceptability of the ITK practice of 'using banana stems in ponds to improve water quality' was assessed by experts as moderate (3.40/5) as compared to other practices, with having moderate scientific value (3.53/5) and it needs to be validated by scientific findings.

According to the table, all the parameters for assessing the appropriateness of the four indigenous practices of fish processing and preservation methods were given very good scores, which means that they are all very cheap, raw materials are also easily available with very high social as well as cultural compatibility, highly environmentally friendly and having a very high scientific value and rationality. The experts suggested that it needs to investigate and validate about how Alocasia micrrohiza helps to preserve Hentak. And turmeric used in the preparation of Nganam imparts a typical flavor to the products and also preservation may be due to reduction in moisture while baking over the chulla and continued keeping near it.

A critical analysis of the findings in the table also enlightened the overall appropriateness of these selected ITK practices. It is noteworthy to mention here that all the twelve ITKs were assessed as more than 72 per cent appropriate", the lowest being 72.23 per cent appropriate for "use of banana stems in pond to improve water quality" and as high as 89.57 per cent appropriate for semi-fermented fish product Ngari.

Conclusion

Many fishermen who have scientific approach and practical knowledge in dealing with various farming systems and technologies can hardly document their experience. This ITK could help the scientific community and fisher farmers to use them to the best advantage of themselves for the sustainable development of the sector. The high scores of appropriateness of experts are an indication that the ITK practices and methods being followed in the Central valley region of Manipur are not only having scientific basis but also they are highly sustainable. The fish farmers had taken quite good considerations of the environment, the culture, the society, the locally available raw material, while evolving these indigenous methods, practices and products in their efforts to achieve sustainable livelihood options. As there are numerous ITKs available, they need to be documented first, and then similar efforts should be made extensively to prioritize and assess the appropriateness, before subjecting them to scientific validation. This will save time in scientific investigation on validation and quicken the amalgamation of ITKs with the modern scientific knowledge base to come out with sustainable aquaculture methods and practices enabling blue revolution.

Acknowledgments: The authors are grateful to the Director/ Vice Chancellor, Central Institute of Fisheries Education, Mumbai, and Dr. C. Shrinivas Rao, Director ICAR-NAARM for constant encouragement and support during the entire period of this work

References

Das, P., Das, S.K., Arya, H.P.S., Reddy, G.S. and Mishra, A. (2002). Inventory of indigenous technical knowledge in Agriculture, Document-1. Mission unit, Division of Agricultural Extension, ICAR, New Delhi.

Das, P., Das, S.K., Arya, H.P.S., Reddy, G.S. and Mishra, A. (2003). Validation of indigenous technical knowledge in Agriculture, Document-3. Mission unit, Division of Agricultural Extension, ICAR, New Delhi.

Devi, B. N., Mishra, S.K., Sharma, A. and Ojha, S.N. (2009). Preserving indigenous technical knowledge in fisheries for sustaining fish production in India. *Intensive Agriculture*, 48(4): 3-7.

Farrington, J. and Martin, A. (1988). Farmer participation in agricultural research – a review of concepts and practices. Occasional Paper No. 9, ODI: 79.

Nirmale, V. (2002). Indigenous knowledge in management of Marine fisheries of Maharashtra. Unpublished PhD. Thesis. CIFE, Mumbai, India.

Rogers, E.M. (1969). Diffusion of innovations. The Free Press, New York, Callier MacMillan Limited, London.

11

Analysis of Food Security of Fish Farmers of Tripura Using Clustering Technique

A.D. Upadhyay; A.B. Patel and D.K. Pandey

Introduction

The Food Security is one of the most important dimensions of livelihood security of household (Rahman and Akter, 2010), because all efforts of the family members are directed towards fulfilling the family need including food. Food security defined as "access by all people at all times to enough food for an active, healthy life". Food security includes at a minimum: (1) the ready availability of nutritionally adequate and safe foods, and (2) an assured ability to acquire acceptable foods in socially acceptable ways (Anderson, 1990). It is also an indicator of sustainability of any vocation be it fish, crop, livestock production or combination of it. In case of Tripura very little study has been made in this aspect. Therefore, in this paper it is attempted to assess the status of food security of fish farmers of Tripura, using Index of Food Consumption (IFC) and Clustering Technique.

Data and Methodology

This study was undertaken in Tripura state and study area comprises 6 Rural Development Blocks viz. Bishalgarh, Melagarh and Mohanpur of West Tripura district; Hrishyamuk, Rajanagar and Matabari of South Tripura District. From these blocks a sample of 105 fish farmers was selected using stratified sampling method for gathering the primary data. The pretested, semi structured survey schedule was administered to collect the data during 2010.

To analyse status of food security of fish farmers, Index of Food Consumption (Anon, 2003) was calculated. In Index of Food Consumption (IFC), ICMR recommended level of intake of ith food item was used as basis, to compute Index for j^{th} food item in i^{th} family using following formula:

Index of Food Consumption (IFC)

$IFC_{ij} = (X_{ij} / X_j ICMR)$

Whereas,

IFC_{ij} = Index of Food Consumption in i^{th} Family for the jth Food Item,

X_{ij}= Per consumer unit per day consumption of jth food item in the ith family in grams

$X_j ICMR$= Per consumer unit per day recommended intake of the jth food item

Food item for which Index of Food Consumption (IFC_{ij}) is 1 or more than one indicates food intake of ith family for j^{th} food item is at par to its recommended level by Indian Council of Medical Research (ICMR). Hence, this index gives an idea about extent of food security and food deprivation of fish farming community.

Classification of Fish Farmers

In order to compare status food security across fish farming community, sampled fish farmers were classified into two groups on the basis of Food Consumption Indices (IFC), using Two Step Clustering Technique (SPSS-15). The clustering is the assignment of a set of observations into subsets (called clusters) so that observations in the same cluster are similar in some sense.

The Two Step Cluster Method is a scalable cluster analysis algorithm designed to handle very large data sets. It can handle both continuous and categorical variables and attributes. It requires only one data pass. It has two steps 1) pre-cluster the cases (or records) into many small sub-clusters; 2) cluster the sub-clusters resulting from pre-cluster step into the desired number of clusters. It can also automatically select the number of clusters. In general, the larger the number of sub-clusters produced by the pre-cluster step, the more accurate the final result is. However, too many sub-clusters will slow down the clustering during the second step. The maximum number of sub-clusters should be carefully chosen so that it is large enough to produce accurate results and small enough not to slow down the second step clustering. The advantage and disadvantage of two-step cluster analysis is that it is not an obligation to know the number of clusters firstly. Two-step cluster analysis tries to determine optimal sub-population number. Especially, latent class and cluster analysis techniques have been identified according to log-likelihood distance measurement (Murat Kayri, 2007).

Status of Food Intake of Fish Farmers

The present level of consumption of different food items and their recommended level by ICMR are given in Table 11.1. The level of per capita per day consumption of cereals (474.97g), fat (22.88g) and fish (58.15g) in fish farming families were at par with recommended level of 420 g/day cereals, 22grams/ day fats and 45 g/day fish. However, the level of consumption of pulses, milk, sugar, egg and meat per capita per day in the study area were below to recommended quantities. This indicates poor level of intake of these food items by the fish farming families of Tripura. The higher intake of fish by fish farming families is an indicative of their better nutritional security specially in terms of protein because in North East in general and Tripura in particular majority of peoples suffers from nutritional insecurity specially from deficiency protein.

Table 11.1: Per capita consumption of different food items by fish farmers and their recommended level by ICMR.

Food Item	Consumption(quantity in g)	ICMR *Recommendation (quantity in g)
Cereals	474.97	420
Pulses	17.52	40
Fat	22.88	22
Sugar	19.45	30
Milk	112.47	150
Fish	58.15	45
Egg	3.46	25
Meat	19.94	45

*(Source: Food Security Atlas of Rural India (2003), MS Swaminathan Foundation, Tamilnadu), Chennai

Clustering of Fish Farmers on the Basis of Index of Food Consumption (IFC)

The fish farming families of which Indices of Food Consumption for different food items were low were assigned in Cluster-1, other group of farmers having higher Indices of Food Consumption was assigned in Cluster-2. Hence, Cluster-1 represent group of fish farmers who have poor access to food, whereas, Cluster -2 represent group of farmers having better access to the food. 73 % of fish farmers belonged to cluster 1 and 27% of fish farmers belonged to Cluster-2 (Table 11.2). The Food Consumption Indices for cereals, oils, milk, fish and meat in Cluster-2 were >1 indicating that intake of these food items was at par to its recommended level. However, the Food Consumption Indices (IFC) for other items like pulses, sugar, milk, egg and meat were less than one in this cluster but intake of these items were greater than cluster one. In cluster-1,

Food Consumption Indices (IFC) for most of the food items were less than 1 showing poor level of intake of these food items by majority of fish farming families (73%). These results reflect that only 27% of fish farmers were having better access to food on the other hand 73% of fish farming families suffered with poor intake of many food items. The reason may be insufficient farm produce and low level of family income of fish farmer due to which they were unable to access to recommended level of foods. Hence, it can be concluded that the fish farmers of the state also suffered with food deficiencies.

Table 11.2: Food Availability Indices of Two Clusters of Fish Farmers

Sl.no.	Food items	Clusters' Centroids for Food Availability Indices		
		Cluster-1	Cluster-2	Combined
1	Cereals	0.96597	1.460476	1.097418
2	Pulse	0.41526	0.561048	0.454013
3	Oils	1.03288	1.267048	1.095127
4	Sugar	0.603483	0.969905	0.700886
5	Milk	0.744655	1.355476	0.907025
6	Fish	1.742259	3.769571	2.281165
7	Egg	0.086759	0.114429	0.094114
8	Meat	0.616983	1.353476	0.812759
	% of Families	73	27	100

Characteristics of two Cluster of Fish Farmers

In order to visualize causes of poor access to food of cluster 1, various socio-economic characteristics such as expenditure patterns, demographic feature, education status, farm and off farm employment, possession of resources in terms of land and water of two categories of fish farmers were mapped out.

Expenditure Pattern of Fish Farmers

The expenditure patterns of two of clusters fish farmers showed that the fish farmers those belonged to cluster-2 spent on an average about Rs. 10581.10 per month on purchase of various foods and non foods items, whereas, the fish farmers of cluster-1 spent about Rs. 8591.83 per month on procurement of similar food and non food items (Table 11.3). These results indicate that in study area living expense of a family (with family size 4.57) with better food access was Rs. 10581.10 per month. However, only 27 % farmers were able to earn from farm and non farm sources of income to meet out this expense. The fish farmers of cluster two spent more on food as well as on non foods including rice purses, milk, fish, meat, health, mobile and transportation etc.

Table 11.3: Monthly Expenditure Patterns of Two Categories of Fish Farmers

Items	Monthly Expenditure(Rs./Family)		
	Cluster-1	Cluster-2	Combined
A. Food Items			
Rice	1026.45	1338.93	1199.75
Pulse	171.38	220.24	182.26
Oils	261.03	291.55	248.46
Sugar	55.48	98.14	60.89
Milk	414.81	610.00	384.04
Vegetable	723.33	998.50	773.26
Fish	829.71	1545.71	1081.53
Dry Fish	143.23	130.20	144.81
Egg	120.44	144.28	107.99
Meat	379.66	732.14	459.66
Miscellaneous Food	220.47	268.33	223.04
Total Food	4345.99	6378.04	4865.70
B. Non Food Items			
LPG	78.96	19.048	54.27
Fuel Wood	381.95	532.94	416.05
Cloth	460.42	607.94	468.29
Education	721.16	332.15	559.67
Health	390.58	563.46	441.02
Electricity	128.58	167.73	122.82
Cable etc	85.63	52.62	73.25
Mobile/telephone	248.60	312.06	239.49
Transportation	289.19	406.28	295.25
Festival	189.88	320.18	192.59
Durable goods	343.89	245.19	321.16
Miscellaneous non Food	422.68	643.52	427.20
Total Non Food	3741.52	4203.1	3611.06
Total Expenditure	8087.509	10581.1	8476.76

The analysis of data on contribution of farm in food and family income showed that 27% of food items were directly met from the farm produce and 31% of family income came from fish production (11%) and agriculture (20%), remaining 69% income earned from nonfarm activities like Government service (36%), private service (5%), labour (12%) and other sources (16%) (Fig 11.1). This shows that even in case of fish farming community of Tripura off farm employment is vital for food and livelihood security.

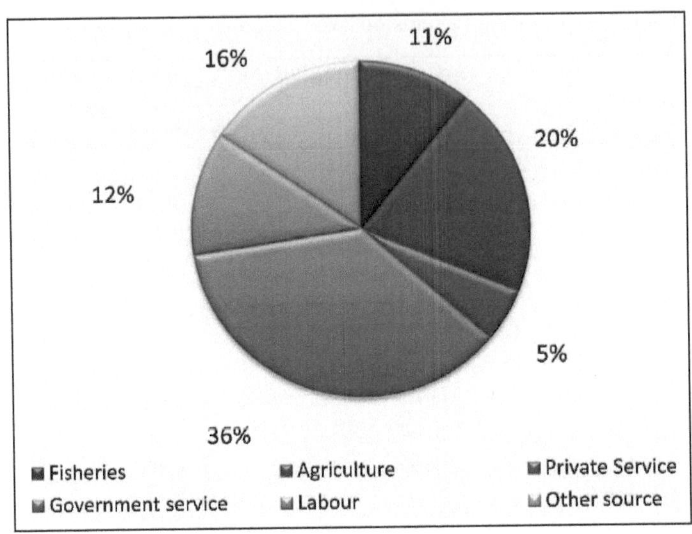

Fig. 11.1: Contribution of Different Sources in of Fish Farmers' Income

Demographic Feature of Two Clusters of Fish farmers

The average family size of fish farmers was 4.74 and male to female ratio was 1.21:1 indicating more number of male as compared to the female. Family size of the fish farmers belonging to the cluster-2 were smaller (4.57) as compared to cluster-1 (4.95). Further, the proportion of male (57.60%) was higher in cluster-2 than cluster-1 (53.62%) (Table 11.4). These results clearly reflect that the smaller families with higher proportion of male population were better in food access.

Table 11.4: Demographic Profile of Two Clusters of Fish Farmers

Gender	Cluster-I		Cluster-2		Overall	
	Population	%	Population	%	Population	%
Male	200	53.62	72	57.60	272	54.62
Female	173	46.38	53	42.40	226	45.38
Total	373	100	125	100	498	100
Family Size	4.95		4.57		4.74	
Sex Ratio	1.16:1		1.36:1		1.20:1	

Education level of Two clusters of Fish Farmers

It was observed from the bar diagram that the majority of the people of fish farming community possesed education between Class five to High school (Fig 11.2). While comparing the education status of two clusters of fish farmers, it was observed that the members of cluster-2 had better secondary or higher education than the cluster-1. This shows that the fish farming families whose education in terms of higher education (intermediate, graduate and post graduate) were higher and belonged to cluster-2 were better in food security.

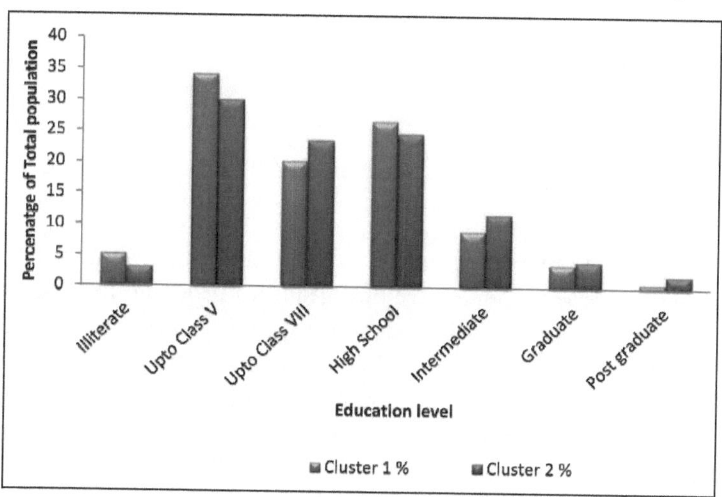

Fig. 11.2: Education in Two clusters of Fish Farmers of Tripura

Farm and off farm Employment in Two Clusters of Fish Farmers

Out of total population of selected fish farming families, 17.75% people were engaged in fish farming and 28% and 31% people were engaged in non income earning activities like students and housewife, respectively (Table 11.5). About 24% peoples were engaged in off farm employments including govt. service, private service, labour and small scale business. It was revealed on comparison of occupations of two categories of fish farmers, that the percentage of population engaged in non income earning activities (Students and House wife) was more in case of cluster-1 (64%) than the cluster-2 (57%). Further, the off farm employments including govt. service, labour, pvt. service and business in cluster-2 was 25% whereas it was 20% in case of cluster-1. Hence, it can be concluded that the members of cluster-2 had better farm and off farm employment opportunities as compared to members of cluster-1. The reason may be education and better level of resource possessions by this category of farmers.

Table 11.5: Occupational structure of two categories of Fish farmers

Occupation	Cluster-1(%)	Cluster-2(%)	Overall (%)
Farming	16.47	18.27	17.75
Government service including pensioner	7.06	5.77	6.49
Unskilled and skilled labour	6.67	8.65	9.09
Private service & tuition	2.35	2.88	3.03
Business	3.92	7.69	4.98
Students	33.33	20.19	27.92
Housewife	30.20	36.54	30.74

Possession of Resources

The possession of resources in terms of land and water area of fish farmers were analysed as these resources are important for food and livelihood security in rural areas. Average size of land holding of fish farmers was 0.61 ha in cluster-2 whereas it was 0.44 ha in cluster-1 (Table 11.6). The fish farmers of cluster-2 had more area (0.2349ha) under fish culture than the fish farmers of cluster-1 (0.1511 ha). This indicates that the farmers who had more land and water area in their possession had better access to food.

Table 11.6: Land holding and pond area of two clusters of fish farmers

Sl No	Area	Cluster-1	Cluster-2
1.	Land area(ha)	0.4361	0.6100
2.	Pond area(ha)	0.1511	0.2349
3.	Ratio of Pond to Land area	0.35:1	0.38:1

Conclusion

This study concluded that the consumption of cereals, fat and fish in fish farming families were at par with its recommended level by ICMR, whereas the consumption of pulses, milk, sugar, egg and meat were below the recommended level. The majority of the fish farmer also suffers from poor access to food. The fish farmers were mitigating nutritional deficiency in terms of protein by taking more fish. Further, the fish farmers whose monthly spending was higher, had better education, family size was smaller, had better farm and off farm employment opportunity and better resource availability (land and water area) were having better level of food security. Though the farming is key to food security as it contributes 27% food and 31 per cent income of fish farmers, but off farm employment is equally important to pursue their livelihood earnings. Hence, it is concluded from the study that for better access to food/food security, a holistic approach that can enhance farm production as well provide better employment opportunities is needed.

References

Anderson, S.A. (1990). The 1990 Life Sciences Research Office (LSRO) Report on nutritional assessment defined terms associated with food access. Core indicators of nutritional state for difficult to sample populations. *Journal of Nutrition*, 102:1559-1660

Anon (2003). Food security atlas of rural India. M.S. Swaminathan Research Foundation, Centre for Sustainable Agriculture and Rural Development, Chennai, pp 162

Murat Kayri (2007). Two step cluster analysis in researches: A case study. *Eurasian Journal of Educational Research*, 28: 89-99

Sanzidur Rahman and Shaheen Akter (2010). Determinants of livelihood security in poor settlements in Bangladesh. International Working Paper Series ,Natural Resources, Agricultural Development and Food Security, International Research Network (NAF-IRN), paper no. 10/01, pp. 30.

12

Traditional Fishing Practices for Livelihood and Income Generation in North Eastern Region of India

A.D. Upadhayay and B.K. Singh

Introduction

Fishing technology is the discipline dealing with the natural sciences and technology for optimizing fish catch and fishing operations, leading to a productive and sustainable capture fishery. The fishes are one of the main exploitable resources of the aquatic ecosystems that provide a cheap source of protein which helps solving the problem of nutritional security particularly in NE region of the country. The rapid development of fishing technology in India during the recent years has paved the way for increased production availability of new synthetic materials, evolution of new designs of fishing craft and gear and avoidance of ignorant harmful fishing techniques etc. It is essential that any developmental activity adopted, should also be base rooted in the objective of uplifting the economy of the fishing communities. This implies that the improvisation of the existing tools and techniques and introduction of new ones to enhance fishing efficiency need a careful study of the traditional technologies.

Further, selection of fishing methods and gear are influenced by various factors such as physiography of the water body, nature of fish stock, characteristics of the material from which gear are fabricated and standard of living. Therefore, variation in application of fishing devices can be observed in different type of water bodies, which have characteristic of their own due to unique nature of the water resources of the region. The success of these fishing techniques depends on various factors like selection of site, time, efficiency of materials used and availability of fish, etc. For successful fishing some attractant as a lure is popularly employed.

The North Eastern Region of the Country is very rich in the indigenous knowledge and techniques of fishing. The fishing gear and craft vary from one state to other states in terms of technical specification, materials and method of fishing. The type and size of water bodies also determine design fishing gear and craft to be used for fishing. The present day fishing methods and fishing devices which are in use in the region are the result of continuous efforts and knowledge inherited through generations of fishing community. Hence, fishing methods and fishing devices in the region was developed by the community according to their need, availability of the material, type of water bodies and target species. In this presentation, an assessment, compilation and documentation of these indigenous fishing technologies of the Tripura which is important for academicians, researchers and other stake holders, is made.

Past Studies on Fishing Technologies of Region

The different aspects of fishing methods and fishing devices in different part of country in general and NE region in particular have been reported by various researchers. A brief account of past studies undertaken and reported for North East region is reviewed here. Detailed aspects of fishing with traps and pots have been described by Slack-Smith (2001). Several researchers have reported on various fishing traps operated in Indian waters Hornell (1938), Job and Pantulu (1953), Prabhu (1954), George (1971), Brandt (1984), Mohanrajan (1993), Nair (1993). Kurup and Samuel (1985) and Kurup et al. (1993) have described the fishing traps used in Vembanad lake, Kerala. Mitra et al. (1987) have described the traps used in upper and middle Hooghly estuary. Details of various traps operated from north eastern India have been reported by Sharma et al. (1993). Traps from the Khachodhara Beel in Assam were reported by Sharma and Ahamed (1998). Some efforts to document the type of traps used in Assam has been carried out by Bhagawati and Kalita (1987), Nath and Dey (1989), Choudhury (1992), Choudhury et al. (1996), Sharma (2001) and Bhattacharya et al. (2004), Gurumayum and Choudhury (2009) and Baruah et al. (2010). Detailed designs of traps of Assam and its operation have been reported by Pravin and Meenakumari (2008). Traditional fishing method of Assam for Catfishes using duck meat as an attractant was studied by Dutta and Bhattacharyya, 2009. Fish Diversity and Fishing Gear used in the Kulsi River of Assam, India was studied and reported by Islam et al, 2013.The use of dug out canoes and different type of fishing gear in Lotak Lake was reported by Hora (1921) Choudhuri & Banerjee (1965), Singh (1983), Devi (1980), Mao (1991) and Singh et.al (1992). Marak et. al. (1998) explained the operational details of the fishing techniques carried out by the Tribal's' in North-Eastern India with reference to Meghalaya and Manipur. The traditional fishing methods and fishing devices of Mizoram have been reported by Lalthanzara and

Lalthanpuii (2009). The traditional knowledge associated with fish harvesting practices of War Khasi community of Meghalaya was studied by Tynsong and Tiwari, 2008. An account of traditional fishing devices of Tripura have been reported by Upadhyay and Singh, 2013. Traditional fishing methods in Central valley region of Manipur and Traditional fish aggregating wisdom of Manipur were studied and reported by Devi et al, 2013 and Devi et al, 2013.

The traditional fishing techniques and fishing devices of Tripura are presented in following section:

1. Tripura

In Tripura, capture fisheries resources constitute about 37 per cent in total water resources, whereas its contribution in total fish production is only 5.50 per cent, the reason being poor management of common fisheries resources, indiscriminate fishing and also due to inefficient fishing devices. However, the sector is important because livelihoods of about 22,373 fishermen families solely depend on it. The family income of fishermen are mainly dependent on daily catch which itself depends on fishing devices, fishing efforts and availability of fishes. Generally the age old traditional fishing gear is being used by the fisher folk of Tripura without realizing the strength and weakness in terms of its impact on the fish diversity. For the sustainable and judicious fishing, it is very important to understand the existing fishing practices and devices in the state.

Traditional Fishing Devices

A wide array of fish catching device have been observed in capture fisheries of Tripura. These fishing devices used in lentic and lotic water bodies showed variations with regards to their fabrications and operations. Fish catching devices based on design, materials used, technical details and operational methods may be broadly categorized as 1.Fishing gear with netting, 2.Hooks and Spears and 3. Traps

I. *Fishing Gear with Netting*

a. Gill Nets (Fash Jal)

Though gillnetting is banned in the state, still it is very common fishing technique in shallow and moving water bodies of Tripura. Both selective and non selective types of gill nets with different mess size were found in operation in the Tripura. Gillnets are generally fixed against the flow of water with bamboo stakes and the catch is collected after 6 to 8 hours. The details of the net and their local names are as follows-

Fig. 12.1. Gillnets

i. ***Fash Jal / Kanke Jal/ Chat jal/ Current net:*** Gillnets are widely used fishing gear during the rainy season especially in sallow moving water bodies in Tripura. It is also known with various names like fash jal, kanke jal, chat jal and current net. There are single walled nets with mesh size 2-12 cm (mostly 5.2 to 6.2 cm) and length of the gear varies from 10 m to 50 m depending upon the width and depth of water bodies. It has been observed that gillnets operated in the study area were mostly made up with polyamide monofilament. Head rope used with this gear is made up with poly propylene. The small stones or gravels are used as sinker. Single man with small kusha/dingi boat is used in fixing the net with bamboo poles across the flow of water. Gill net is operated during night time from evening to morning. After 4-6 hours, fishes are collected from the net. The target species of fish are puntius, tengra, glassogobuis, loaches, chanda, Indian major carps and prawn.

ii. **Drift gill net:** Drift gill net without foot rope is commonly practiced in Tripura. Drift gillnets are operated in the surface layer and drift with the current either separately or with the boat to which it is attached. But, it is not desirable to operate such gill net in reservoir and lake. Catches mainly composed of catfish and carps.

b. Drag Nets

i. **Laitya Jal**: It is a surrounding net which is mostly used to catch Channa spp. One end is hold by fishermen on shore and a boat speeds the net and encircles the areas. Its mesh size is 2.5 to 3.75 cm. It is used to catch Gudusia chapra, Chanda nama in Rudrasagar Lake. It is made up of nylon.

ii. **Fy-Jal**: It is a surrounding net, operated by two boats. First the boat goes to the centre of lake and then they spread the net by carrying each boat each end of the net and encircling the area. The two ends are pulled up and catch the entire lot of fishes. Generally it is made up of very small mesh size for catching pelagic small size fish.

iii. **Ber-Jal:** It is the most popular dragnet of Tripura. Mesh size varies from 0.5-13 . About 60% of resources are exploited by this gear. The length of net varies from 25-150 meter. Its breadth is 4-25 meters. It is mainly used for catching carps. It is operated from boats in deeper lakes. It is one type of surrounding gear mostly used in domestic ponds. During operation, the net is spread horizontally and vertically one side of tanks by covering the whole breadth of the tanks with float and sinkers. Sometimes it is also called as "Gharra jal".

iv. **Tana-Jal:** It is two men operated drag net mesh size of the net varies from 1 to 8 mm. Mouth is kept open to 4.5meters and its breadth is 1.8m Mouth is kept open by bamboo sticks of 1.2m height. It is dragged in shallower water. Small and medium sized fishes are caught during daytime. It is made of mosquito net hence it also called "Masari jal".

v. **Para-Jal:** Its mesh size is 2 - 2.5 3 . It is used only for demersal big-sized fish. A boat will spread the net and 2 or 3 fishermen press the net sinker onto bottom by leg and after surrounding the areas, they will pull the net and harvest the fish.

c. Bag nets

i. **Pelni Jal:** It is a scoop net, having a round or triangular frame. The mesh size of pelni jal varies from 4 to 5 mm and bamboo strip used to form Small Square. The diameter of bamboo frame is 50-60 cm. It may be operated by boat or by walking. Puntius spp., Amblypharyngodon mola etc. are caught by it.

ii. **Phaloon/ Khara JAl/ Bel Jal:** It is a traditional boat operated dip net. It can be operated by moving bag up and down or by pushing forward. The netting is hung to triangular frame. The two bamboo poles of length 3-5 meter were tied together to form acute angle. The net is fastened on two poles of bamboo while at the free end, it is supported by rope. The mesh size of the phaloon gear is ranged from 5 mm to 8 mm. The fabrication cost of the gear is Rs. 1200/ and expected economic life is 3 years. In good season a fisherman with small dingi boat catches on an average about 7-10 kg of fish per day. During rainy season it is fixed in canals facing the current of water to collect fishes came through drifting.

iii. **Chapila Jal:** It is used to catch Gudusia chapra mainly. It is operated from two boats. During spreading the net, the fishermen make sound by leg over boat. It attracts the fishes to aggregate and the bag like net is pulled up and the fishes are harvested. Length of the net is 100-150 m and breadth 3 m. Mesh size varies between 6-10 mm.

iv. **Maiya / Aain-Jal:** It is mainly to catch catfishes specially Mystus spp. in rivers. Float has a bit lesser weight than sinkers. During rainy season, it is most frequently used.

Fig. 12.2. Fishing Operation with Phaloon Jal

d. Falling Gear

(i) Koni JAl / Uran Jal

It is a cast net operated by one fishermen. The principle is only to throw the net to cover the fish. Lower edge is provided with series of sinkers. It may be cast from bank of the rivers, beels, and ponds or from boat. It is provided with a line to haul it.

i. Polo Jal

It is used in shallower water. Its length 0.5-5 meters. Its frame made up off bamboo surrounded with mosquito cloth. A 15-25 cm hole is kept at the upper side of gear. A fisherman pushes it over the fish and takes it out from the hole.

Fig. 12.3. Fishermen Operating Kuni Jal for fishing

Fig. 12.4. A village women knitting Kunijal

e. Lift Net

(i) Dharma Jal

It is used in running water especially in rivers. Frame is made up of bamboo. Net is made up of cotton twine. Its mesh size varies 1-3 cm. It is also widely used in beels, Chera river lets. Mainly used for catching weed fishes. The length of bamboo frames is 5-7feet.

Fig. 12.5. Operation of Dharma Jal for fishing in river system

In case of Rudrasagar Lake of Tripura, Kunijal (cast net) was used by 45.37 per cent fishermen, followed by faloon jal (21%), ber jal (10%), gill net (9.24%) and chapila jal (8.40%) as reported by Upadhyay (2013).

II. Spears and Hooks

Fishing hooks and lines are simplest and the most important fish catching devices of Tripura. They are simple, easy to operate, cost effective and selective. The hooks serves the functions of holding bait, enticing the fish to it and ensuring that the fish shall be unable to spit out the bait after swallowing it. It is usually penetrates into the mouth of the fish when the bait is taken or when hook is pulled. The different types of lines and fishing hooks found in Tripura were described below-

a. Spears

i. **Juitya:** It is a long bamboo stick (70-80cm long). The tip of bamboo stick is equipped with 5-8 hooks and extreme end of stick is attached with a nylon rope. When the fishermen recognized that fish is passing nearby to him, he threw the "Juitya" that pierced to the fish. These are practiced by experienced or trained persons only. This practice used to catch mainly big size fishes like silver carp and some other carps.

ii. **Gathynna/Guthia**: Gathynna a steel or wooden stick with pointed end. This type of spear mainly used to locate *Monopterus cuchia* in swampy areas and marginal embankments. A fisherman first seal hole and then inserts the stick at soft ground randomly and when he feels the movement of eel, he tans out the eel by digging with hand.

iii. **Ek-Kaithia:** It is a type of spear its length is varying from 1.5 to 2.5meter. It has 4-10 sharp prongs made up by splitting seasoned bamboo pole at one end. Sometime this type of wounding spear referred as "Chal" which is used by the tribal fishermen in catching of or fishing of large fishes. They pushed it in shallow areas and the fish cannot come out of it. Specially, *Cyprinus carpio* is caught by spearing with the above. It is useful only in clear water. Sometime it may cause injury to the fishes, which may lower the market value of them. But strike and injured fish which escape may die afterwards. It is mostly used in beels, "Charra" and swampy areas.

b. Hooks (Barshi)

"Barshi Fishing" in Tripura is very common and old age practice. This type of fishing is specially used in shallow water bodies with weeds or standing crops and also to catch bottom fishes. It is operated throughout the year. A Nylon line is tied on the tip of bamboo pole. It is a type of hand line. The sinker in it is locally called as "Bhar". The hook is provided with lure/bait like earthworm, forage fishes, small prawns, bread and rice paste. The following types of Barshi are being used by fishing community to catch different type of fishes.

i. **Boal Barshi:** A Nylon line is tied on the tip of bamboo pole. Length varies from 1-3 meters. The hook is provided with small live bait. As the name of the fishing hook, this used specially to catch riverine fish like Boal fish (*Wallago attu*).

ii. Tang Barshi: This type of hooks used in large scale during the flood. A Nylon line is tied on the bamboo stick. Length varies from 1-1.5 meters (Fig 7). The hook is provided earthworm, small prawns, bread, and rice paste. This type of fishing hook is used to catch puntius, chana and tilapia fish.

iii. **Lait Barshi:** This type of hooks used in large scale during the flood. A Nylon line is tied on the bamboo stick. Length varies from 1-1.5 meters. The hook is provided with earthworm, small prawns, bread, and rice paste. This fishing hook, used to catch Tilapia, common carp and medium size IMCs, Anabus and Kanla fish.

iv. **Gucha Barshi:** This type of hooks used in large scale during the flood. A Nylon line is tied on the bamboo stick. The hook is specially designed and made up with wooden sticks. The hook is provided earthworm, small prawns, bread, and rice paste. This fishing hook, used to catch snake fish Monopterus cuchia generally found in the mud/soil of water bodies.

v. **Baira:** This is an arc like hook made up with flexible bamboo stick. In the middle of the stick a Nylon line is tied and on other end line is tied with bamboo round stick of 9-12". The hook is specially used in either paddy field or weedy shallow water body to catch Koi fish (*Anabas testudinous*), Magur, Singhi, kanla etc. The hook is provided with an insect found in paddy field. Putting both end of hook together and insect is put into it which attracts the fish, as soon as the fish attack on insect, both ends of the hook get free inside the mouth of the fish.

Fishing Traps: Anta trap

It is a rectangular box-shaped trap made up of bamboo wire mesh or iron or polythene strips. There is a small mouth opening which only opens in its inner side by water pressure. Earthworms, rice bran mixed with dry fish are given inside the box to attract prawns. These are placed in a series marking each with bamboo stick. These traps are placed in the evening hours. Mouth opening is against current. Prawn can enter in the device which open by water pressure but cannot come out from it. In the morning these are collected. The crabs and small fishes were also collected by this trap.

i. **Benta:** This trap is like Anta with little difference that it has wide front and round or narrow back. It is also made up with bamboo strips and tightly woven with steel wire or plastic threads. It is fixed in moving water especially on the dyke of the paddy field. This trap is mainly used to catch fish like chanda, putti, baim and gutom etc.

ii. **Singchai:** It is tubular trap with circular mouth and tapered back portion of trap. It is made with narrow bamboo strips, bamboo sticks and woven with plastic threads. The Singchai is used to catch bottom feeder fish like magur, singhi, baim and baila, crab, prawn etc. in small shallow water bodies. It is fixed at the bottom, mouth opening across the flow of water.

iii. **Pharam:** It is a rectangular box-shaped trap which made up with bamboo strips. There is a small mouth with flexible opening which only opens in its inner side. Mouth opening is against current. It is set in stagnant weedy water body to catch medium and small size fish including IMCs. These are placed in a series marked with bamboo stick. These are placed in the evening hours. The fish entered in the trap but cannot come out from it. In the morning these fishes are collected

iv. **Deur:** This trap is just like carry bag made up with bamboo strips, sticks and woven with plastic thread. This trap is set in the moving water to catch fish such as putti, baim, tengra, chanda and gutam etc.

v. **Phala:** This trap is funnel shaped with circular bottom and narrow at upper side. The trap is made up with bamboo strips, sticks and woven with plastic or iron wires. This trap is very common in Bangladesh and some part of Tripura state. The small and medium size fishes were caught by this device.

Jak: It is sometimes referred to fish aggregating device. It is mainly practiced in Rudra Sagar Lake as well as most of wild water body of Tripura. The fishermen keep bushes, Eichhornia, Bamboo poles with leaf in the water body (Fig. 12.6). They also provide foods in it. Fish consider the "Jak" as their habitat, starts to congregated there and breed there. After 2-3 months the fishermen harvest the whole plot by mosquito net. They take out the bushes and are placed in a new location again. They harvest all the fishes residing at that "Jak".

It appears from the study of fishing devices of Tripura that though it is small state of the country but very rich indigenous know how about capture fishing. The indiscriminate fishing, fishing in breeding season, poisoning and fishing with gillnets are some of the ill practices of fishing observed in the study area. This is threat to the fish biodiversity of Tripura, which should be discouraged. The mesh size regulation is very much required to protect the young fishes and juveniles to be caught. The fishermen are using nylon net for gillnetting which is harmful in case ghost fishing. Therefore, only bio-degradable netting should be allowed for gill netting. If the synthetic netting is used then proper care should be taken for positioning of gillnets and collecting fishes must be taken. Knotless netting should be introduced to reduce the manpower for operations in case of larger nets particularly drag nets. It is also need of the hours to stock hatchery produced seeds to the river and reservoirs to improve the catch and income of the fishermen.

Fig. 12.6. Jak fishing

References

Biramani N. Singh (1977). An account of the fishing gear of Manipur with special reference to Loktak Lake Dissertation Submitted Towards Partial Fulfilment for Post Graduate Diploma in Fisheries Science Volume 58 of D.F. Sc Dissertation, Central Institute of Fisheries University Dept. of Fisheries. Central Institute of Fisheries Education, p.40.

Brandt, A.V. (1984). Fish catching methods for the world (3rd Ed.). Fishing News Books Ltd., London, p. 418.

Chakravartty, Pranjal, and Sharma, Subrata (2013). Different types of fishing gear used by the fishermen in Nalbari District of Assam. *International Journal of Social Science & Interdisciplinary Research*, 2 (3):177-191.

Chatia, Riyajul Islam and Seydur Rahman (2007). Traditional riverine fish catching devices of Assam. Fishery Technology, 44(2):137-146.

Dutta, R. and Bhattacharyya B.K. (2009). Traditional fishing method of Assam for catfishes using duck meat as an attractant. Indian Journal of Traditional Knowledge, 8(2): 234-236

Devi, B Nightingale; Mishra, SK; Das, Lipi; Pawar, NA; Chanu, Th.Ibemcha (2013).Traditional fishing methods in Central valley region of Manipur, India, *Indian Journal of Traditional Knowledge*, 12(1): 137-143

Devi, B Nightingale; Mishra, S K; Pawar, N A; Das, Lipi; Das, Soma (2013). Traditional fish aggregating wisdom of Manipur, North-eastern India. *Indian Journal of Traditional Knowledge*, 12(1): 130-136

George, V. C, (1971). Fishing techniques of riverine and reservoir systems: Present status and future challenges, *Riverine and Reservoir fisheries*, SOFT (I), 192-196.

Hickling, C. F. (1961). Tropical inland fisheries Publication. Longman, Green and Co. Limited London, 287 pp.

Hornell, J, (1923). The fishing methods of the Ganges. Mem. Asiatic Soc. Bengal, 8(3):199-237.

Hornell, J. (1938). Fishing methods of Madras Presidency, Part II: The Malabar Coast, Madras Fish. Bull., 27 1: 69 pp.

Islam M.R., Das B., Baruah, D., Biswas, S.P. and Gupta A. (2013). Fish Diversity and Fishing Gear used in the Kulsi River of Assam, India. *Annals of Biological Research*, 4 (1):289-293

Jhingran, A. G. and Natarajan, A. V. (1969). Study of the fisheries and fish population of the Chilka Lake during the period 1957-65. *J. Inland Fish. Assoc. India*, 1:49-63.

Job, T. J. and Pantulu, V. R. 1953. Fish trapping in India. *J. Asiatic.Soc. Sci.*, 19(2): 175-196.

Jones, S. and Sujansingani, K. H. 1952. Notes on the crab fishery of Chilka Lake. *J. Bombay Nat. Hist. Soc.*, 51(1): 119-34.

Kar, D and Dey, S.C, (1993). Variegated encircling gear in Lake Sone of Assam. *J. of Applied Zool. Res.*, 4(2): 171-175.

Kurup, B. M. and Samuel, C. T. (1985). Fishing gear and fishing methods in Vembanad Lake. In Harvest and Post-harvest Technology of Fish, SOFT (I), Cochin, 232 pp.

Kurup, B. M., Sebastian, M. J. Sankar, T. M. and Ravindranath. P. (1993). An account of Inland fishing gear and fishing methods of Kerala. In: Low Energy Fishing, Fish. Technol. (Special issue), Society of Fisheries Technologists (India), Cochin, p. 145-151.

Lalthanzara, H. and Lalthanpuii, P. B. (2009). Traditional fishing methods in rivers and streams of Mizoram, North-East India. *Research Note Sci Vis* 9 (4), 188-194.

Mitra, P. M., Ghosh, K. K. Saigal, B. N. Sarkar, N. D. Roy, A. K. Mondal, N. C. and Paul, A. R. (1987). Fishing gear in the upper and middle Hooghly estuary. Bullein. Central Inland Fisheries Research Institute, Barrackpore, No.49: 1-22.

Miyamoto, H. (1962). A field manual suggested for fishing gear surveys (In Mimeo), CIFT, Cochin, 15 pp.

Mohanrajan, M. (1993). Fish trapping devices and methods of southern India, Fish.Technol. 36: 85-92.

Varghese, M.D. George, V.C., Gopalkrishna Pillai, A.G. and Radhalakshami, K, (1997). Properties and Performance of fishing hooks. *Fishery Technology*, 34:39-44.

Nair, P. R. (1993). Fishing with traps. In: low energy fishing. Fish. Technol. (Special Issue) Society of Fisheries Technologists (India), Cochin, p. 207-209.

Nath, P. and Dey, S. C. (1989). Fish and fisheries of North-eastern India Vol. 1 (Arunachal Pradesh), 194 pp.

Prabhu, M.S. (1954). Trap fishing. *Indian. J. Fish.*, 1: 94-129.

Pravin, P. and Meenakumari, B. (2008). Fishing traps of Assam, CIFT Special Bull. No. 14, 162 pp.

P. Pravin, B. Meenakumari, M. Baiju, J. Barman, D. Baruah and B. Kakati (2011). Fish trapping devices and methods in Assam - a review. *Indian J. Fish.*, 58(2) : 127-135, 2011

Remesan, M. P., Pravin, P. and Meenakumari, B. (2007). Collapsible fish traps developed for inland fishing ICAR News. 13(1): 117-118.

Sainsbury, J.C. (1996). Commercial Fishing Methods (an introduction to vessel and gear). Blackwell Science Ltd., Oxford. 368pp.

Saly N. Thomas, Gipson, Edappazham, B. Meenakumari and P.M. Ashraf (2007). Fishing Hooks: An Overview. *Fishery Technology*, 44(1) 1-16.

Sathinshas, R.; Panikkar, K.K.P.and Salini, K.P. (1993). Economic of traditional gill net fishing using wind energy along with Tamilnadu coast. Proc. Nat Workshop on low energy fishing *Soc. Offish Technol.* India: 272-277.

Sharma, P. and Ahamed, S. (1998). Relative efficiency of fish capturing devices in Kachodhara Beel of Morigaon district, Assam. *Environ. Ecol.*, 16(1): 123-126.

Sharma, P., Kalita, K. K. and Dutta, O. K. (1993). Low energy fishing techniques of the North eastern India. In: George, V. C., Vijayan, V. Varghese, M. D., Radhalakshmi, K., Thomas, S.N. and Jose Joseph (Eds.), Low Energy Fishing, Fish. Technol., Society of Fisheries Technologist (India), Cochin, p. 163-167.

Sharma, R. (2001). Traditional fishing methods and fishing gear of Assam. *Fishing Chimes*, 20 (12): 23-26.

Slack-Smith R. J. (2001). Fishing with traps and pots. FAO training series 26: 66 pp.

Tynsong, H and Tiwari B.K. (2008). The traditional knowledge associated with fish harvesting practices of War Khasi community of Meghalaya. *IJTK* 7 (4): 618-623.

Upadhyay, A D; Singh, B. K. (2013). Indigenous fishing devices in use of capture fishing in Tripura. Indian Journal of Traditional Knowledge, 12(1): 149-156.

Verma, Anand Mohan (2007). Fishing gear used by riparian fisheries posing threat to the conservation of rare fish fauna in North Bihar. Fishery Technology, 44(2):147-152.

Wilimovsky, N .J. and Alverson, D. L. (1971). In: Kristjonsson, H. (Ed.), Modern Fishing Gear of the World, Vol. 3, Fishing News (Books) Ltd., London, 509 pp.

Yadava, Y. S. and Choudhury, M. (1986). Banas fishing in beels of Assam. J. Bombay Nat. Hist. Soc., 83(2): 452.

Yadava, Y. S., Choudhury, M. and Kolekar, V. (1981). Fishing methods of flood plain lakes in North Eastern region. *J. Inland Fish. Soc. India*, 13(1): 82-86.

13

Fish Marketing Practices of Kombuthurai Fishing Village, Thoothukudi District: A Case Study

T. Umamaheswari and G. Sugumar

Introduction

Fisheries play an important role in our national economy by providing a quality source of protein thus ensuring nutritional security of our people, foreign exchange and employment for about 14 million people. The ocean is continuously subjected to enormous stress through utilization of the marine resources and preservation of biodiversity is often ignored. Precautionary use of living aquatic resources in harmony with the environment leads to responsible fisheries and moreover it's our responsibility to make fishery wealth as perennial as possible and adoption of responsible method of fishing can take us to a prosperous future. Though various fishing techniques are in practice, line fishing using baited hooks or artificial jigs is an effective method to capture predatory fishes. The catch obtained by line fishing is generally of high quality and commercial value with minimal environmental impact. The supply of fish is highly seasonal, leading to price fluctuations across regions and seasons, even within a day and being exploited by middlemen resulting in reduced welfare of fishermen. The existence of traditionally been highly unorganized and unregulated marketing system is the prime reason for its inefficiency. Various attempts have been made to overcome this problem by fishermen group as well as government agencies in India. But, these kinds of efforts have largely been confined to a few small locations and were highly scattered. Unlike poultry or dairy industry, innovations in fish marketing have not been on a macro level (Ganesh, et al., 2008). Against this background, the study has attempted to have a better understanding the innovative fish marketing practices being adopted by people institutions called "Sangams" in Kombuthurai fishing village and to draw lessons for upscaling and to promote such successful institutions in a similar socio-politico-economic scenario in other parts of the country as well for efficient fish marketing system.

Background

Compared to the achievements in fish production, the fish marketing system is very poor and highly inefficient in India (Ganesh Kumar et al., 2008). Unlike conventional marketing systems of agricultural products, fish marketing is characterized by heterogeneous nature of the product regarding species, size, weight, taste, quality and price. Certain other problems in fish marketing include high perishability and bulkiness of material, high cost of storage and transportation, no guarantee of quality and quantity of commodity, low demand, and high price spread (Ravindranath, 2008).

Gupta (1984) and Srivastava and Kant (1985) had analyzed the price variations among the fish species across states and had identified infrastructural bottlenecks in efficient marketing system of India. The fishermen could be saved from exploitation by encouraging group marketing, cooperative marketing, contract marketing, etc., which would increase the marketing efficiency and improve their profit (Chahal et al., 2004; Ali et al., 2008). Katiha et al. (2004) studied the governance, institutions and policies for fisheries of floodplain wetlands, wherein they revealed that the stakeholders include fishers, government agencies, lessee, fishery co-operatives, village authority, community leaders, market agents, political leaders and NGOs and their interactions and transparency are essential to increase the efficiency of the markets. Acharya (1997) had suggested encouraging the local farmers, traders and processors to market the produce under their brands and promoting such local brands based on graded produce to improve the marketing efficiency.

Ganesh Kumar et al. (2010) studied the fish marketing by various institutions in the southern states of India, viz. Tamil Nadu, Kerala, Karnataka and Andhra Pradesh with the hypothesis that the institutional arrangements in the marketing of fish and fishery products reduce the transaction cost and improve the market access and its efficiency. The study reported the primary activities of the institutions in the efficient fish marketing and through these advantages, the fishermen have been found to achieve economies of scale, technological innovations, capacity development, linkage among activities, degree of vertical integration, timing of market entry, product differentiation, market access, credit access, etc. It has been suggested to promote such successful innovative institutions in marketing of fish and fishery products through appropriate policies and programmes.

Brinson et al. (2011) stated that through Community Supported Fishery (CSF) Programs, an arrangement between fishermen and consumers provide upfront payments to fishermen in exchange for scheduled seafood deliveries. It is a form of direct-to-consumer-marketing where fishermen receive higher prices for fish, a guaranteed stable income and can activate political and regulatory support through direct interaction with consumers. He concluded that direct marketing through a CSF is not likely to completely replace traditional markets for fishermen, but could be a valuable supplement to their operations.

Design of the Study

Kombuthurai, a fishing village in Thoothukudi district of Tamil Nadu was purposively selected owing to the existence of typical fish marketing system through people institutions. From a total population of 709 (as per Tamil Nadu Marine fisher folk census 2010), 87 fisher folk respondents were randomly selected and primary data was collected through structured interview schedule during 2016-17. Simple tools of analyses like percentage and descriptive statistics analyses were used for this study.

Study area

It is a traditional fishing village (Lat. 8°34'50.49" N and Long. 78°08'12.91" E) originated in late 20th century (1979-1980) and situated 30 km away from Thoothukudi (on the way to Tiruchendur) mainly characterized by line fishing (Fig. 13.1). Fishermen, in general, mainly depend upon fish catch that is influenced by resource potentiality, physical characteristics of the coast and the type of fishing craft and gear used for their livelihood. Thus, the fishermen skilled in the fabrication and operation of hook and lines, long lines, troll lines and squid jigs operate Fiber Reinforced Plastic (FRP) boats. Since the line fishing excels in capturing high market value fish species namely seer fish, carangids, barracudas, lethrinids, cuttle fishes, tuna etc., the use of other fishing gear is self restricted and regulated. Besides, these gears (line fishing) are eco-friendly and help in sustainablity of reasources.

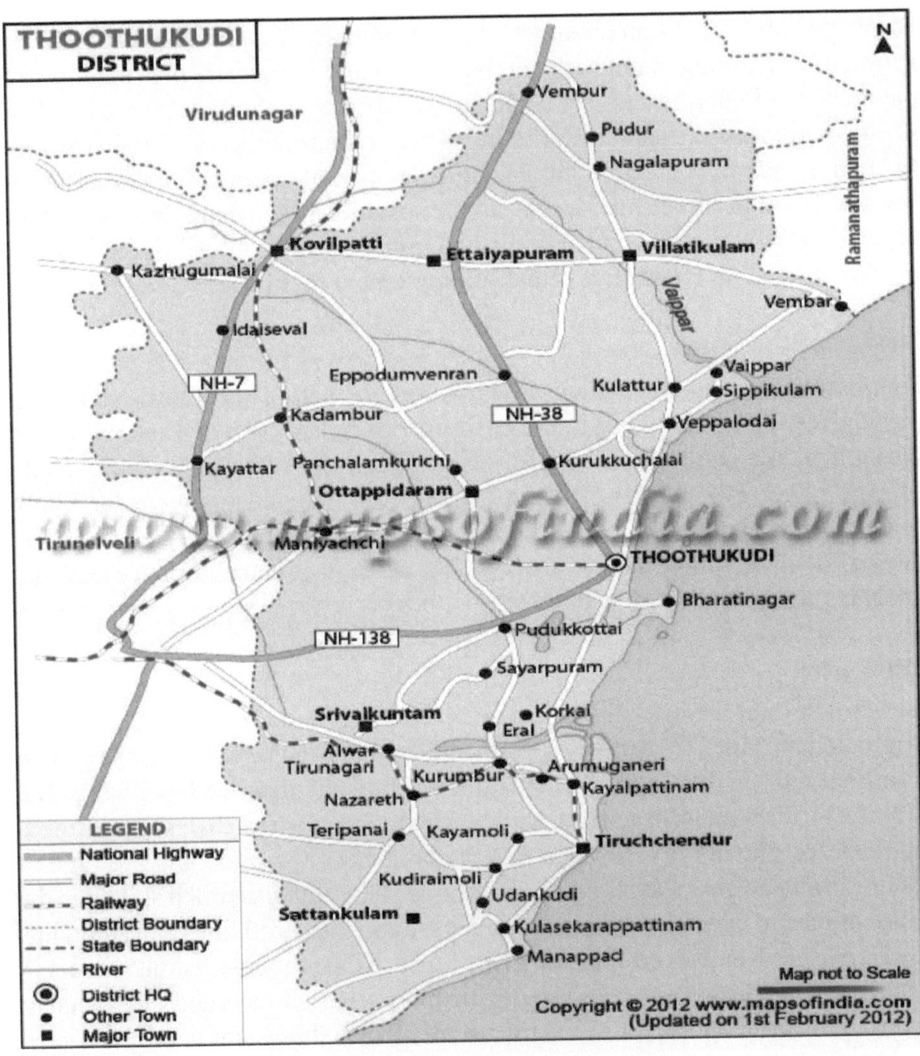

Fig. 13.1: Map showing the study area

Socio-Economic Characteristics of Fishers

The socio-economic profile of fishers in Kombuthurai fishing village is given in Table 13.1. The average age and fishing experience of the fishermen is about 40.10 years and 18.68 years, respectively. The literacy level measured on five point rating scale viz., illiterate (i), schooling up to 5th (2), 6th to 10th (3), 11th to 12th (4) and graduate (5) reveals that the literacy level is found lowr with a mean value of 2.20.

Table 13.1: Socio-economic profile of fishermen of Kombuthurai

(n=87)

S.No.	Variables	Values
1.	Age (years)	40.10±10.62 23-62 26.48
2.	Literacy level (scores)	2.20±1.01 1-5 45.91
3.	Experience in fishing (years)	18.68±8.30 5-40 44.43

(The values in first, second and third rows indicate mean and S.D, range and C.V, respectively)

Consumption Pattern

The consumption pattern of Kombuthurai fishers is depicted in Fig 13.2. The data on consumption expenditure of fishers was collected under two heads viz., family expenditure and expenditure towards welfare and village administration. Family expenditure accounts to 93% of the total expenditure while remaining (7%) for welfare and village administration. Family expenditure is further classified under various heads viz., food (42%), clothing (5%), savings through chits (4%), education (9%), social obligation (2%), transport (7%), medical (8%), recreation (1%), loan repayment (13%) and mobile recharging (2%).

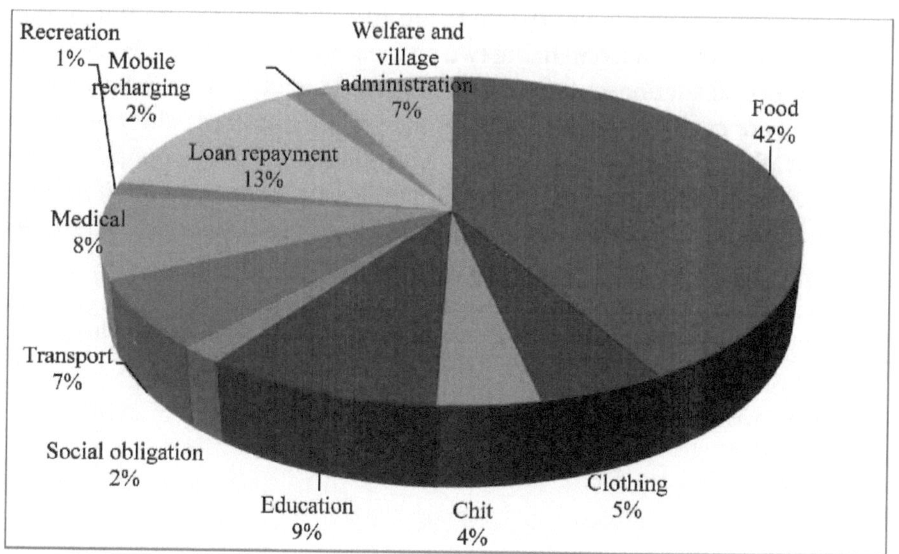

Fig. 13.2: Consumption Pattern of Fishers in Kombuthurai Fishing Village

Marketing Practices in Kombuthurai

Fish marketing in Kombuthurai village of Thoothukudi district was similar to the other landing centres of the region initially, wherein fishes were sold to retailers through open auction at low price. Later, the fishes were sold based on Tender system. To administer and look into their daily transactions of fishing activities in Kombuthurai, the fisher folk formed two marketing societies (later called as **"Sangam"**) namely St. Xavier Fishermen Sangam operated by Thoothukudi Multipurpose Social Service Society (TMSSS), a Non-Governmental Organisation (NGO) and St. George Fishermen Sangam by fishermen themselves for sale of fish. The two sangams invite quotations for fishes in a sealed cover once in two weeks and the highest bidder will be offered the entire lot of fish. But, the demand for premium varieties of fish like seer fish, barracudas, carangids, etc. landed in fresh condition and in good quality by line fishing at Kombuthurai attracted many wholesalers to purchase fish at a better price compared to other landing centres of the region, where fishes from gill netting and trawling are landed. Later during 2005, the marketing method followed by the fishermen of Kombuthurai was changed to daily auction, the major difference of which from the other landing centres was that in Kombuthurai the fishes were segregated species-wise and auctioned on per-kg-basis and then sold on weight-basis. The fishes are then transported in trucks and insulated vans by buyers.

In the landing center, auctioneers have the responsibility of collecting the money from the wholesalers and distribute it to the fishermen through sangams. The fishermen have a common agreement that all of them should become member in one of the sangams and presently, there are five sangams, among which three operate under TMSSS and remaining two on their own (Table 13.2). Each sangam has an authorized auctioneer to auction the fishes landed by its members and the auctioneers are paid on monthly basis for the service rendered. The wholesalers from other places such as Thoothukudi, Kanyakumari and Manappad of Tamil Nadu and from major markets in Kerala purchase fish through auction either directly or through their agents and the price is fixed based on its size and species.

Table 13.2: Sangams of Kombuthurai fishing village

Sl.No.	Name of the Sangam	Total members (nos.)	No. of years in existence
Under TMSSS			
1	Saint Xavier	20	16
2	Saint Christhuraja	45	12
3	Sahaya Matha	30	9
On their own			
4	Mudiyappar	15	18
5	Soosaiyappar	2	6
	Total	112	

Crew Share Pattern

Crew share pattern of Kombuthurai fishing village is well defined. Four or five members including owner-cum-crew head or crew head are engaged in fishing. Oil cost include starting oil (Petrol), fuel oil (Kerosene) and lubricant oil for the fishing trip and cost of the fishing materials lost is deducted from the actual sale proceeds of the fish landed. The balance of proceeds is shared between crew members, engine and vallam in the ratio of 4 or 5: 2: 1. No share is entitled to fishing gear. While the owners of the craft meet the cost of maintenance of engine and vallam, maintenance of fishing gear is shared by the crew members.

Role of TMSSS

TMSSS plays a major role in governing the sangams under their control and normally involved in conducting sangam's monthly meeting at sangam hall (1st Friday of every month), checking of sangam's accounts once / twice in a month, extending marketing services, settlement of savings to fishermen once in 5 years, maintenance of records like member, cash, bank, fuel, credit, old age pension (OAP), savings and administration ledgers, extending need based financial assistance and getting working capital from commercial banks and offering special schemes for the benefit of fishermen and elders called "Pension Scheme" and "Savings Scheme". The charges rendered over the services by the society are as follows (Table 13.3).

Table 13.3: Distribution of Service Charges by TMSSS

Sl.No.	Distribution of services	Old (%)	New (%)
1.	Marketing service	2	—
2.	Bank loan premium	10	—
3.	Pension scheme	1	1
4.	Savings scheme	1	3
5.	General fund	1	—
6.	Salary to staff / Sangam administration	1	3
7.	Welfare activities	1	1
	Total	17	8

Extent of Adoption of Hygienic Practices by Fishermen

Fishing harbours and fish landing centres play an important role in determining the quality of sea food produced as it is the main area where fish is handled after their landing at shore (Thomas et al., 2015). The extent of adoption of hygienic practices by fishermen on board and on beach measured on a three-point scale rating viz., adopted, partially adopted and not adopted, with the scoring pattern of 3, 2 and 1, respectively at Kombuthurai fish landing centre is presented in Table 13.4. A total of eight hygienic practices were taken into account for measuring the extent of adoption of hygienic practices. The results revealed that the overall adoption index of hygienic practices is found as 2.82 ± 0.13 with coefficient of variation of 5.54%. While on board hygienic practices like use of clean seawater for washing and cleaning, use of ice for preserving fish, cleaning of insulated ice box and on beach hygienic practices like use of clean water for cleaning fish and cleaning of fishing vessels and nets periodically are being adopted (3.00±0) by all the fishers of Kombuthurai, use of hygienic materials like gloves, gumboots, apron, mouth guard and head gear on beach is being adopted at a lower rate (1.78±0.42). Balasubramanium et al. (2009) reported that low level of scores was obtained for use of adequate clean water for washing fish (39.49%) and prompt disposal of waste (40%) in fish landing centres of Vishakapattinam.

Table 13.4: Adoption of hygienic practices by fishermen of Kombuthurai

S.No	Hygienic practices	Values
Onboard		
1	Use of clean seawater for washing and cleaning	3.00±0 3 0
2	Use of ice for preserving fish	3.00±0 3 0
3	Cleaning of fish hold	2.90±0.30 2-3 10.34
4	Cleaning of insulated ice box	3.00±0 3 0
5	Use of hygienic materials for fish handling like gloves	2.90±0.30 2-3 10.34
On beach		
1	Use of clean water for cleaning fish	3.00±0 3 0
2	Use of hygienic materials like gloves, gumboots, apron, mouth guard and head gear	1.78±0.42 1-2 23.60
3	Cleaning of fishing vessels and nets periodically	3.00±0 3 0
	Overall	2.82±0.13 - 5.54

(The values in first, second and third rows indicate mean and S.D, range and C.V, respectively)

Prevalence of Unhygienic Practices at Fish Landing Centre

Besides the hygienic practices, unhygienic practices like spitting, urination, walking with slippers / barefoot, throwing liquor bottles and plastic wastes at fish landing centre are also documented (Table 13.5). The table clearly reveals that all the fishermen (100%) are walking either with slippers or bare foot at fish landing centre. While urination and the prevalence of liquor bottles (100%) are completely unnoticed spitting and plastic wastes disposal at fish landing centre are observed at 12% only.

Table 13.5: Adoption of unhygienic practices at Kombuthurai fish landing centre

Sl.No.	Unhygienic practices	Practiced	Not Practiced
1	Spitting	12	88
2	Urination	-	100
3	Use of slippers / bare foot	100	-
4	Throwing liquor bottles	-	100
5	Throwing plastic wastes	12	88

(The figures indicate percentage)

Fish Catch and Behavior of Fish Prices

Data on fish landings and price bahaviour of fishes landed at Kombuthurai during the study period is shown in Fig 13.3. It is clearly inferred that except *Rastelliger kanagurta* (being used as live bait), no other fish varieties landed at Kombuthurai fish landing centre were of less than 1 kg weight. Also, the average price of different varieties of fishes marketed at Kombuthurai depicts the price advantage the fishermen get on account of selective fishing methods, grading practices and auctioning on per kg basis.

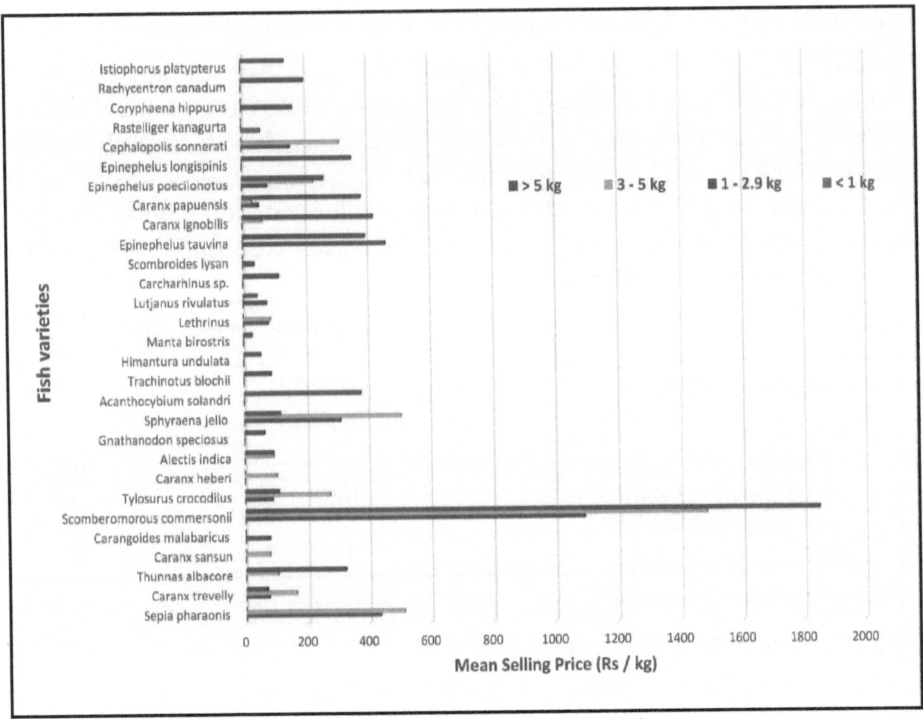

Fig. 13.3: Landings and price of fish varieties

Economics of Marine Capture Fisheries at Kombuthurai

The economics of marine capture fisheries in Kombuthurai fishing village during the study period was calculated and estimated the profitability. For a comparative approach economic estimation was done simultaneously for Therespuram, one of the major fish landing centres of Thoothukudi (Table 13.6). While the number of fishing trips was accounted as 155 for Kombuthurai, it was 180 for Therespuram, on an average. Comparatively, the calculated $BCR_{(TC)}$ indicate the higher returns on investment for Kombuthurai fishermen (1.44:1) than the Therespuram fishermen (1.27:1) and the price advantage clearly reflects the profitability in Kombuthurai fishing village.

Table 13.6: Comparative economics of marine capture fisheries of Kombuthurai and Therespuram fishing villages

Sl.No.	Particulars	Kombuthurai		Therespuram	
		Total value (Rs. in lakhs/ annum)	% to Total cost	Total value (Rs. in lakhs/ annum)	% to Total cost
a. Fixed Cost (FC)					
i.	Depreciation				
	Fishing craft @ 10%	14.57	1.99	12.82	1.57
	Engine @ 8%	8.76	1.20	6.61	0.81
ii.	Repairs and maintenance	23.12	3.16	11.03	1.35
iii.	Interest on capital cost @ 10%	29.15	3.99	22.42	2.74
	Total Fixed Cost (TFC)	75.60	10.34	52.87	6.47
b. Variable Cost (VC)					
i.	Fuel	302.25	41.34	94.72	11.59
ii.	Ice	8.49	1.16	5.08	0.62
iii.	Bait fish	—	—	227.16	27.79
iv.	Food	9.15	1.25	9.39	1.15
	Crew share	321.67	44.00	428.13	52.38
v.	Other expenses	13.95	1.91	—	—
	Total Variable Cost (TVC)	655.51	89.66	764.47	93.53
c.	Total Cost (TC)	**731.11**	**100.00**	**817.34**	**100.00**
d.	Total Returns (TR)	**1054.22**	—	**1036.08**	—
e.	Total Net Returns (NR)	**323.11**	—	**218.74**	—
f.	Net Returns / Head	**3.72**	—	**2.51**	—
g.	Marketing charges / Head @ 8%	**0.30**	—	—	—
Cash in hand as profit / Head		3.42	—	2.51	—
$BCR_{(TC)}$		**1.44**	—	**1.27**	—
$BCR_{(TVC)}$		**1.61**	—	**1.36**	—

Conclusion

The existing system of fish marketing practices and service charge system of Kombuthurai fishing village is well-known for its innovativeness. From the study, it is concluded that the adoption of eco-friendly fishing methods and innovative marketing model exclusively managed through people institutions of Kombuthurai ensuring sustainability and maximum producer's share with price advantage is a worthy model which may be replicated in a similar socio-politico-economic scenario of other parts of the country as well with suitable modifications. Segregating the fish catch species-wise and auctioning on weight basis through fishermen associations, combined with proper market information and required infrastructure facilities would definitely result in maximizing income to the fishermen and minimizing exploitation by the middlemen.

References

Acharya, S.S. (1997). Agricultural marketing in India: Policy framework, emerging issues and needed initiatives.*The Bihar Journal of Agricultural Marketing*, 5 (3): 253-269.

Ali, E.A., Gaya, H.I.M. and Jampada, T.N. (2008) Economic analysis of fresh fish marketing in maiduguri gamboru market and kachallari alau dam landing site of North-eastern Nigeria. *Jouranl of Agriculture and Social Science*, 4: 23-26.

Ayeisha A. Brinson, Min-Yang A Lee, Barbara Rountree (2011). Direct Marketing Strategies: The Rise of Community Supported Fishery Programs. *Marine Policy* 35(4): 542-548.

Chahal, S.S., Singh, S. and Sandhu, J.S. (2004). Price spreads and marketing efficiency of inland fish in Punjab: A temporal analysis. *Indian Journal of Agricultural Economics,* 59(3): 498.

Ganesh Kumar, B., Datta, K.K., Joshi, P.K., Katiha, P.K., Suresh, R., Ravisankar, T., Ravindranath, K. and Muktha Menon (2008). Domestic fish marketing in India – Changing structure, conduct, performance and policies, *Agricultural Economics Research Review* (Conference Issue.), 21 : 345 - 354.

Ganesh Kumar, B., Datta, K.K., Vidya Sagar Reddy, G and Muktha Menon, 2010. Marketing System and Efficiency of Indian Major Carps in India, *Agricultural Economics Research Review*, Vol. 23: 105-113.

Gupta, V.K., 1984. Marine Fish Marketing in India (Volume I – Summary and Conclusion). IIM Ahmadabad & Concept Publishing Company, New Delhi.

Katiha, P.K. and Barik, N.K., (2004). Governance, institutions and policies for fisheries of floodplain wetlands. *Indian Journal of Agricultural Economics*, 59 (3): 490.

Ravindranath, K. (2008). Domestic marketing of fish and fishery products in India – Opportunities and challenges. In: National Workshop on Development of Strategies for Domestic Marketing of Fish and Fishery Products held at College of Fisheries Science, Nellore, India, 7-8 February: 43-48.

Srivastava, U.K. and Kant Uma (1985). Inland Fish Marketing in India (Volume I – Overview: Summary and Conclusion), IIM Ahmadabad & Concept Publishing Company, New Delhi.

14

Meta Data Analysis on Production and Marketing Constraints in Freshwater Fish in Asian Countries

Pradip C. Bhuyan

Introduction

Production of consumer preferred varieties of fish and fish products have been facing several constraints. Identification of potential constraints is important for growth and development of fisheries sector. The constraints of production and marketing have been examined by a number of researchers in different time and place. In this chapter, it is tried to explain an overview of the different constraints of production and marketing of fish and value added fish.

Constraints of production of freshwater fish

Constraints of aquaculture production in developing countries have been identified by Lee (1997) and categorized them as:

i) natural and environmental (inequitable allocation of land resources, insufficient quantity and degraded quality of water, highly seasonal variation in temperature and natural disaster);

ii) socioeconomic (insufficiency of infrastructure for production and marketing, variation of prices of inputs, ageing and poor training of aqua-farmers); and

iii) Institutional constraints (inefficiency of extension services, lack of a better organization of producers, shortage of rural finance).

The study has provided some strategies to overcome these constraints which includes structural adjustment; better market management and effective institutional programmes intervened by the Government. Structural adjustment through establishment of cultivated areas, strengthening of early monitoring systems and acceleration of technological changes were suggested to promote aquaculture development. For effective market management the study suggested shortening of marketing margins and transmitting market information to the producers. For institutional improvement, better organization of producers, better extension services and effective aid of rural credit were suggested.

Steinbronn *et al.,* (2005) studied the constraints in fish production in Yen Chau district of Son La Province, Vietnam. The typical pond system in the study area was utilized for polyculture of grass carp, other carp species and tilapia. The study revealed that the main problem of pond farming were lack of training or extension services in the field of aquaculture, frequent outbreak of disease, poor quality of the seed, application of pesticides in paddy fields (which ultimately come to fish pond), shortage of water while irrigating the paddy fields, low water temperatures during the winter, and limited supply of feed resources in the cold dry season.

Liao and Chao (2009) studied the constraints in the aquaculture industry of Asia-Pacific region. Constraints faced by aquaculture industry included competition for land and water with other industrial sectors, insufficient aquaculture engineering for land-based and off-shore aquaculture, unpopularity of automatic devices for super intensive aquaculture and post-harvest processing, high prevalence of disease outbreaks and natural disasters, and complete dependency of farmers on government aids.

Martin and Ross (2011) reviewed the constraints that poor people faced in accessing markets in Cambodia and analyzed implications for proper domestic aquaculture development. Analysis of secondary data confirmed that the potential for poor aquaculture producers to interact with urban markets in Cambodia was currently low but the potential of aquaculture to interact with rural markets was high. The study concluded that aquaculture had the potential for attaining self-sufficiency in rural food fish and could provide farm income diversification through rural market development.

Common constraints faced by freshwater fish farmers, especially in Bangladesh and India, are plurality of ownership, lack of credit facilities, lack of technical know-how, illegal poaching, deliberate poisoning, inadequate marketing opportunities, non-recognition of aquaculture as a land-based activity, absence of long-term leasing policies, and non-assurance of seed supplies at appropriate times (FAO 2001, v). Dey *et al.,* (2005) reported that freshwater fish farming is generally profitable in Asia but fish culture practices in most of the Asian countries have some constraints.

Alam and Thomson (2001) identified the problems against the fuller utilization of potential of Bangladesh fisheries sector. They reported that resource limitations, poor implementation of fisheries laws, limited spread of fish farming technology, low financial capacities and ineffective extension practices were the main factors responsible for the under-utilization of fishing areas. Lack of reliable fisheries statistics in the country was identified as a basic problem for fisheries required for analysis, planning and evaluation. The study further revealed that rich farmers and other affluent people enjoyed most of the benefits of easy credit but the small and medium sized fish farmers found it difficult to get credit. There was lack of proper training and field experience among Fisheries Officers. The Fisheries Officers were often constrained by lack of conveyance for which they found it difficult to visit farm in remote areas. Mohsin and Haque (2009) studied the constraints of carp production in Rajshahi district of Bangladesh. The study revealed that 34% farmers perceived financial crisis as the prime constraints for the carp farming followed by adequate availability of seed (25%), feed (14%), high mortality rate of fry (11%), poaching (6%), poisoning of pond (4%), and scarcity of sufficient water (4%), and disease incidence (2%). The study found that farmers did not adopt recommended carp culture technology completely. Illiteracy was one of the major problems of non-adoption of fish culture technology. Sarkar, Chowdhury and Itohara (2006, 68-73) analysed entrepreneurship barriers of pond fish culture in Mymensing district of Bangladesh. Their study revealed that lack of technical knowledge on pond management, unavailability of credit, poor extension service and lack of information were the potential barriers of pond fish culture entrepreneurships.

The constraints for growth of small-scale freshwater fish culture in India are lack of basic inputs, poor fisheries extension mechanism, poaching, conflicting interests with regard to water use between agriculture and aquaculture, short lease period, inadequate institutional finance, lack of infrastructure facilities like cold storage, good approach roads from production sites to marketing centers, and quick transport facility etc. (Sinha and Ranadhir).

Occurrence of trash fishes and weeds, fish disease, and poaching are the major constraints of fish production in Tirunelveli district of Tamilnadu (Selvaraj). V. Kumar and Selvaraj (1988) conducted a socio-economic study on composite fish culture in five districts of Tamil Nadu and categorized the constraints as production, management, and marketing constraints. Production constraints were related to availability and dearness of inputs like seed, fertilizer and labour. Untimely supply of fish seed was a major constraint faced by majority of the respondents. The management constraints were associated with predators, weeds, trash fish and poaching. Unremunerative price, lack of transportation, tied sale and spoilage were included in marketing constraints.

Padhy (1994) identified constraints of fish culture in Birbhum district of West Bengal and categorized them as environmental and situational constraints, lack of technological intention, and socio-economic and infrastructural constraints. Environmental and situational constraints included occurrence of flood, drought and weeds. Inadequate availability of inputs such as feed, credit, transportation cost and returns, management, trained extension services, marketing, storage facility etc. were included under socio-economic and infrastructural constraints. The study indicated that socio-economic and infrastructural constraints accounted for 53.85 percent, technological constrains 27.22 per cent, and environmental and situational constraints were 18.92 percent. Influence of these constraints individually on productivity of fish culture and functional relationship between them were ascertained by multiple regression analysis, which indicated that out of 14 independent variables (constraints), only 5 were statistically significant determinants in productivity scenario of fish culture.

Chakraborty (1991) identified technological constraints of inland fish cultivation in 24 Parganas (North) district of West Bengal. The study examined and identified the gaps between potential and actual yield and real problems so as to formulate future programmes for increased fish production in inland sector. The yield rate of beneficiaries belonging to the Fish Farmers Development Agency (FFDA) schemes was significantly higher (1650 kg/ha) than that of non-FFDA farmers (613 kg/ha), although this was less than the potential yield (2500 kg/ha/yr) for FFDA schemes. The study further revealed that about 73 percent beneficiaries of the FFDA were satisfied with the support of the scheme which uplifted the economic condition of the fish farmers through proper technology adoption.

Perceived problems of composite fish culture in 4 districts of West Bengal studied by Bhaumik and Saha (1995) revealed that the major perceived problem in adoption of composite fish culture was high cost of inputs followed by poaching, poisoning, high rent of water body, lack of follow- up action, marketing of harvested fish, non availability of subsidy, non- availability of finance, multi ownership of water body, stagnancy of capital, and non-achievement of expected results. The study stressed upon development of low cost package of practices on the lines of single stocking- multiple harvests or multiple stocking-multiple harvest.

Goswami *et al.*, (2010) studied the factors influencing adoption behaviour of fish farmers towards scientific fish culture practices in the purposively selected Dakshin Dinajpur district of West Bengal during 2005-2008. The study revealed that majority of the fish farmers of the study area adopted scientific fish culture practices at a medium to high extent as most of the fish farmers had correct information and knowledge about scientific fish culture practices. The study indicated a positive and significant co-relation between the extension agency and adoption of scientific fish culture practices. It recommended planning more knowledge building activities like meeting, discussion, mass media etc. by fisheries extension personnel to increase knowledge base of farmers for adoption of scientific technologies. The level of knowledge base among fisher folks of Dakshin Dinajpur district, West Bengal was investigated by Goswami and Samajdar (2011) which revealed that majority of the fish farmers (50.8%) had a medium level of knowledge about scientific fish culture practices. 90.8 per cent knew about the need of manure application in the fish pond, while 16.7 per cent were aware of the recommended stocking rate in composite fish culture. Insufficient awareness on scientific fish culture was the factor responsible for low fish production in West Bengal. The study suggested increase in innovative proneness, extension agency contact and mass media participation by means of organizing awareness campaigns, field days, demonstrations, exhibitions, krishan mela etc. to enable farmers to accrue latest knowledge on scientific fish culture practices. It emphasized introduction of a system of evaluation at apex government level in order to improve the process of reorienting the fishery extension system and to provide technical and input support to the farmers to enhance knowledge level on fish culture.

Srivastava (2000) identified some of the constraints and problems faced in freshwater aquaculture development in India. Non-availability of quality fish seed of commercial species in adequate quantities at the right time, absence of cheap and acceptable supplementary feeds, difficulties in mobilizing institutional finance and credit for small fish farmers, low price realization by the producer due to the poor market structure and absence of uniform leasing policy in different states were reported as main constraints in the study.

The SWOT analysis carried out by Radheshyam (2001) with participatory efforts of farmers revealed some important weaknesses in community based aquaculture. The major constraints were poor organizational capacity among rural farmers due to personal disputes, non existence of capable community leader, lack of infrastructures, weak research-extension support, low technical awareness, and dual leasing policy with short leasing period. The majority of rural farmers did not own ponds and over 67% of freshwater fish farming were in leased out ponds in certain areas of the Country. For the short duration leasing policy, fish farmers were reluctant to invest more resulting underutilization of the resources. The impact assessment suggested intervention of research/ extension/policy planners for setting directions and priorities for further improvement in the sector.

Investigations of Sasmal et al. (2006) in Dharsiwa Block of Raipur District revealed constraints perceived by the fish farmers for adoption of recommended composite fish culture technology such as high cost of pond preparation, eradication of weeds, lack of knowledge, lack of efficient marketing structure, and restriction posed by the village community regarding the use of some of aspect of recommended technology. Maximum fish farmers were adopting the traditional practices of fish farming instead of recommended technology.

Meena, Prasad and Singh (2009) investigated the constraints perceived by rural agro-processors of Punjab to adopt post-harvest technologies and categorized the constraints as socio-economic, technological, farming, marketing, and extension aspects. Socio-economic, technological and farming constraints were more important than extension and marketing constraints. The suggested measures for removing the constraints were appropriate policy interventions for boosting-up the rural agro-processing sector. A national intervention for adopting the post-harvest technologies towards rural poverty eradication programmes facilitating formation of SHGs/social capital at cluster level and transfer of technology through these SHGs were also suggested. They also recommended conduction of need-based and skill oriented training, long term institutional credit support, market-driven and decentralized extension system, orientation towards high-value enterprises through technology based entrepreneurship development programmes, and awareness creation about post-harvest technologies and government schemes through mass media.

Abraham *et al.,* (2010) studied the aquaculture practices of Andhra Pradesh and West Bengal and revealed that majority of the respondent farmers of the two states cultured carps. But there were differences in farm holdings, size of the pond/farm, species cultured, stocking rate and stocking density, fish seed procurement policy, nursery management, feed and feeding rates, pond fertilization, harvesting frequency, mode of fish marketing, source of information on aquaculture, fish seeds and disease treatment, and perceptions on aquaculture practices. The major constraint faced by farmers of Andhra Pradesh and West Bengal was incidence of fish diseases. Other constraints included fluctuation of market price, irregular electricity supply, poaching, declining production, poor seed quality, floods, financial problem, siltation etc. The magnitude and impacts of these problems were different among the farmers of both the states. The study concluded that in order to enhance fish production from culture systems a strong commitment from Government organizations and research institutions in the form of more training and extension services were urgently needed.

Mohanty et al. (2011) identified the major constraints in adopting/developing participatory agri-aquaculture in three different watershed sites in Orissa, India through preferential ranking technique and delineated as many as nine constraints.

Those were lack of awareness and technical knowledge, high feed cost, low water depth in summer, lack of interest, priority to domestic use, non-availability of fingerlings in time, etc. The study suggested putting efforts to improve marketing of produce through information dissemination on prices and nutritional value among vulnerable groups; improving road access to urban markets to ensure better price, formation of marketing groups, and providing information on preservation and storage.

The status of freshwater aquaculture resources in Boudh District of Orissa was investigated by Chattopadhyay et al. (2012) which identified some constraints faced by fish farmers such as unavailability of desired quality and quantity of fish seed, poor water retention capacity (6-7 months) of the pond, high rate of evaporation, high lease value, lack of technical knowledge among fish farmers, aquatic weed infestation, and presence of predatory fishes.

Unavailability of quality seed, inadequate technology transfer, lack of private entrepreneurship, lack of infrastructure facilities, low temperature regime, complex land ownership patterns, small fragmented land holdings etc. were the main constraints of development of fish culture in North East India (Munilkumar and Nandeesha 2007). Major constraints in fishery sector of Tripura were identified as genetic degradation in fishes due to inbreeding in hatcheries, lack of diversification of culture fisheries research facilities, soil and water quality mapping, and recurrent flood (Barman and Mandal, 2011). Singh *et al.*, (2009) assessed the technical efficiency level and its determinants in small-scale fish production units of West Tripura district. Primary information collected from 101 fish farmers of three blocks through a multi-stage random sampling method revealed that farmers were not getting quality fish seed. The middlemen were the source of fish fingerling supply to the farmers who made it available as a mixture of different species and different size. Farmers had no access to other assured sources of quality fish seed. All these constituted a low technical efficiency level, whereas those farmers who purchased fingerlings from the government firms enjoyed better technical efficiency. The study suggested that the State Government needs to play a role to ascertain the supply of quality fish fingerlings adequately and timely to ensure the technical efficiency of the culture systems.

Non-availability of inputs, disease outbreaks, inadequate financial and extension support, and frequent flood problems were some of the constraints limiting the productivity of fish in Assam (Goswami and Sathiadhas 2000, Goswami *et al.*, 2002) conducted a study on socio-economic dimension of fish farming in two districts of Assam, *viz.*, Darrang and Nagaon during the period 1998-2000. The study revealed that, on an average, 57 per cent of the respondents were engaged in agriculture followed by fishery (21%), business (17%) and service (5 %) as other occupations. Only 16.67 per cent of the respondents of Darrang and 25 per cent of Nagaon had fishery as a major occupation. Majority of the respondents did not receive training on fish culture practices. The percentage of trained respondents in Darrang and Nagaon were 25 and 20, respectively.

The SWOT analysis of fishery enterprises, carried out by Agricultural Technology Management Agency (ATMA), in 2006 revealed some weakness of culture fishery in Nagaon district of Assam. They have been identified as low water retention capacity of the soil, occurrence of flood, dominance of aquatic macrophytes, unregulated retail fish markets, lack of storage and preservation facilities, non-streamlined institutional finance, exploitation by market intermediaries, subsistence nature of fish farming, imbalance use of organic and chemical fertilizers, non availability of quality fish seeds, non-availability of large size fingerlings, and poor soil and water management (Agricultural Technology Management Agency, Nagaon, 2006). In addition to these constraints, poor extension machineries, low pH value of soil and water, lack of proper marketing channels, lack of credit, lack of entrepreneurship, social taboo, natural calamities etc are some of the problems of fish farming in Assam (Kalita, et al., 2007).

The adoption behaviour of composite fish culture practices was positively influenced by the factors like extension participation, economic motivation, cosmopolitanisms, scientific orientation and knowledge of fish farmers, and negatively influenced by their age (Talukdar and Sonataki, 2005). The study recommends that efforts should be made by extension agencies through various programmes to highlight the economic benefits of composite fish farming to promote large-scale adoption of this technology. Study tours, exposure visits, participation in fairs and exhibitions were recommended as the ideal methods for promoting adoption of composite fish culture. This study was carried out in Sonitpur district of Assam.

There is a good market demand for endemic fish species like magur (*Clarias batrachus),* singi (*H. fissilis*), koi (*Anabus testudineus*), Pabda (*N. notopterous*) etc. in North-Eastern parts of India (Sugunan, 2009). But the culture practices of these species have not received much attention due to lack of standardized package of practice of culture of these varieties. Das (2002) revealed that inadequate supply of seed and proper feed hinders the culture of magur in the area. Again, there is good demand for the snakehead, *Channa striatus*, commonly known as striped murrel and locally known as '*sol*' in Assam. But the culture of murrels in Assam is still not common due to the lack of seed supply (Marimuthu *et al.*, 2001). Proper technologies for captive breeding of such other alternate potential commercial fish species are necessary to diversify the culture systems for better economic returns.

From the above review of literature major constraints of freshwater fish production are summarized below:

i) Lack of technical know-how
ii) Lack of training
iii) Lack of extension services in the field of aquaculture
iv) Illegal poaching
v) Frequent outbreak of disease
vi) Poor quality of the seed
vii) Shortage of water while irrigating the paddy fields
viii) Low water temperatures during the winter
ix) Limited supply of feed resources in the cold dry season etc.
x) Deliberate poisoning and poaching
xi) Multiplicity of pond ownership
xii) Absence of long-term leasing policies
xiii) Poor implementation of fisheries laws
xiv) Non-assurance of seed supplies at appropriate times
xv) Social taboo against stocking community ponds
xvi) Limited spread of fish farming technology
xvii) Poor fisheries extension mechanism
xviii) Lack of reliable fisheries statistics planning and evaluation
xix) Conflicting interests with regard to water use between agriculture and aquaculture
xx) Short lease period
xxi) Inadequate institutional finance
xxii) Inadequate availability of seed, feed
xxiii) Lack of infrastructure facilities like cold storage,
xxiv) Good approach roads from production sites to marketing centers, and
xxv) Lack of technical knowledge on pond management
xxvi) Incidence of disease

Constraints of Marketing of Freshwater Fish

Several constraints related to distribution of fish and fish products have been identified by different studies at different times and places. Fish marketing is not an easy task as it has to face many peculiar and special problems at different stages of production and marketing management. Some of the specific problems of marketing of fish are greater uncertainties in fish production, the high perishability of fish, collection of fish from too many scattered landing centers, too many varieties of fish and Therefore, too many demand patterns, wide fluctuation in prices, lack of proper transportation of fish etc. (Rao 1997). According to FAO Fisheries Circular No. 973 (FAO 2001) major constraints of fish marketing were bad transportation system, poor bargaining power, high marketing margins, low institutional credit for production and marketing of fish etc. The study reported that credit was provided by market intermediaries to the marketers as well as to the producers and force them to sell their produce. But the credit supplier often paid less than the market price. The report concluded that sustainable development policies are needed that could address issues related to use of natural resources, research, pricing, credit, trade, investment, and exchange rates.

Ahmed et al. (2012) carried out a study in order to develop sustainable tilapia marketing systems in Bangladesh. The study revealed that almost all the tilapias produced in Bangladesh were marketed internally for domestic consumption. Constraints in marketing of tilapia, as perceived by the farmers were inadequate knowledge of marketing systems, low market prices (24% of respondents), exploitation by intermediaries (42% of respondents), and lack of infrastructure (mainly poor road and transport facilities, 34% of respondents). Other constraints were higher transport costs, insufficient supply of ice, unhygienic conditions, lack of financial support, lack of credit facilities, and poor markets infrastructure (i.e., inadequate drainage systems, poor supply of water, limited ceiling, and flooring space), lack of standard practices for handling, washing, sorting, grading, cleaning and icing of tilapia. The strategies formulated in this study were-provision of capacity building for the development of stakeholder organization, government institutions for technical advice and support on marketing, proper market infrastructure, encouraging involvement of appropriate NGOs, and the implementation of a management plan to address existing constraints. The study developed conceptual framework which consists of three basic components: market awareness, market access, and marketing facilities.

The major constraints for both domestic and export markets in Cambodia were inadequate facilities for handling, sorting, weighing and packing fish, and lack of storage facilities and preservation equipment or materials (e.g., ice, ice-crushing machines, ice boxes, freezers, salt) at landing sites (Mohammed et al.

2012). According to them due to lack of modern equipment or production methods, small and medium-scale fish processing operations were unable to adopt quality control measures and hygiene standards.

The study carried out by Chea and McKinney (2003) on fish marketing from Great Lake to Phnom Penh (South East Asia) revealed that most fishers are in a weak price negotiation position since they are compelled to sell their produce to the trader with whom they are in debt. The lack of transparent interest rate on loans for fishers is another constraint identified in the study area. Lenders are likely to take advantage of this lack of transparency to increase returns on their loans. The study further indicated that fish marketing is affected by a number of other constraints such as high financing costs, spoilage and weight loss, monopolistic control of distribution, high transportation and ice costs, and fees charges along the road during transport.

Different marketing constraints of fish marketing system of Swarighat, Dhaka were lack of modern hygienic fish landing centers, shortage of adequate ice-plants with sufficient capacity, cold and freezer storage, lack of handling and preservation facilities, inadequate transportation and distribution facilities, lack of insulated and refrigerated fish vans, etc. (Alam et al.2010). The study further revealed that the consumers had to pay higher price due to the participation of too many intermediaries in the marketing channel, but the actual fishers never got the actual price for their products and major portion went to the intermediaries. The study suggested establishment of more ice-plants, cold-storage and preservation facilities, introduction of insulated and refrigerated fish vans and fish carriers to maintain cold-chain during transportation, improvement of existing fish market structure, and establishment of modern wholesaling facilities. The study further stressed upon the government intervention at national level for formulation of proper price policies, arrangement of training and extension works in marketing and management of relevant market research. It was also stated that present marketing system could be improved through strengthening the bargaining power of the farmers/fishers by providing actual information about the present market status, pricing policy, credit facilities, and formation of association.

There are no organized fish marketing policies that cover price structure and marketing outlets among others which are related to both export and domestic markets (Mohite and Mohite 2008). The domestic fish marketing system in India is neither efficient nor modern and is mainly carried out by private traders with a large number of intermediaries between producer and consumer. This leads to reduction in the fisherman's share in consumer's rupee (G. Kumar *et al.*, 2008). Hence, efforts are necessary to convey the prices prevailing at the nearby fish markets for various species daily through appropriate media. Better hygienic conditions of fish markets can not only attract more consumers to the markets, but also build confidence among buyers to consume fish. The study suggested that modern retail outlets have to be promoted vigorously through public-private partnership in every major city so that fish consumption becomes an easier proposition in days to come.

Due to lack of proper shelter and fish dressing platforms, drainage system, drinking water etc. in most of the markets in Mumbai, most retailers used to sell fish at roadside without maintenance of quality or hygiene (Mugaonkar *et al.*, 2011). Market infrastructure was very poor in Dakshin Dinajpur district of West Bengal. No grading, sorting, standardization, certification, etc. were found in either rural or urban fish markets in the district. There was no proper regulatory mechanism of the government over the market (Roy 2008).

A study was carried out by Upadhya *et al.*, (2011) in one wholesale market and four retail markets in Agartala, Tripura. The study revealed that there were inadequate infrastructural facilities in terms of auction/selling platforms, market sheds, power supply, drainage facilities, water supply in both wholesale and retail markets. The study emphasized on intensive investigation on dry fish marketing covering entire Northeast region to bridge the gap on available information on demand of dry fish, seasonal variability in prices and species availability, source of supply of dry fish and employment opportunities in Northeast region. The study stated that though much progress has been made in Indian fisheries marketing system, especially in the private sector, much remain to be done with regard to improving the performance of the fish marketing system in India.

There exist some marketing constraints in fish marketing systems of Assam too. Lack of adequate transport and communication facilities in Assam has constrained the sale of fish to limited outlets and prevents the growth of specialized marketing (Goswami *et al.*, 2012). In addition to this constraint it was reported that insufficient credit and differential pricing policies were emerging as hindrance to the market development. The study emphasized on infrastructure development by means of providing ice plants, storage and processing facilities and improvement in transportation system for improvement of marketing system in the State. The authors also suggested motivating fish farmers/fishermen to start fish production including seed production and marketing through cooperative system.

Shil and Bhattacharjee (2009) reported that fish markets in Barak Valley are not well organized and there is need to reform the markets by introducing proper marketing techniques. Absence of proper transportation, insufficient parking facilities, inadequate storage facilities, poor power supply, lack of proper drainage and water supply, lack of credit facilities, seasonal differentiation of price in the markets, lack of assistance from Municipal Board are some other bottleneck for development of fish marketing in the study area. The study suggested some measures to be adopted for development of retail marketing in the area such as improvement of road condition, provision of storage facilities, marketing information service, credit provision to retailers, provision of parking facilities etc.

According to Kumar *et al.,* (2008), fishery is a state subject under the Constitution of India. Only a few states have a policy specifically aimed at fish marketing. The only legislation for fish marketing is the West Bengal Fish Dealer's Licensing Order, 1975. The Act has a variety of legal procedures to control the process of supply of fish to other states from West Bengal. Every fish merchant has to get a license to conduct business by paying an annual fee. All the fish commission agents and wholesaler-cum-retailers are to be registered with the Directorate of Fisheries under this Order. All state Fisheries Departments, State Fish Development Corporations and Apex Fishermen Cooperative Societies have schemes to help fishermen to market their catch efficiently. The schemes include provision of vehicles for transporting fish from landing centres to markets, fish kiosks and marketing implements like insulated boxes, utensils, dressing knives, etc.

Several organizations have been set up at the national level to promote the fisheries sector and help the fishermen. These include organizations such as the National Cooperative Development Corporation (NCDC), the National Federation of Fishermen's Cooperatives Ltd. (FISHCOPFED) and the National Fisheries Development Board (NFDB). NCDC's fisheries related activities include creation of infrastructural facilities for fish marketing, ice plants, cold storages, retail outlets, etc. FISHCOPFED promotes fishery cooperatives and assists fishermen to market their produce efficiently through hygienic retail fish centres in metropolitan cities thereby providing remunerative prices to fish farmers. NFDB is promoting domestic fish marketing through modernization of wholesale markets, establishment of cold chains, popularization of hygienic retail outlets and technology up gradation. Fish is not a notified commodity under the APMC Act of 1966, leading to the exploitation of fishermen by commission agents. Unlike other agricultural commodities, where commission charges are paid by the traders, in fisheries, all commission charges are paid by fishermen. This reduces the share of fishermen in consumer's rupee and makes fishing a non-viable venture. Suitable modifications are to be introduced in the Act to overcome this situation.

Different studies suggested different measures to overcome the constraints of marketing of fish. Some of the important measures suggested for overcoming constraints and problems of marketing in India by the FAO report (FAO 2008: Fisheries Circular. No.1033) are as follows:

- Modern fish markets should be established in major urban centers
- Development of a legal framework for the establishment and management of fish markets
- Culture of suitable species/ new species of fish should be encouraged
- The development of value-added products from low value fish species should be urgently promoted
- Women self-help groups should be promoted
- Proper training in the techniques of production and marketing should be provided;
- Fish as a healthy food needs to be popularized among consumers. A special campaign, similar to the campaign currently undertaken for dairy and poultry products, to promote eating of fish is necessary
- It is necessary to formulate a nation-wide fish marketing strategy with the specific objectives of helping fishers to market their products at a remunerative price and to supply safe and quality fish and fish products to consumers. The example of the cooperative structure of the small-scale dairy industry in India should be followed;
- In order to promote the marketing of frozen fish products, the excise duty on these products should be waived. Value-added tax (VAT) should also be reduced;
- The improvement of fishery statistics, especially with regards to inland fisheries landings and marketing of fish from various sources are necessary.

Though there are a number of organizations and policies relating to promotion of fish marketing in the country, there is a need to formulate a uniform market policy for fishes so that it becomes easier in operation and regulation, which will not only improve the level of country's fish production but also availability to the consuming population, ensuring a remunerative price to the fishers at the same time (Kumar *et al.,* 2008, 2010) suggested promoting institutions like SHGs, producer/fishermen associations, cooperatives, etc. and allowing the entry of private agencies with appropriate regulatory mechanism to improve the efficiency of fish marketing in the country. Recommendations to improve fish marketing by the organized sectors in India have been provided and policy implications have been discussed in this study.

In areas where aquaculture has developed to a significant level, the general trend is to educate the public on the quality of farmed products and use this as a selling criterion (Pillay and Kutty 2005). Many countries have established specialized sales federation, cooperatives or similar organizations to reduce the number of intermediaries involved, harmonize marketing within the country and compete effectively in export markets. Such organizations are able to undertake useful promotional and publicity programmes and thus improve sales.

Major constraints of freshwater fish marketing in Asian countries (as revealed from review of literature) are summarized below

1. Inadequate infrastructural facilities in terms of auction/selling platforms, market sheds, proper shelter and fish dressing platforms, power supply, drainage facilities, water supply, drinking water etc. in both wholesale and retail markets
2. Lack of proper refrigeration facilities both in urban and rural markets
3. Inadequate facilities for handling, sorting, weighing and packing fish
4. Lack of storage facilities and preservation equipment or materials (e.g., ice, ice-crushing machines, ice boxes, freezers, salt) at landing sites
5. Shortage of adequate ice-plants with sufficient capacity, cold storage
6. Lack of modern hygienic fish landing centers
7. Lack of suitable containers
8. Poor transportation system, no means of cooling during transportation by fishers
9. Fees charges along the road during transport
10. Inadequate transportation and distribution facilities, lack of insulated and refrigerated fish vans, etc.
11. Lack of market information
12. Lack of assistance from Municipal Board
13. Low institutional credit for marketing of fish etc.
14. Lack of access to credit
15. Lack of transparent interest rate on loans for fishers
16. High financing costs
17. Differential pricing policies

18. Monopolistic control of distribution
19. Unorganized fish marketing
20. No organized fish marketing policies that cover price structure and marketing outlets among others which are related to both export and domestic markets
21. No proper regulatory mechanism of the government over the market
22. A large number of intermediaries between producer and consumer, thereby reducing the fisherman's share in consumer's rupee
23. No grading, sorting, standardization, certification, etc.
24. Lack of knowledge to prepare and preserve fish etc.
25. Lack of business and management skills
26. Seasonality and perishable nature of fish
27. Seasonal differentiation of price in the markets
28. Poor bargaining power
29. High marketing margins

References

Abraham, T. Jawahar, S. K. Sil, and P. Vineetha (2010). A comparative study of the aquaculture practices adopted by fish farmers in Andhra Pradesh and West Bengal. *Indian Journal of Fisheries,* 57(3): 41-48

Agricultural Technology Management Agency (ATMA) (2006) Nagaon. District Agricultural Development Strategy of Nagaon District, Assam.

Ahmed, Nesar, James A. Young, Madan M. Dey, and James F. Muir. (2010). From production to consumption: a case study of tilapia marketing systems in Bangladesh. *Aquaculture International* 20 (2012): 51–70.

Alam, Ferdous, and Kenneth J. Thomson. (2001). Current constraints and future possibilities for Bangladesh Fisheries. *Food Policy,* 26 : 297–313

Alam, Jobaer, Rumana Yasmin, Arifa Rahman, Nazmun Nahar, Nadia Islam Pinky and Monzurul Hasan (2010). A study on fish marketing system in Swarighat, Dhaka, Bangladesh. *Nature and Science* 8(12):96-103

Barman, Debtanu, and Sagar C. Mandal (2011). Achieving self-sufficiency in fish production in Tripura state of India some policies and suggestions. *Aquafind:* Web.19 July 2011<http://aquafind.com/index.php>

Bhaumik, U. and S. K. Saha (1995). Need for modification of composite fish culture technology in West Bengal as perceived by the fish farmers. *Workshop Proceeding on Current and Emerging Trends in Aquaculture*, Ed. Thomas, P. C., Daya Publishing House, New Delhi-110035: 348-59.

Chea, Yim, and Bruce McKenney (2003). Domestic fish trade: a case study of fish marketing from the Great Lake to Phnom Penh. Working paper no. 29. Phnom Penh: Cambodia Development Resource Institute.

Chakra borty, S. (1991). Constraints to technological progress in inland fish cultivation - a case study in 24 Parganas (North) District of West Bengal." *Proceeding of the National Workshop on Aquaculture Economics, November 20-22*: Ed. S. D. Tripathy, M. Randhir, and C. S. Purushothaman. Asian Fisheries Society, Indian Branch, Mangalore, India.

Chattopadhyay, D. N, H. K. De, G.S. Saha, Radheyshyam, Soumi Pal, and T. S. Satpati (2012). Fresh water Aquaculture in Boudh District of Odisha: Prospects and Potentials. *Fishing Chimes,* 31(11): 20-23.

Das, S. K. (2002). Seed production of Magur (*Clarias batrachus*) using a Rural Model Portable Hatchery in Assam, India – A Farmer Proven Technology." *Aquaculture Asia,* 7(2) : 19-21.

Das, S. K., and U. C. Goswami. (2002). Current status of culture fisheries in Nagaon and Morigaon Districts of Assam. *Applied Fisheries and Aquaculture,* 11(2): 33-36.

Dey, Madan Mohan, Mohammed A. Rab, Ferdinand J. Paraguas, Ramachandra Bhatta, Ferdous Alam, Sonny Koeshendrajana, and Mahfuzuddin Ahmed (2005). Status and Economics of Freshwater Aquaculture in selected Countries of Asia. *Aquaculture Economics and Management,* 9(1-2): 11-37.

Directorate of Extension Education, Assam Agril. University. Web. 21 March 2013 <http://www.aau.ac.in/dee/annexture6.php>

Economic Survey, Assam. Guwahati (2011-12): Government of Assam, Directorate of Economics and Statistics.

FAO. (2001) Production, accessibility, marketing and consumption patterns of freshwater aquaculture products in Asia: A cross-country comparison. Fisheries Circular, No. 973 (2001).

FAO (2008). Present and future markets for fish and fish products from small-scale fisheries – Case studies from Asia, Africa and Latin America. Fisheries Circular. No. 1033. Rome, FAO. 2008. 87p. <http://www.ibcperu.org/doc/isis/9819.pdf>

FAO (2001). Production, accessibility, marketing and consumption patterns of freshwater aquaculture products in Asia: a cross-country comparison. *FAO Fisheries Circular.* No. 973. Rome, FAO. 2001. http://www.fao.org/DOCREP/004/Y2876E/y2876e02.htm

Goswami, Biswajit, and Tanmay Samajdar (2011). Knowledge of fish growers about fish culture practices. *Indian Research Journal of Extension Education* 11(2): 25-29.

Goswami, Biswajit, Golam Ziauddin, and S. N. Datta (2010). Adoption behaviour of fish farmers in relation to scientific fish culture practices in West Bengal. *Indian Research Journal of Extension Education,* 10(1): 24-28.

Goswami, Mukunda, R. Satbiadbas, and U. C. Goswami (2012). Market flow, price structure and fish marketing system in Assam - a case study. n.d. 146-155.

Goswami, Mukunda, and R. Sathiadhas. (2000). Fish farming through community participation in Assam. *Naga, The ICLARM Quarterly* 23(3): 29-32. Print.

Goswami, Mukunda, R. Sathiadhas, U.C. Goswami, and S. N. Ojha (2002) Socio-economic Dimension of Fish Farming in Assam. *Journal of the Indian Fisheries Association,* 29(2002): 103-10.

Kumar, B. Ganesh, K. K. Datta, P.K. Joshi, P.K. Katiha, R. Suresh, T. Ravisankar,K. Ravindranath and Muktha Menon (2008). Domestic fish marketing in India changing structure, conduct, performance and policies. *Agricultural Economics Research Review* 21(2008): 345-54.

Kumar, B. Ganesh, K. K. Datta, G. Vidya Sagar Reddy, and Muktha Menon (2010). Marketing system and efficiency of Indian major carps in India. *Agricultural Economics Research Review* 23(2010):105-13.

Kumar, B. Ganesh, T. Ravisankar, R. Suresh, Ramachandr Bhatta , D. Deboral Vimala, M. Kumaran, P. Mahalakshmi, and T. Sivasakthi Devi (2010). Lessons from innovative institutions in the marketing of fish and fishery products in India. *Agricultural Economics Research Review* 23 (2010): 495-504.

Kumar, Vasanta, and Selvaraj (1988). A socio-economic constraints to composite fish culture in Tamilnadu.*Aquculturel Tropical,* 3(1988): 63-69.

Lee, Chaur Shyan. (1997). Constraints and government intervention for the development of aquaculture in developing countries. *Aquaculture Economics and Management,* 1(1-2): 65-71.

Liao, Chiu, and Nai-Hsien Chao. (2009). Aquaculture and food crisis: opportunities and constraints. *Asia Pacific Journal of Clinical Nutrition,* 18(4): 564-69.

Marimuthu, K., M. A. Haniffa, M. Muruganandam, and A. J. Arockia Raj (2001). Low Cost Murrel Seed Production Technique for Fish Farmers. *Naga,* the ICLARM Quarterly, 24(1- 2)

Martin, L. van Brakel, and Lindsay G Ross (2011). Aquaculture development and scenarios of change in fish trade and market access for the poor in Cambodia. *Aquaculture Research,* 42: 931-42.

Meena, M. S., M. Prasad, and Rajbir Singh (2009). Constraints perceived by rural agro-processors in adopting modern post-harvest technologies. *Indian Research Journal of Extension Education,* 9(1): 1-5.

Mohammed A., Rab Hap Navy, Seng Leang, Mahfuzuddin Ahmed, and Katherine Viner (2012). Marketing infrastructure, distribution channels and trade pattern of inland fisheries resources Cambodia: An exploratory study. The World Fish Centre, Batu Maung, Penang, Malaysia, Web. 23Aug.2012 a. <www.http.//worldfishcenter.org >

Mohanty, Rajeeb, K., A. Mishra, S. Ghosh, and D. U. Patil (2011). Constraint analysis and performance evaluation of participatory agri-aquaculture in watersheds. *Indian Journal of Fisheries* 58(4): 139-45.

Mohite, S. A., and A. S. Mohite (2008). Marketing of Fish and Fish Products- Dominant Role of Fisherwomen. *Fishing chimes,* 28(8): 35-36.

Mohsin, A. B. M., and Emadadul Hague (2009). Effect of Constraints on Carp Production at Rajshahi District, Bangladesh. *Journal of Fisheries International,* 4(2): 30-33.

Munilkumar, S., and M. C. Nandeesha. Aquaculture Practices in Northeast India: Current Status and Future Directions. *Fish Physiology and Biochemistry,* 33 (2007): 399–412.

Mugaonkar, Pankaj kumar Hanmantrao, P.S. Ananthan, Suman Sekhar Samal and Biswajit Debnath (2011). A Study on Consumer Behaviour at Organized Fish Retail Outlet. *Agricultural Economics Research Review,* 24:133-40.

Njai, Sirra E (1994). Traditional fish processing and marketing of the Gambia. UNU- Fisheries Training Programme, Final Project 2000, 1-28. Web.4 May 2011

Padhy, M. K. (1994). Problems and prospects of pond fisheries in Birbhum District of West Bengal." *Fishing Chimes* (1994): 9-10.

Pillay, T. V. R., and M. N. Kutty. *Aquaculture Principles and Practices.* 2nd ed. Blackwell Publishing Limited P, 2005.

Radheyshyam (2001). Community-based aquaculture in India- strengths, weaknesses, opportunities and threats. *Naga,* the ICLARM Quarterly 24. 1-2 (2001): 9-12.

Rao, P. S., and S. Surapa Raju (2006). Fish as a food security in India and the World. *Journal of Fisheries Economics and Development,* 6(1): 1 -14.

Rao, P S. (1983) Fishery economics and management in India. Bombay: Pioneer Publishes and Distributors.

Rao, P. S., and S. Surapa Raju (2006). Fish as a food security in India and the World." *Journal of Fisheries Economics and Development,* 7(1): 1 -14.

Roy, Tuhin Narayan. Analysis of Marketing of Fish Fingerlings and Environmental Awareness Level of Fishermen in Dakshin Dinajpur District of West Bengal. *Agricultural Economics Research Review* 21 (2008): 425-32.

Sarker, A., Chowdhury A. H., and Itohara, Y. (2006). Entrepreneurship barriers of pond fish culture in Bangladesh- a case study from Mymensingh District.'*Journal of ocial Sciences*, 2(3): 68-73.

Sasmal, S, H. K. Patra, J. D. Sarkar, and S. R. Gaur (2006). Constraints of technology transfer in adoption of composite fish culture at rural level. *International Journal of Agricultural Sci*ence, 2(1): 134-42.

Selvaraj (1987). An economic analysis of inland fish culture in Tirunelveli District of Tamilnadu. *Fishing Chimes* 25-30.

Shil, P., and J. Bhattacharyya (2009).Fish production and retail market scenario in Barak Valley. *Banijya* 2(1): 80-88.

Sinha, V. R. P., and M. Randhir (2010). Potential and Constraints of Small scale Freshwater Fish Culture Enterprise in India." 526-38.Web.24 Nov. 2010 <http://www.apfic. org. dt.24.11.10>

Singh, Kehar, Madan M. Dey, Abed G. Rabbani, Pratheesh O. Sudhakaran, and Ganesh Thapa (2009). Technical efficiency of freshwater aquaculture and its determinants in Tripura, India.*Agricultural Economics Research Review* 22 (2009): 185-95.

Srivastava, U. K. *Aquaculture Research Needs for 2000 AD.* New Delhi: Oxford and IBH Publishing Co. Pvt. Ltd. P, 2000.

Srivastava, U. K. Aquaculture: marketing and economics in India 310-325.

Steinbronn, Silke, Nguyen Ngoc Tuan, Ulfert Focken, Klaus Becker, and Nguyen Thi Luong Hong (2005). Limitations in fish production in Yen Chau/ Son La Province/Northern Vietnam." *Conference on International Agricultural Research for Development,* Oct. 11-13.

Sugunan, V.V. (2009). Domestic marketing and post–harvest management in inland fisheries. *Fishing Chimes,* 29(9): 7.

Talukdar, P. K., and B. S. Sontaki (2005). Correlates of adoption of composite fish culture practices by fish farmers of Assam, India." *Journal of Agricultural Sciences* 1(1): 12-17.

Upadhyay, A. D., A. K. Roy, and J. R. Dhanze (2013). Dry fish marketing at Agartala, Tripura state- infrastructure and pattern of marketing. *Fishing Chimes* 31(8): 15-19.

Vasanta, K., and Selvaraj (1988). A socio-economic study of constraints to composite fish culture in Tamilnadu. *Journal of Aquaculture Tropical* 3 (1988): 63-69.

15

Linear Programming Approach in Search of an Efficient Supplementary Fish Feed for Composite Fish Culture System

A.D. Upadhyay adn A.B. Patel

Introduction

Aquaculture, the farming of aquatic organisms, has been the agro industrial activity with the highest growth rate worldwide in the last four decades. From 1970 to 2008 the production of aquaculture organisms grew at a rate of 8.3% per year, compared to less than 2% of fisheries, and 2.9% of livestock (Luchini et al. 2008). In India, aquaculture feed market was valued at USD 1.20 billion in 2017 and is expected to register a CAGR of 10.4% during the forecast period (2018-2023). Indian feed mills have the capacity to produce 2.88 million metric ton. Andhra Pradesh is the largest feed consuming state in India (Anon. 2017). The growth in production has been faster for fed species than for non fed species indicating that adoption of feed based aquaculture as the major driver of aquaculture production growth (FAO, 2016). However, the demand for aquaculture feed in the country will touch to 7 million tons by 2017-18. Hence, there is wide gap in demand and supply of feed. The feed cost is a major production cost in aquaculture in India yet feeding of fish is still mostly guesswork, each fish producer following different guidelines (feed charts) or adopting different practices. Feeding too much leads to feed wastage, a pure economic loss, and greater waste output. Feeding too little results in less growth and this also represents an economic loss. In intensive and semi intensive aquaculture, natural foods are not sufficient to meet our overall nutritional requirement of the fishes for their survival and faster growth. Hence, the good quality supplementary diet which contains all required nutrients in recommended quantity were recommended by many aquaculture scientists for the better production. It is further opined by the researchers that it is economical to prepare supplementary feeds by using locally available ingredients. The dietary carbohydrates are usually quantitatively the largest constituent of diet (40-55% W/W) particularly in feeds for carps, the most dominant group of fishes with respect to not only world aquaculture but also Indian Aquaculture. The rice bran, corn, and wheat are the conventional carbohydrate sources in India. In North East Region of the country none of these items are produced locally in sufficient quantity (Kumar et al. 2018). Due to less of local production of these ingredients, cost of feed which generally constitute about 40-50% of total cost of production, increase tremendously in NEH region due high transportation cost. Hence, for development of ideal feed for carp based aquaculture in North East Region, optimization of feed cost is very important to make the fish production cost effective and more profitable.

Basic Concepts of Linear Programming

Linear programming: is a mathematical tool that enables a decision maker to arrive at the optimum solution. In linear programming a linear objective function composed of a set of decision making variables, subject to a set of restrictions and presented as a mathematical equations.

Decision Variables: The variables in a linear program are a set of quantities that need to be determined in order to solve the problem; i.e., the problem is solved when the best values of the variables have been identified. The variables are sometimes called decision variables because the problem is to decide what value each variable should take. Typically, the variables represent the amount of a resource to use or the level of some activity. Frequently, defining the variables of the problem is one of the hardest and/or most crucial steps in formulating a problem as a linear program. Sometimes creative variable definition can be used to dramatically reduce the size of the problem or make an otherwise non-linear problem linear.

***Objective Function*:** The objective of a linear programming problem is to maximize or to minimize some numerical value. Linear programming is an extremely general technique, and its applications are limited mainly by our imaginations and our ingenuity. The objective function indicates how much each variable contributes to the value to be optimized in the problem. The objective function takes the following general form:

Maximize or Minimize $Z=\sum_{i=1}^{n} ciXi$

Where

Z is the objective function which can be maximized or minimized

c_i = the objective function coefficient corresponding to the ith variable, and

X_i = the ith decision variable.

The coefficients of the objective function indicate the contribution to the value of the objective function of one unit of the corresponding variable.

Constraints: Define the possible values, the variables of a linear programming problem may take. They typically represent resource constraints, or the minimum or maximum level of some activity. They take the following general form:

Subject to $\sum_{i=1}^{n} ajiXi \geq bj$

Where,

j=1,2,3,4,.....m

$a_{j,i}$ = the coefficient on X_i in constraint *j*, and

X_i =the ith decision variable,

b_j = the right-hand-side coefficient on constraint *j*.

Note that *j* is an index that runs from 1 to *m*, and each value of *j* corresponds to a constraint. Thus, the above expression represents *m* constraints (equations, or, more precisely, inequalities) with this form. Resource constraints are a common type of constraint. In a resource constraint, the coefficient $a_{j,i}$ indicates the amount of resource *j* used for each unit of activity *i*, as represented by the value of the variable X_i. The right-hand side of the constraint (b_j) indicates the total amount of resource *j* available for the project.

Non-negativity Constraints: For technical reasons, the variables of linear programs must always take non-negative values (i.e., they must be greater than or equal to zero). In most cases, where, for example, the variables might represent the levels of a set of activities or the amounts of some resource used, this non-negativity requirement will be reasonable even necessary. In any case, the non-negativity constraints are part of all LP formulations, and should always include. They are written as follows:

$X_i \; 0 \; i = 1, 2, \ldots, n$

where X_i = the ith decision variable.

Assumptions of linear programming

Linearity: There must be a linear relationship between the output and the total quantity of each resource consumed. If the objective function is not linear, the technique will not be applicable. The variable cost and resource requirements per unit do not alter over the relevant range of output.

Additively: This means that the sum of resources used by different activities must be equal to the total quantity of the resources used by each activity for all the resources i.e. the total units of a constrained resource used is the sum of all the resource units needed to produce a given quantity of output.

Divisibility: Perfect divisibility of outputs and resources must exist. Output can be in fractions and resources also can be in fraction. It should be noted that this condition however doesn't hold for integer programming.

Non-negativity: Decision variables cannot be added to the final objective function in a negative way. That is each of the decision variables must either be positive or zero.

Simple objective: The objective in a linear programming model can either be maximization or minimization of one activity.

Finiteness: The constraints and the variables must be finite so that it can be programmed. Hence, a finite number of activities and constraints must be employed.

Certainty: All values and quantities must be known with certainty. That is, the input/output coefficients are known and thus give the model the property of being deterministic.

Proportionality: This implies that the contribution of each variable to the final objective function is directly proportional to each variable. If we want to double the output then all decision variables must be doubled.

External factors: All external factors not taking into consideration in the model but can affect the final output must be unchanging e.g. in the formulation of feeds, cost of milling and bagging the ingredients.

Application of LP in Minimization of Feed Cost of Grow out Carp Fish

Conceptualization of LP Model

The structural model for minimization of feed cost formulation is given below:

Minimize

$\Lambda = 14.5X_1 + 23X_2 + 22X_3 + 22X_4 + 30X_5 + 35X_6 + 178X_7$ (Costfund)

Subject to

$12.6X_1 + 13.5X_2 + 13.5X_3 + 7X_4 + 27.8X_5 + 44X_6 + 0X_7 >= 22$ (Protein Constraints)

$3.1X_1 + 2.9X_2 + 2.9X_3 + 1.5X_4 + 11.4X_5 + 11.3X_6 + 0X7 >= 6.26$ (Lipid Constraint)

$50.5X_1 + 77.7X_2 + 77.7X^3 + 85.3X_4 + 39.6X_5 + 16.9X_6 + 0X_7 <= 45$ (NFE Constraint)

$26.7X_1 + 3.3X_2 + 3.3X_3 + 4.4X_4 + 12.5X_5 + 3X_6 + 0X_7 <= 12$ (Fiber Constraint)

$15X_1 + 2.5X_2 + 2.5X_3 + 1.7X_4 + 9.2X_5 + 24X_6 + 0X_7 <= 10$ (Fiber Constraint)

$0X_1 + 0X_2 + 0X_3 + 0X_4 + 0X_5 + 0X_6 + 100X_7 >= 1.5$ (Mineral Constrain)

$X_1, X_2, X_3, X_4, X_5, X_6, X_7 >= 0$ (non negativity constraints)

Whereas

λ = Total cost per unit of supplementary feed

X_1 = quantity of rice bran

X_2 = quantity of broken wheat

X_3 = quantity of wheat bran

X_4 = quantity of maize

X_5 = quantity of MOC

X_6 = quantity of fish meal

X_7 = quantity of mineral mixture

The data was analysed using MS EXCEL Solver 2026.

Results and discussions

In this study feed cost for grow out composite fish culture were optimized. For minimization of feed costs, first nutritional requirement of the grow with size 100 mm major nutritional requirement with respect to protein, lipid, crude fiber, NFE, ash and minerals in the ideal fish feed is shown in table- 15.1

Table 15.1: Nutritional Requirements of Fishes under Grow out Composite Fish Culture

Sl.NI.	Nutrients	Nutritional requirement in Feed (%)
1.	Protein	22
2.	Lipid	6.26
3.	Mineral	1.5
4.	NFE	45
5	Fiber	12
6	Ash	10

Further, the proximate composition of the locally available fish feed ingredients like rice bran, mustard oil cake, fish meal, broken wheat, wheat flour, broken maize grain were adopted from PG research work carried by the student in Aquaculture Lab of College of Fisheries, CAU(I), Lembucherra, Tripura. Proximate composition is presented in Table-15.2:

Table 15.2: The Market Prices and Proximate Composition Locally Available Fish Feed Ingredients

Sl. No	Feed ingredient	Market Prices (Rs. /kg)	Protein	Lipid	Mineral	NFE	Fibre	Ash
1.	Rice Bran	14.50	12.6	3.1	0	50.5	26.7	15
2.	Broken Wheat	23	13.5	2.9	0	77.7	3.3	2.5
3.	Wheat Bran	22	13.5	2.9	0	77.7	3.3	2.5
4.	Maize	22	7	1.5	0	85.3	4.4	1.7
5.	MOC	30	27.8	11.4	0	39.6	12.5	9.2
6.	Fishmeal	35	44.7	11.3	0	16.9	3	24
7.	Mineral Mixture	178						

Sources: 1.Sinha, 2017,

2. Current market price for feed ingredients was collected ELP programme of Aquaculture Dept. COF, CAU, Tripura

As structural model of LP given in methodology, all the data were entered into the MS excel spread sheet (Fig 15.1 & Fig 15.2). In the spread sheet (fig 15.2) objective function which is to be minimized is called target variable is given in cell E14, the objective function is optimized by changing in the quantities of ingredients with fixed prices is called changing variables shown in cell F7 to L7. All the constraints functions were given in cell D17 to N23. Thereafter, with the help of solver objective function was minimized with respect to given constraints in terms of minimum requirement of the nutrients in the ideal artificial diet for grow out composite fish culture. The Initial and final solution of the LP is given Fig 15.1& Fig 15.2.

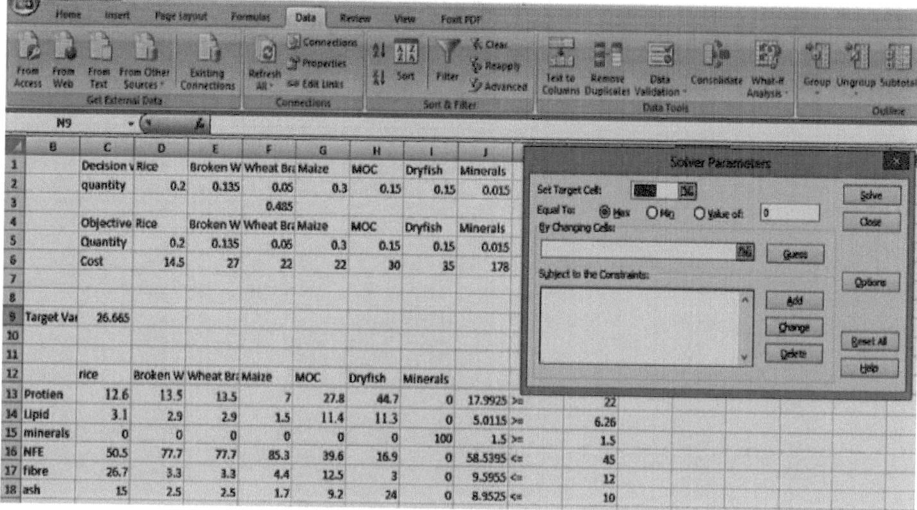

	Decision v	Rice	Broken W	Wheat Br	Maize	MOC	Dryfish	Minerals
	quantity	0.2	0.135	0.05	0.3	0.15	0.15	0.015
				0.485				
	Objective	Rice	Broken W	Wheat Br	Maize	MOC	Dryfish	Minerals
	Quantity	0.2	0.135	0.06	0.3	0.15	0.15	0.015
	Cost	14.5	27	22	22	30	35	178
Target Va	26.665							

	rice	Broken W	Wheat Br	Maize	MOC	Dryfish	Minerals			
Protien	12.6	13.5	13.5	7	27.8	44.7	0	17.9925	>=	22
Lipid	3.1	2.9	2.9	1.5	11.4	11.3	0	5.0115	>=	6.26
minerals	0	0	0	0	0	0	100	1.5	>=	1.5
NFE	50.5	77.7	77.7	85.3	39.6	16.9	0	58.5395	<=	45
fibre	26.7	3.3	3.3	4.4	12.5	3	0	9.5955	<=	12
ash	15	2.5	2.5	1.7	9.2	24	0	8.9525	<=	10

Fig. 15.1: Initial solution of LP

The obtained results indicate that for preparation of 1 kg of supplementary feed for grow out carp fishes costs Rs. 26.78 per kg. The composition of ingredients in the artificial diet was obtained and it indicated that out of seven ingredients included it the initial step of the LP five ingredients were entered into the final solution with composition to 20.5% rice bran, 19.7% wheat bran, 44.8% MOC, 9.7% fish meal and 1.5% minerals. This diet meet out all the recommended level of nutritional requirement that is protein 22%, lipid 7.4%, NFE 45% crude fiber 12% ash 10% and mineral 1.5% (Fig 15.2). The feed cost per hectare per cycle of production of composite fish culture is given in Table-15.3. Feed cost is estimated to be about Rs. 1.77 Lakh/ha for the production of 3 tones/ha of fish in 26 weeks.

Decision Variab	Rice bran	Broken Wheat	Wheat Bran	Maize	MOC	Fishmeal	Minerals
quantity	0.205	0	0.197	0	0.448	0.097	0.015
Objective Funct	Rice	Broken Wheat	Wheat Bran	Maize	MOC	Fishmeal	Minerals
Quantity	0.205	0	0.197	0	0.448	0.097	0.015
Cost	14.5	23	22	22	30	35	178

Optimized Cost of Feed	26.78

	Rice Bran	Broken Wheat	Wheat Bran	Maize	MOC	Fishmeal	Minerals	Sumproduct(E		MinimumR
Protien	12.6	13.5	13.5	7	27.8	44.7	0	22.00	>=	22
Lipid	3.1	2.9	2.9	1.5	11.4	11.3	0	7.40	>=	6.26
Mineral	0	0	0	0	0	0	100	1.50	>=	1.5
NFE	50.5	77.7	77.7	85.3	39.6	16.9	0	45.00	>=	45
Fibre	26.7	3.3	3.3	4.4	12.5	3	0	12.00	>=	12
Ash	15	2.5	2.5	1.7	9.2	24	0	10.00	<=	10

Fig. 15.2: Final solution of LP

Table 15.3: Per Hectare Fish Feed Requirements in Composite Fish Culture System

Week	Body wt (gms.)	No. of Fish	Total body wt. (kg)	App % of feed against body weight	Feed rate (kg/day)	Weekly feed quantity (Kg)	Feed Cost (Rs.)
1	10	10000	100	0.06	6	42	1124.76
2	15	9700	145.5	0.06	9	63	1687.14
3	20	9400	188	0.06	11	77	2062.06
4	25	9100	227.5	0.06	14	98	2624.44
5	30	8900	267	0.05	14	98	2624.44
6	40	8700	348	0.05	17	119	3186.82
7	50	8500	425	0.05	20	140	3749.2
8	60	8300	498	0.05	25	175	4686.5
9	75	8100	607.5	0.04	25	175	4686.5
10	90	7900	711	0.04	28	196	5248.88
11	105	7750	813.75	0.04	33	231	6186.18
12	120	7600	912	0.04	36	252	6748.56
13	145	7500	1087.5	0.04	44	308	8248.24
14	170	7500	1275	0.035	45	315	8435.7
15	195	7500	1462.5	0.035	51	357	9560.46
16	220	7500	1650	0.035	58	406	10872.68
17	250	7500	1875	0.035	66	462	12372.36
22	280	7500	2100	0.035	74	518	13872.04
23	310	7500	2325	0.035	81	567	15184.26
24	340	7500	2550	0.035	89	623	16683.94
25	370	7500	2775	0.035	97	679	18183.62
26	400	7500	3000	0.035	105	735	19683.3
						6636	1,77,712.1

Summery and Conclusion

The supplementary feed is essentially important in almost all livestock enterprises and aquaculture production system because the levels of output on per unit basis, of these enterprises are highly dependent on the quality of supplement diet. Hence, the good quality supplementary diet which contains all required nutrients in recommended quantity were advised by many aquaculture scientists for the better production in intensive and semi-intensive aquaculture system. For the formulation of the ideal diet for composite fish culture has been attempted this study using linear programming technique. The analysis of the data indicates that for preparation of 1 kg of supplementary feed for grow out carp fishes costs Rs. 26.78 per kg. The composition of ingredients in the artificial diet was obtained and out of seven ingredients included it the initial step of the LP five ingredients were entered into the final solution with composition to 20.5% rice bran, 19.7% wheat bran, 44.8% MOC, 9.7% fish meal and 1.5% minerals. This diet meet out all the recommended level of nutritional requirement that is protein 22%, lipid 7.4%, NFE 45% crude fiber 12% ash 10% and mineral 1.5% . The feed cost per hectare per cycle of production of composite fish culture is given in Table-3. Feed cost is estimated to be about Rs. 1.77 Lakh/ha for the production of 3 tones/ha of fish in 26 weeks. Hence, this study can helpful in formulation of ideal supplementary diet for different fish culture and seed production systems of North East region of the country where feed cost is very high as compared to the other region of the country because of the poor supply of local ingredients in the market.

References

Anon (2017) India Aquaculture Feed Market - Segmented by Type - Growth, Trends, and Forecast (2018 - 2023). https://www.mordorintelligence.com/industry-reports/india-aquaculture-feed-market

FAO. The State of World Fisheries and Aquaculture. Rome. 2016; 190.

Kumar Manmohan, Patel Arun Bhai, Keer Naresh Raj, Mandal Sagar C, Biswas, Pradyut and Das, Satyajit (2018) Utilization of unconventional dietary energy source of local origin in aquaculture: Impact of replacement of dietary corn with tapioca on physical properties of extruded fish feed. *Journal of Entomology and Zoology Studies*, 6 (2)2324-2329.

Luchini L, Panné-Huidobro S. *Perspectivas en Acuicultura: Nivel Mundial, Regional y Local.* Buenos Aíres, Argentina: Dirección de Acuicultura. Subsecretaría de Pesca y Acuicultura; 2008.

Mukhopadhyay, P.K. Present status of carp nutrition and feed development studies in India.

Singh, S.K.; Singh, M.k. and Prasant Kumar (2007) Preparation and application of supplementary feed. Fishing Chimes, 26(11):16-22.

Htun, Moc Sanda; Thein Tin Tin and Pyke Tin. Linear programming approach to diet problem for black tiger shrimp in shrimp aquaculture.

Oscar J. Cacho, Henry Kinnucan and Upton Hatch (1991). Optimal Control of Fish Growth. *American Journal of Agricultural Economics*, 73(1):174-183.

Sinha, Alok (2017). Evaluation of mixed feeding schedule using artificial feed and fresh wolffia arrhiza for Osteobrama belangeri (Valenciennes, 1984). M.F.Sc. thesis College of Fisheries, Central Agricultural University (Imphal), Lembucherra Tripura. Pp 59. (unpublished).

16

Methodological Approach for Attitude Scale Construction for Members of Fishery Co-operatives Societies

Narendra Kumar Verma and Shyam Sundar Dana

Introduction

The word cooperation is of Latin origin and it means to work together (Filley, 1929).Cooperatives are founded on the values of self-help, self-responsibility, democracy, equality, equity and solidarity. Based on the founding principles of co-operatives, members believe in the ethical values of honesty, openness, social responsibility and caring for others (Canadian Business Service Centre, 2004). A fishery co-operative society can be defined as "An autonomous association of people united voluntarily to meet their common economic, social and cultural needs and aspirations through a jointly-owned and democratically controlled fishery enterprise or fishery activities."

Attitude implies that the individual is no longer neutral toward the referent psychological objects. The person would be positively inclined or negatively disposed in some degree towards the referents (Campbell, 1963; Allport, 1966; Newcomb, 1966; Zanden, 1977; Burr, 2000). The response in this connection is a lasting one, as long as the attitude in question is operative. Attitude refers to a psychological individual's stands about objects, issues, persons, groups, or Institutions. The definition of attitude proposed by Triandis (1971) suggests that attitude has three components: (a) a cognitive component (the idea), (b) an affective component (the emotions), and (c) a behavioural component (the action).

In attitude scale, the examinee/respondent simply reflects his/her opinions with reference to the statements given. As a whole, it indicates his/her attitude in that context. For the laboratory fishery many scales and techniques are available for fishery related activities but there are very few scales and techniques available for the field fishery/fish farmers. As fish farmers play an important role in fish production (from production till the consumption), there are very few scales available to measure the socio-economic and psychological variables of fish farmers. Keeping in view these points at attempt was made to construct the scale for measuring the attitude of members of fishery co-operative society. Different types of methods are available for scale construction like Likert, Thurstone and Guttman. Some of the methods have some limitations or some difficult and lengthy processes, i.e. in the method of paired comparison it is very difficult to get judges because the number of judges required, increases with the number of statements. Mostly the choice of method to use for the development of attitude scale lies between Thurston and Likert procedures. "The Thurston method lacks good indices of validity of items. For this reason, some investigators recommend that an item analysis of the usual kinds be made of the items. Such analysis tends to select items of the more extreme scale positions, which is to be expected. If one is going to use Thurston method of scale construction, neutral items would have to be retained in spite of their invalidity."Hence, for constructing an attitude scale "Likert's" summated ratings method was attempted.

Methodology

Following procedure or steps were followed for construction of scale:

i. Collection of items/statements
ii. Selection of items/statements
iii. Scoring of items/statements with experts
iv. Item evaluation with "t" test
v. Scoring techniques
vi. Reliability of the scale
vii. Validity of the scale
viii. Administering the scale

For the construction of the scale, above mentioned steps were followed. All respondents were selected by using simple random sampling from fishery co-operative societies from Kolkata district of West Bengal. Data was collected by using simple random sampling. A checklist of 30 statements was scrutinized with the help of 26 experts by following 1st and 2nd step i.e. collection of items/ statements and selection of items/statements. Further, statements were administered to 30 respondents (fishery cooperative members) for the evaluation of the selected statements. T-test was administered to calculate the significance between the statements and to select the statements. Further, reliability was tested on 25 respondents by means of split half method/regression test.

Result and Discussion

Collection of items/statements

All possible statements which will discriminate the positive and negative attitudes of the fishery cooperative members towards cooperative society were collected from relevant literature and also by having discussion with experts working in the field of fishery. Proper care was taken to include various levels of statements which have direct consequences on co-operative members. The statements were edited by following the procedure suggested by Edwards (1969). A total of 70 statements were collected. Further, 44 statements were selected from above collected 70 statements with the help of experts working in the relevant field.

Selection of items/Statements

After the formation of 44 statements for the attitude scale, it was essential to scrutinize them to select the most effective statements and to reduce the number of statements, looking at the effectiveness to assess the attitudes of fishery co-operative members. For this, opinion was taken from learned educationists, researchers and psychologists. A list of such experts was prepared with the help of different sources for their consultation in the selection of effective statements for the final form of attitude scale. In this regard proper contact was made, either personal contact, telephonic contact or contact through email. Further, a checklist prepared by the researcher was given to 26 experts to score the selected 44 statements based on their relevancy. The response was collected on a five point continuum of the most relevant, more relevant, relevant, less relevant and least relevant. The scores were assigned as 5,4,3,2 and 1respectively. The total score for each statement given by the experts was calculated. Further, the statements were ranked in descending order based on their scores. From those statements, 30 statements with highest scores were selected and subjected to item analysis. The procedure suggested by Edwards (1969) was followed.

Table 16.1: Statement Wise Total Score Given by Experts

Sl.No.	Statements	Total Score by Experts
1	More participation can make fishery cooperative more successful.	117
2	Friendliness of manager makes fishery cooperative more successful.	105
3	Price and profit loss information affects fishery cooperative working.	90
4	Fishery cooperatives should publish the financial report on a regular interval.	113
5	There should be the provision of competition and awards for members of fishery cooperative to be judged it in terms of attendance, active participation and innovative suggestions.	105
6	Fishery cooperatives can uplift poor sections of society.	115
7	Active participation can build the capacity of members in a cooperative.	117
8	Full-time enrolment of members in a cooperative can increase the livelihood security of members.	98
9	Cooperatives are a way to develop "we" feeling among group members.	111
10	Active participation of members will make fishery cooperatives economically and politically strong.	115
11	Involvement of more women in fishery cooperatives will broaden the scope of cooperatives and improves their social role by empowering them to decision-making level.	116
12	Lack of awareness about principles, values and the by-laws of fishery cooperatives hinder the growth of fishery cooperatives.	107
13	Fishery cooperatives generate awareness among members about government development programmes.	98
14	Fishery cooperative promotes self-confidence.	100
15	Up-to-date information about operation plays a vital role in the success of fishery cooperatives.	105
16	Training is an integral part of successful fishery cooperatives.	115
17	Income generating activity of cooperative through fisheries is a viable option.	106
18	Fishery cooperatives help members to adopt new or improved technology more rapidly because of group influence.	107
19	Fishery cooperative members can avail loan more easily through banks.	95
20	New members view should be encouraged.	93
21	Entrepreneurial qualities of members make fishery cooperative more effective.	107
22	Fishery cooperative works as a powerful tool for socio-economic empowerment of poor in rural areas.	107
23	Fishery cooperatives can be a way to eradicate the poverty and unemployment.	99
24	Fishery cooperative improves the saving behavior of the members.	90
25	Fishery cooperatives should take up business activities according to market demand.	107
26	To make fishery cooperative successful, information of market is necessary.	115
27	Risk taking is the important characteristics of a successful cooperative.	87
28	Meeting of different cooperatives contributes in exchange of their experiences.	105
29	Banks are more eager to sanction loan to groups or cooperatives compared to an individual.	93
30	Cooperatives are good source to solve the agricultural and household related problems.	89

Contd.

Sl. No.	Statements	Total Score by Experts
31	Cooperative helps to generate family feeling in between group members that helps to work more efficiently in the group.	92
32	Cooperative promotes group leadership.	98
33	Cooperative members posses more knowledge than their counter-part non-member farmers.	82
34	Financial assistance is essential to set up a new venture.	109
35	Group goal achievement is facilitated in the cooperative.	98
36	Groupism among the members of the cooperative badly affects the functioning of cooperatives.	107
37	Lack of co-operation among the members adversely affects the functioning of the cooperative.	109
38	Products prepared by fishery cooperative members have no market.	65
39	Cooperative improves the social behavior of its members.	86
40	Cooperative tries to tap social capital within the group for meeting collective needs.	84
41	Cooperatives are helpful to break social, cultural and religious barriers.	81
42	Cooperative promotes mutual cooperation among members.	102
43	Cooperative is a collective effort approach.	109
44	Group rules and regulations are based on democratic principle.	100

Table 16.2: Top 30 Statements with Highest Score for Scale Construction by Experts

Sl. No.	Statements	Total Score
1	More participation can make fishery cooperative more successful.	117
2	Active participation can build the capacity of members in a cooperative.	117
3	Involvement of more women in fishery cooperatives will broaden the scope of cooperatives and improves their social role by empowering them to decision-making level.	116
4	To make fishery cooperative successful, information of market is necessary.	115
5	Fishery cooperatives can uplift poor sections of society.	115
6	Active participation of members will make fishery cooperatives economically and politically strong.	115
7	Training is an integral part of successful fishery cooperatives.	115
8	Fishery cooperatives should publish the financial report on a regular interval.	113
9	Cooperatives are a way to develop "we" feeling among group members.	111
10	Financial assistance is essential to set up a new venture.	109
11	Lack of co-operation among the members adversely affects the functioning of the cooperative.	109
12	Cooperative is a collective effort approach.	109
13	Lack of awareness about principles, values and the by-laws of fishery cooperatives hinder the growth of fishery cooperatives.	107
14	Fishery cooperatives help members to adopt new or improved technology more rapidly because of group influence.	107
15	Entrepreneurial qualities of members make fishery cooperative more effective.	107

Contd.

Sl. No.	Statements	Total Score
16	Fishery cooperative works as a powerful tool for socio-economic empowerment of poor in rural areas.	107
17	Fishery cooperatives should take up business activities according to market demand.	107
18	Groupism among the members of the cooperative badly affects the functioning of cooperatives.	107
19	Income generating activity of cooperative through fisheries is a viable option.	106
20	Friendliness of manager makes fishery cooperative more successful.	105
21	There should be the provision of competition and awards for members of fishery cooperative to be judged it in terms of attendance, active participation and innovative suggestions.	105
22	Up-to-date information about operation plays a vital role in the success of fishery cooperatives.	105
23	Meeting of different cooperatives contributes in exchange of their experiences.	105
24	Cooperative promotes mutual cooperation among members.	102
25	Fishery cooperative promotes self-confidence.	100
26	Group rules and regulations are based on democratic principle.	100
27	Fishery cooperatives can be a way to eradicate the poverty and unemployment.	99
28	Fishery cooperatives generate awareness among members about government development programmes.	98
29	Group goal achievement is facilitated in the cooperative.	98
30	Full-time enrolment of members in a cooperative can increase the livelihood security of members.	98

Scoring of Items/Statements

The statements were administered to 30 fishery cooperative members of non sample area, selected randomly. They were asked to respond to each statement in the terms of their own agreement or disagreements on a five point continuum, namely, strongly agree, agree, undecided, disagree, and strongly disagree. The scores were assigned respectively as 5,4,3,2, and 1 for positive statements and reverse for the negative statements. The total score for each of the respondents were the sum of all the items.

Table 16.3: Total Score of Selected Items for Cooperative Members

Sl.No.	Statements	Score by Cooperative members
1	Involvement of more women in fishery cooperatives will broaden the scope of cooperatives and improves their social role by empowering them to decision-making level.	139
2	Active participation can build the capacity of members in a cooperative.	89
3	More participation can make fishery cooperative more successful.	86

Contd.

Sl.No.	Statements	Score by Cooperative members
4	Fishery cooperatives can uplift poor sections of society.	76
5	To make fishery cooperative successful, information of market is necessary.	138
6	Fishery cooperatives should publish the financial report on a regular interval.	145
7	Active participation of members will make fishery cooperatives economically and politically strong.	139
8	Cooperatives are away to develop "we" feeling among group members.	148
9	Training is an integral part of successful fishery cooperatives.	142
10	Fishery cooperative works as a powerful tool for socio-economic empowerment of poor in rural areas.	79
11	Financial assistance is essential to set up a new venture.	138
12	Lack of co-operation among the members adversely affects the functioning of the cooperative.	76
13	Cooperative is a collective effort approach.	143
14	Lack of awareness about principles, values and the by-laws of fishery cooperatives hinder the growth of fishery cooperatives.	140
15	Entrepreneurial qualities of members make fishery cooperative more effective.	89
16	Groupism among the members of the cooperative badly affects the functioning of cooperatives.	122
17	Income generating activity of cooperative through fisheries is a viable option.	143
18	Friendliness of manager makes fishery cooperative more successful.	144
19	There should be the provision of competition and awards for members of fishery cooperative to be judged it in terms of attendance, active participation and innovative suggestions.	141
20	Fishery cooperatives help members to adopt new or improved technology more rapidly because of group influence.	145
21	Fishery cooperatives should take up business activities according to market demand.	146
22	Meeting of different cooperatives contributes in exchange of their experiences.	142
23	Up-to-date information about operation plays a vital role in the success of fishery cooperatives.	141
24	Fishery cooperatives can be a way to eradicate the poverty and unemployment.	129
25	Cooperative promotes mutual cooperation among members.	140
26	Fishery cooperative promotes self-confidence.	135
27	Group goal achievement is facilitated in the cooperative.	141
28	Group rules and regulations are based on democratic principle.	136
29	Fishery cooperatives generate awareness among members about government development programmes.	144
30	Full-time enrolment of members in a cooperative can increase the livelihood security of members.	71

Evaluation

The statements were then arranged in an array based on the total score obtained by them. 25 per cent of respondents with higher total scores and 25 per cent of respondents with lower total score were selected from among the respondents. These two groups formed the criterion groups. Whether the individual statements are varying among the two groups, t-statistic has been worked out.

Table 16.4: Arrangement of Statements Based on Score

Sl.No.	Statements	Score by Cooperative members
1	Cooperatives are away to develop "we" feeling among group members.	148
2	Fishery cooperatives should take up business activities according to market demand.	146
3	Fishery cooperatives should publish the financial report on a regular interval.	145
4	Fishery cooperatives help members to adopt new or improved technology more rapidly because of group influence.	145
5	Friendliness of manager makes fishery cooperative more successful.	144
6	Fishery cooperatives generate awareness among members about government development programmes.	144
7	Cooperative is a collective effort approach.	143
8	Income generating activity of cooperative through fisheries is a viable option.	143
9	Training is an integral part of successful fishery cooperatives.	142
10	Meeting of different cooperatives contributes in exchange of their experiences.	142
11	There should be the provision of competition and awards for members of fishery cooperative to be judged it in terms of attendance, active participation and innovative suggestions.	141
12	Up-to-date information about operation plays a vital role in the success of fishery cooperatives.	141
13	Group goal achievement is facilitated in the cooperative.	141
14	Lack of awareness about principles, values and the by-laws of fishery cooperatives hinder the growth of fishery cooperatives.	140
15	Cooperative promotes mutual cooperation among members.	140
16	Involvement of more women in fishery cooperatives will broaden the scope of cooperatives and improves their social role by empowering them to decision-making level.	139
17	Active participation of members will make fishery cooperatives economically and politically strong.	139
18	To make fishery cooperative successful, information of market is necessary.	138
19	Financial assistance is essential to set up a new venture.	138
20	Group rules and regulations are based on democratic principle.	136
21	Fishery cooperative promotes self-confidence.	135

Contd.

Sl.No.	Statements	Score by Cooperative members
22	Fishery cooperatives can be a way to eradicate the poverty and unemployment.	129
23	Groupism among the members of the cooperative badly affects the functioning of cooperatives.	122
24	Entrepreneurial qualities of members make fishery cooperative more effective.	89
25	Active participation can build the capacity of members in a cooperative.	89
26	More participation can make fishery cooperative more successful.	86
27	Fishery cooperative works as a powerful tool for socio-economic empowerment of poor in rural areas.	79
28	Fishery cooperatives can uplift poor sections of society.	76
29	Lack of co-operation among the members adversely affects the functioning of the cooperative.	76
30	Full-time enrolment of members in a cooperative can increase the livelihood security of members.	71

For calculation of t-statistics, the following formula was used.

$$t = \frac{\bar{X}_H - \bar{X}_L}{\sqrt{\dfrac{\sum(X_H - \bar{X}_H)^2 + \sum(X_L - \bar{X}_L)^2}{n(n-1)}}}$$

Where,

$$\Sigma(X_H - \bar{X}_H)^2 = \Sigma X^2{}_H - (\frac{\Sigma X_H}{n})^2$$

$$\Sigma(X_L - \bar{X}_L)^2 = \Sigma X^2{}_L - (\frac{\Sigma X_L}{n})^2$$

$\bar{X}_H$ = mean score of a given statement for the high group

$\bar{X}_L$ = mean score of a given statement for the low group

n = number of subjects

The statements with highest "t" value (i.e., more than 1.75) were selected for the attitude items. Thus, the attitude scale consisted of 13 items, which were finally included in the study.

Table 16.5: Distribution of 't"Ttest According to Farmer´s Score

Sl.No.	Statements	"t' value	Selected/ Rejected
1	Involvement of more women in fishery cooperatives will broaden the scope of cooperatives and improves their social role by empowering them to decision-making level.	1.13	Rejected
2	Active participation can build the capacity of members in a cooperative.	4.31	Selected
3	More participation can make fishery cooperative more successful.	4.22	Selected
4	Fishery cooperatives can uplift poor sections of society.	-0.15	Rejected
5	To make fishery cooperative successful, information of market is necessary.	1.05	Rejected
6	Fishery cooperatives should publish the financial report on a	0.00	Rejected
7	Active participation of members will make fishery cooperatives economically and politically strong.	1.66	Rejected
8	Cooperatives are away to develop "we" feeling among group members.	1.00	Rejected
9	Training is an integral part of successful fishery cooperatives.	0.00	Rejected
10	Fishery cooperative works as a powerful tool for socio-economic empowerment of poor in rural areas.	5.05	Selected
11	Financial assistance is essential to set up a new venture.	1.53	Rejected
12	Lack of co-operation among the members adversely affects the functioning of the cooperative.	2.50	Selected
13	Cooperative is a collective effort approach.	1.53	Rejected
14	Lack of awareness about principles, values and the by-laws of fishery cooperatives hinder the growth of fishery cooperatives.	1.13	Rejected
15	Entrepreneurial qualities of members make fishery cooperative more effective.	5.16	Selected
16	Groupism among the members of the cooperative badly affects the functioning of cooperatives.	1.66	Rejected
17	Income generating activity of cooperative through fisheries is a viable option.	2.05	Selected
18	Friendliness of manager makes fishery cooperative more successful.	1.66	Rejected
19	There should be the provision of competition and awards for members of fishery cooperative to be judged it in terms of attendance, active participation and innovative suggestions.	2.05	Selected
20	Fishery cooperatives help members to adopt new or improved technology more rapidly because of group influence.	0.61	Rejected
21	Fishery cooperatives should take up business activities according to market demand.	1.53	Rejected
22	Meeting of different cooperatives contributes in exchange of their experiences.	2.65	Selected

Contd.

Sl.No.	Statements	"t' value	Selected/ Rejected
23	Up-to-date information about operation plays a vital role in the success of fishery cooperatives.	1.82	Selected
24	Fishery cooperatives can be a way to eradicate the poverty and unemployment.	2.08	Selected
25	Cooperative promotes mutual cooperation among members.	1.53	Rejected
26	Fishery cooperative promotes self-confidence.	2.05	Selected
27	Group goal achievement is facilitated in the cooperative.	1.27	Rejected
28	Group rules and regulations are based on democratic principle.	0.30	Rejected
29	Fishery cooperatives generate awareness among members about government development programmes.	2.05	Selected
30	Full-time enrolment of members in a cooperative can increase the livelihood security of members.	2.84	Selected

Scoring Techniques

The items on the attitude scale were provided with five point continuum namely, strongly agree, agree undecided, disagree, and strongly disagree with scores of 5,4,3,2 and 1 respectively for the positive statements and 1,2,3,4, and 5 respectively for negative statements. The attitude score of the respondents could be obtained by summing up the scores for all the items in the scale.

Reliability of the Scale

A scale is said to be reliable when it consistently produce the same or similar result when applied to the same sample at different time. Here the reliability was tested by means of split half method.

The scale was administered to 25 non-sample respondents and was divided into two halves based on odd and even number of statements. Thus total 12 statements from 13 were selected for final study. The total obtained for odd and even numbered items were subjected to correlation analysis. The correlation coefficient (r) was 0.884and found to be significant at one per cent level of probability. Since the "r' value was more than 0.8, the scale was considered to be reliable.

Table 16.6: Distribution of Total Score of Respondents for Scale Construction.

Sl.No. of respondents	Total score of Even Statements	Total score of Odd Statements	"r" Value
1	26	21	0.884
2	28	27	
3	30	29	
4	25	24	
5	29	28	

Contd.

Sl.No. of respondents	Total score of Even Statements	Total score of Odd Statements	"r" Value
6	22	20	
7	22	18	
8	24	27	
9	30	30	
10	25	25	
11	29	26	
12	27	26	
13	27	23	
14	30	30	
15	30	30	
16	30	30	
17	29	29	
18	24	24	
19	26	24	
20	30	29	
21	29	29	
22	30	29	
23	23	18	
24	27	27	
25	25	24	

Validity of the scale

The developed scale was tested for content validity. The main criterion of content validity is how well the contents of the scale represent the subject matter under study. Since the items selected were from the universe of the content, it was ensured that the items covered all aspects of cooperative.

Administering the scale

The final scale which measured the attitude of members towards cooperative society consist 12 statements. Each statement was noted on a five point continuum as strongly agree, agree, undecided, disagree and strongly disagree with score of 5,4,3,2 and 1 respectively for positive statements. The scoring was reversed in the case of negative statements. The score was obtained for each statement and summed up to get the attitude score of a cooperative member. The maximum score was 60 and the minimum was 12.

Table 16.7: Final Statements for Scale Construction.

Sl.No.	Statement	Strongly Agree(5)	Agree (4)	Undecided (3)	Disagree (2)	Strongly Disagree (1)
1	Active participation can build the capacity of members in a cooperative.					
2	More participation can make fishery cooperative more successful.					
3	Fishery cooperative works as a powerful tool for socio-economic empowerment of poor in rural areas.					
4	Lack of co-operation among the members adversely affects the functioning of the cooperative.					
5	Entrepreneurial qualities of members make fishery cooperative more effective.					
6	Income generating activity of cooperative through fisheries is a viable option.					
7	There should be the provision of competition and awards for members of fishery cooperative to be judged it in terms of attendance, active participation and innovative suggestions.					
8	Meeting of different cooperatives contributes in exchange of their experiences.					
9	Up-to-date information about operation plays a vital role in the success of fishery cooperatives.					
10	Fishery cooperatives can be a way to eradicate the poverty and unemployment.					
11	Fishery cooperative promotes self-confidence.					
12	Full-time enrolment of members in a cooperative can increase the livelihood security of members.					

Conclusion

The scale was constructed by using"Likert's" summated ratings method. Considerable measures were taken during the construction of the scale. At the first stage 70 statements were selected and after following the procedure 12 statements were finally selected for the attitude scale construction. All the measures like evaluation, reliability and validity was considered during the sale construction.

Acknowledgement

As this is a work of Ph.D. thesis, I would like to express my special thanks of gratitude to my advisor and guide(Prof. Shyam Sundar Dana, Registrar & HOD, Dept. of Fishery Extension, Faculty of Fishery Sciences (WBUAFS), Kolkata), my teachers, Dr. Ajit Kumar Roy, Ex. Consultant (Statistics), College of Fisheries, (CAU), Tripura, Dr. Anil Datt Upadhyay, Assistant Professor (SS), College of Fisheries, (CAU), Tripura who gave me the golden opportunity to do this work on the topic (A Multidimensional Study on Fisheries Co-Operatives in Some Selected Districts of Uttar Pradesh, India), which also helped me in doing a lot of research and I came to know about so many new things for which I am really thankful to them.

References

Campbell, D., (1963). Social attitude and other acquired behavioral dispositions. McGraw Hill, New York

Allport, G.W., (1966). Attitudes in the history of social psychology. *In*: M. Jahoda and N.l Warm (eds.). Attitudes–Selected readings. Pp.15-21 Penguin Book Inc., USA

Newcomb, T.M., (1966). On the definition of attitude. *In:* M. Jahoda and N. Warm (eds.), Attitudes–Selected Readings. pp. 22-24. Penguin Book Inc: USA

Zanden, J.W.V., (1977). Social psychology. 4 th edition, McGraw-Hill, Inc., U.S.A.

Filley, H. Clyde (1929). Cooperation in agriculture, New York; Wiley.

Triandis, H.D. (1971). Attitude and attitude change. New York: John Wiley & Sons, Inc.

J. P. Guilford, Psychomatric Methods. New Delhi: Tata Me Graw - Hill Publishing Co. Ltd., 1975, P. 456.

17

Tips on Application of Statistical Methods in Social Science Research

Ajit Kumar Roy

Introduction

This chapter outlines recently developed and traditional statistical techniques, which are increasingly being applied in social science research. The social sciences cover diverse phenomena arising in society, the economy and the environment, some of which are too complex to allow concrete statements; some cannot be defined by direct observations or measurements; some are culture-or region specific, while others are generic and common. Statistics, being a scientific method as distinct from a 'science' related to any one type of phenomena is used to make inductive inferences regarding various phenomena. The chapter addresses both qualitative and quantitative research a combination of which is essential in social science research and offers valuable updates for advanced level for researchers. In this chapter, I will attempt to addresses the gaps between the research and data analysis faced the professionals and researchers working under different environment with multitude of problems particularly in the social science sector. It provides students with theoretical perspectives and methodological tools to explore the social and cultural systems that influence policy interventions.

The Social Science: Social science is the field of study concerned with society and human behaviors. Social science covers a broad range of disciplines like demography and social statistics, methods and computing; education, social anthropology, and linguistics apart from anthropology, archaeology, criminology, economics, education, history, linguistics, communication studies, political science and international relations, sociology, geography, law, and psychology. Further, it covers development studies, human geography and environmental planning; law, economic and social history; politics and international relations; psychology and sociology; and social policy and social work. Social science is, in its broadest sense, the study of society and the manner in which people behave and influence the world around us. Social science tells us about the world beyond our immediate experience, and can help explain how our own society works from the causes of unemployment or what helps economic growth or what makes people happy. It provides vital information for governments and policymakers, local authorities, non-governmental organisations and others. **Economics** seeks to understand how individuals interact within the social structure, to address key questions about the production and exchange of goods and services. **Management studies** explores a wide range of aspects relating to management of fisheries activities, such as strategic and operational management, organizational psychology, employment relations, marketing, accounting, finance and logistics.

Role of Economics in Fisheries Research

An overview of recent social science research pertaining to fisheries management is presented particularly on the relevance of social science information for successful management regimes. This section provides an introduction to social issues in planning and R&D activities in the fisheries sector. ***Economics*** is a social science. It studies about a particular aspect of human behavior that is full of complexity. It is not easy to study it. So, economic science is not as precise and exact as the physical sciences. But economics is to be considered as an important science than other social sciences like politics or history because in economics we make use of money as a measuring indicator of utility. Economics is the study of social behavior guiding in the allocation of scarce resources to meet the unlimited needs and desires of the individual members of a given society. Economics seeks to understand how those individuals interact within the social structure to address key questions about the production and exchange of goods and services. Production is about the conversion of scarce resources into desired goods and services. These resources are often referred to as the factors of production like Land, Labor, Capital and Entrepreneurship. It is widely accepted that Economics as a social science deals with human wants and their satisfaction. It is related to other social sciences like sociology, politics, history, ethics, and psychology. The economic development of a nation depends not only on economic factors but also on historical, political and sociological factors. Sociology is the science of society. Social sciences like politics and economics may be considered as the branches of sociology. Sociology attempts to discover the facts and laws of society as a whole. Sociology deals with all aspects of society. But economics deals only with the economic aspects of a society.

Effective fisheries management relies on biological information of the fish resource, as well as economic and social sciences information on commercial and recreational fishermen and other stakeholders. Economic information includes information on market conditions in commercial fisheries in terms of price and value information. Social sciences information is typically broader sources of information specific to commercial and recreational fishermen, their families, and the fishing community in general. For many managed fisheries economic and social sciences information are not available and is provided in an informal manner by fishermen during public interaction periods. At times, this information is viewed as anecdotal and may be difficult to use in the fishery management decision-making process. In managing fisheries, it is humans who must be understood and managed. Social science studies of fisheries indicate that not all members of a given user group operate in the same way, or have the same impact on marine ecosystems. People's behavior is often influenced by family, community, and other socio-cultural variables in addition to economic and ecological considerations. Using the perspectives and methodologies derived from the disciplines such as anthropology and sociology, fisheries managers should be able to develop policies which integrate and balance economic, social and biological objectives. Management systems which are more compatible with broad user group values should result in higher compliance and reduced enforcement costs. If the "social" sphere is concerned with all forms of relations between individuals and groups, sociology is concerned more specifically with the collective behaviour of people. This means understanding the ways "society", as a grouping of individuals, has developed, the way it is organised, how the various groups within a society interact, the norms of behaviour which they observe and how groups and group behaviour affect the individuals which make up those groups. Sociologists must include economic factors in their analysis of development issues, and economists clearly need to take sociological factors into equal account. The term socio-economics therefore, needs to be handled with some care. The inclusion of proper sociological analysis as part of the process of managing fisheries will ensure that a whole range of social impacts are taken into account which might otherwise have been missed. These social impacts can often jeopardize the success or sustainability of fisheries interventions. In fact, sociological analysis in fisheries can be focused on a few key levels. These levels can be used to provide a basic framework to sociological analysis for fisheries. Five "levels" of analysis are 1. Gender 2. Age 3. Community 4. Household and 5. Production-unit. These five elements can generally be considered the "building blocks" of most social systems the world over so an understanding of how they are constituted and understanding their significance is needed. Few studies on economics related to fisheries and aquaculture is reported (Roy, 2008; 2010; 2010; 2018).

The United Nation proclaims that sustainable development comprises environmental, economic, and social sustainability. Fisheries contribute to livelihoods, food security, and human health worldwide. The relationships between environmental and economic sustainability, as well as between economic and social sustainability, continue to receive attention. Sea has since ancient time been used for navigation, fishing, and hunting, and there is a long social science tradition of studying fisheries and fishery communities. Fisheries economists included the economic aspects of harvesting and promoted the concept of maximum economic yield (*Schaefer, M.B. 1957;_Ricker, W.E. 1954*). Based on bio-economic models, they sought to stipulate the level of effort that would create the largest difference between Aquaculture, which accounts for an increasing share of global seafood production, has also attracted more attention from social science, and focus has partly shifted from managing fisheries to managing coastal zones and ocean areas with their multiple activities and interests involved (Johnsen and Hersoug 2014; *OECD. 2016*). In recent past fisheries social science developed and widened in terms of both topics and approaches (Urquhart et al. 2014; Bavinck et al. 2018). With roots in fishing communities and common property dynamics, fisheries social science learned to look up and study these issues relative to larger institutions, management systems, regional economies, and knowledge hegemonies. But fisheries social sciences remain dedicated to revealing how such larger processes play out locally. During the past decades, the marine social sciences have developed a vast knowledge about the development of fisheries and these insights and the critical approaches that have been applied in the context of fisheries, are highly relevant to the wider agenda raised by the UN Sustainable Development Goal 14 and the visions of conservation and blue growth (*Silver, J.J et. al. 2015)*. The role of social science in blue growth facilitating sustainable economic growth, job creation, provide food and energy, and even reduce poverty through the extension of land-based growth policies. Presently, society is moving out to sea through processes of ocean industrialization and the privatization and marketization of its resources (Knott and Neis 2017; Soma et al. 2018;_*Knott, C., and B. Neis. 2017*). As sustainable development is fundamentally about societal transitions, there is a central role for the social sciences in the formulation of governance alternatives, the anticipation of future trends, the imagination of desirable futures, and the facilitation of socially just processes and outcomes (Bennett et al. 2017). The combination of critical social science approaches studies not only emergent processes in the marine environment, but also practices, enactments and discourses that shape them. It is envisaged that further research should focus on the governance instruments that are deployed and their outcomes in terms of rules, norms, and shared conceptions (Lascoumes and Le Galès 2007; Knol 2011; Song et al. 2018). The contributions of fisheries to achieving the Sustainable Development Goals of the United Nations (UN) system has affirmed its commitment to putting equality and nondiscrimination at the heart of the implementation of the 2030 Agenda (FAO, 2018). In fisheries and aquaculture, the commitment to leave no one behind is a call to focus action and cooperation on achieving the core ambitions of the 2030 Agenda for the benefit of all fish workers, their families and their communities as mentioned in the "Fisheries and the Sustainable Development Goals: meeting the 2030 Agenda" in Part 2). Achieving the SDGs is the collective responsibility of all countries and all actors. It will depend on collaboration across sectors and disciplines, international cooperation and mutual accountability, and requires comprehensive, evidence-based and participatory problem solving, financing and policy-making.

Statistical Thinking about the Social Sciences*:* As because Social Sciences study human societies and human behaviors, we are concerned with complex issues and subtle problems. Human motivations are many and complex. Our societies can be very complex. Sometimes, we feel that artists capture and express society's issues and people's problems much better than social scientists. Their contributions are immense and essential. But to design policies we need to analyze and study. Trying to understand these issues, we require thinking. To quantify things in the social sciences requires even more thinking. The original idea of "Statistics" was the collection of information about and for the State/ Country. During the 20th Century statistical thinking and methodology has become the scientific framework for literally dozens of fields including education, agriculture, economics, biology, and medicine, and with increasing influence recently on the hard sciences such as astronomy, geology, and physics. Social Science requires thinking about what you need for your analysis. It is about thinking clearly about the social sciences. It is about thinking through complex issues. So, let us start on a journey of thinking through the fundamental concepts from the basics to the frontiers of current work.

Conducting Research on **Social Science:** In most research studies the following steps are taken for sampling, data collection, and data analytics. A high level of clarity in research methods is needed to ensure that the findings are not biased by the researcher's preconceptions. In writing the report, the researcher should describe very clearly the detailed process used for sampling, data collection, data analysis, and hypotheses development, so that readers can independently assess the reasonableness, strength, and consistency of the reported inferences

i. Define research questions.

ii. Select case sites.

iii. Create instruments and protocols.

iv. Select respondents.

v. Start data collection.

vi. Conduct within-case data analysis.

vii. Conduct cross-case analysis.

viii. Build and test hypotheses.

ix. Write case research report.

In social science, qualitative versus quantitative research refers to empirical or data-oriented considerations about the type of data to collect and how to analyze them. ***Qualitative research*** relies mostly on non-numeric data, such as interviews and observations, in contrast to ***quantitative research*** which employs numeric data such as scores and metrics. Hence, qualitative research is not amenable to statistical procedures such as regression analysis, but is coded using techniques like content analysis. Sometimes, coded qualitative data is tabulated quantitatively as frequencies of codes, but this data is not statistically analyzed. Interpretive research should attempt to collect both qualitative and quantitative data pertaining to their phenomenon of interest, and so should positivist research as well. Joint use of qualitative and quantitative data, often called "mixed mode designs", may lead to unique insights and are highly prized in the scientific community (Berkowitz, S. 1996).

Quantitative Research Methodology

Quantitative Research: Quantitative research studies the size or extent of particular issues or trends in society. Quantitative research can measure and describe whole societies, or institutions, organizations or groups of individuals that are part of them. The strength of quantitative methods is that they can provide vital information about a society or community, through surveys, examination or records or censuses that no individual could obtain by observation.

Quantitative methodologies: Some of the most common quantitative research methodologies are described here. These methodologies are widely-used across various research disciplines (Howe, K., and Eisenhart, M. 1990).

i. ***Cross-sectional studies:*** Cross-sectional studies are surveys undertaken at one point in time, rather like a photo taken by a camera. If the same or similar survey is repeated, we can get good measures of how society is changing.

ii. *Longitudinal studies:* Longitudinal studies follow the same respondents over an extended period of time. They can employ both qualitative and quantitative research methods, and they follow the same group of people over time.

iii. *Opinion polls*: An opinion poll is a form of survey designed to measure the opinions of a target population about an issue, e.g. support for political parties, views about crime and justice, the economy or the environment.

iv. *Questionnaires:* Questionnaires collect data in a standardized way, so that useful summaries can be made about large groups of respondents, such as the proportion of all young people of a given age who are bullied. Usually most questions are 'closed response', where respondents are given a range of options to choose from.

v. *Social attitude surveys*: Social attitude surveys ask more general questions about beliefs and behaviour: e.g. how often people go to church; how much trust they have in the police force; whether they think children need a strict upbringing; how content they are with their life; how often they see other family members; whether they are in employment.

vi. *Surveys and censuses*: A census is a survey of everyone in the population. Because of the vast number of respondents they are very expensive to organise. Governments now depend much more on administrative records like hospital records or tax returns and sample surveys. Surveys use a questionnaire to investigate respondents in a sample. Samples are chosen in such a way that they can represent a much larger population. A precise calculation can be made of how accurate the information from any sample is likely to be.

Qualitative Research Methodology

***Qualitative Research*:** Social scientists often want to understand how individuals think, feel or behave in particular situations, or in relations with others that develop over time. They use in-depth interviews, participant observation and other qualitative methods to gather data. Researchers might watch a school playground to observe and record bullying behaviours, or ask young people about exactly what they understood by being bullied, and how they thought it affected them. Qualitative methods are scientific, but are focused more on the meaning of different aspects of peoples' lives, and on their accounts of how they understand their own and others' behaviour and beliefs. Some of the most common qualitative research methodologies are outlined here.

***Qualitative methodologies*:** Most common qualitative research methodologies are described here. These methodologies are widely-used in research.

i. ***Semi-structured interviews:*** In semi-structured interviews the researcher has a small core of questions or areas they wish to explore, but will then take the questions in different directions, depending on the answers they receive. Flexibility is important with this type of interview. This method is used when seeking richly descriptive information, e.g. what makes a good teacher?

ii. ***Unstructured interviews:*** Unstructured interviews are open-ended and informal. The researcher is seeking a detailed picture and tries to bring no preconceptions. This type of interview is often used in narrative research. Generally the researcher asks one question and then leaves the interviewee to talk or 'tell their story'.

iii. ***Observation:*** Observation relies heavily on the skills of the researcher to understand and interpret what they are seeing in an unbiased way. It might be used, for example, in education research to see how much time young people spend 'on task' and what they do when distracted. In this method, the researcher observes what is happening and makes field notes either at the time or soon afterwards.

iv. ***Open questionnaire survey:*** Unlike questionnaires in quantitative research, which offer a limited range of choices, open surveys seek opinion and description in response to open-ended questions. They may be used to gather information and ideas from more people than one-to-one interviewing would allow.

v. ***Keeping logs and diaries:*** Researchers and participants can keep logs or diaries as a way to collect details about daily life. Participants are asked to keep detailed records of some aspect of their life, such as social activities or exercise, so the researcher later can analyse this material. Researchers also keep diaries during the period of data collection on aspects of the research, such as the context in which interviews or observation takes place. This is then used alongside other data to help them to broaden their understanding of the research findings.

vi. ***Case studies:*** Researchers examine a small number of specific examples and narratives where researchers study respondents' stories in depth. Case studies can help researchers to explore life in different families, cultures and communities. However, in order to examine how far we can generalize the specific cases for wider society, some form of quantitative methods are often needed.

vii. ***Content analysis:*** This is a popular method which involves counting up the prevalence and sequencing of certain words, sentences, expressions, metaphors, and similar, in texts such as newspaper articles or transcripts of interviews. It can also be used to identify the types of explanations people give for their own behavior or use in order to persuade people to support them or agree with their argument. It is predominantly a quantitative method.

viii. ***Discourse analysis:*** Another popular method is discourse analysis. This is a qualitative method that provides detailed analyses of exactly what language is used and how it is used. Discourse analysts do not aim to find an absolute truth about how people use language. They are more interested in the processes whereby people construct meanings socially and individually. Most discourse analysts are interested in subjectivity people's own sense-making and often include an analysis of the researcher's own subjective understandings as part of the analysis of data, thus using a mixture of insider and outsider viewpoints.

Sampling Methods for Data Collection:

In most of the research the use of sampling is common. A sample is selected, evaluated and studied in an effort to gain information about the larger population from which the sample was drawn. As for an example, a very large sample consisting of millions of households can be selected to respond to a lengthy questionnaire that is part of a census. A sample represents a population, and information obtained from a sample is generalized to be true for the entire population from which it was drawn. The validity or accuracy of generalizations from samples to populations depends on how well a sample represents its population. A well-selected sample can provide information comparable to that obtained by a census.

Advantages of sampling: Studying a sample instead of a population, can have the following advantages.

i. ***Cost*:** Samples can be studied at much lower cost. The smaller number of units or individuals involved in a sample requires less time and money to evaluate. Samples can provide affordable, accurate, and useful information in cases where a census would cost more than the value of the information obtained.

ii. ***Time*:** Samples can be evaluated more quickly than a population. If a decision had to wait for the results of a census, a critical advantage might be missed, or the information might be made obsolete by events or changes that took place while the data were being collected and analyzed.

iii. ***Accuracy***: Any time data are collected, there is a chance for errors to occur. Errors of measurement, incorrect recording of data, transposition of digits, recording of information in the wrong area of a form, and errors in entering data into a computer can all influence the accuracy of results. In general, the larger the data set, the more opportunity there is for errors to occur. A sample can provide a data set that is small enough to monitor carefully and can permit careful training and supervision of data gatherer and handlers.

iv. ***Feasibility***: In some research situations, the population of interest is not available for study. A substantial portion of the population might not yet exist or might no longer be available for evaluation. In other cases, evaluation of an item requires its destruction. For example, a manufacturer interested in how much pressure could be applied to a part before it cracked, could not perform a census without destroying the entire production run.

v. ***Scope of information***: In a sample survey, there are greater varieties of information that can be considered which may be impracticable in a complete census due to constraints such as blood sample from human being. Here entire blood cannot be taken.

Sampling Designs: There are two categories of sample designs, namely, *probability* or random sampling and *non-probability sampling.*

1. ***Probability Sampling***: The major principle of these designs is to avoid bias in the selection procedure and to achieve the maximum precision for a given outlay of resources. The main types of probability sampling designs are: simple random sampling, systematic sampling, stratified sampling, cluster sampling and multi-stage sampling.

2. ***Non-probability sampling:*** Non-probability sampling designs select samples with features not embodying randomness. The selection of the elements in the sample lies solely on personal judgment. The chance of selecting an element cannot be determined. For this reason, there is no means of measuring the risk of making erroneous conclusion desired from non-probability samples. Thus the reliability of results i.e. sampling errors cannot be assessed and also used to make valid conclusion about the population. The main methods of non-probability sampling are Convenience, Judgmental and Quota Sampling

Process of Collection of Data: The actual study begins with the collection of data. Once the survey instrumentation is completed then the collection of data is a critical step in providing the information needed to answer the research question. Every study includes the collection of some type of data whether it is from the literature or from subjects to answer the research question. As mentioned earlier, data can be collected in the form of words on a survey, with a questionnaire, through observations, or from the literature.

Operationalization and Measurement: This is the most important part to keep in mind three basic questions at this stage

- What do you measure?
- How do you measure?
- How well do you measure?

What is a concept? A concept is a mental image that summarizes a set of similar observations, feelings, or ideas. All theories, ideas, are based on concepts. A key difference is that normal science deals with concepts that are well defined and to a great extent standardized measures e.g. speed, distance, volume, weight, size, etc. On the contrary, ***social sciences*** often use concepts that are more abstract and therefore, the standardization in measurement varies or there is little agreement e.g. social class, development, poverty, etc. Thus, our goal is that our measurements of the different concepts are valid or *match* as much as possible the "real" world. Care must be taken about creating variables and their qualities as described below.

Creating Variables: First goal is to create measurable variables out of our concepts.

- We first must nominally define our concepts.
- We are moving from the abstract to the concrete.
- We must be able to observe our variables.
- We link our variables to data.
- When we link our variables to data, this is operationalization.

Qualities of Variables

- ***Exhaustive***: Should include all possible answerable responses.
- ***Mutually exclusive***: No respondent should be able to have two attributes simultaneously for example, employed vs. unemployed. It is possible to be both if looking for a second job while employed.

Transferring Concepts into Something Measurable

Variable: A representation of concept in its variation of degree, varieties or occurrence. A characteristic of a thing is such matter that can assume varying degrees or values.

Example: Concept and Variable

- **Concept**: Political participation
- **Variables:** Voted or not; How many times a person has voted; What party a person votes for

Selection of Variables: Scientists try to figure out how the natural world works. In doing so, they use experiments to search for cause and effect relationships. Cause and effect relationships explain why things happen and allow you to reliably predict what will happen if you do something. In other words, scientists design an experiment so that they can observe or measure if changes to one thing cause something else to vary in a repeatable way. The things that are changing in an experiment are called variables. A variable is any factor, trait, or condition that can exist in differing amounts or types. An experiment usually has three kinds of variables: independent, dependent, and controlled. The independent variable is the one that is changed by the researcher. The dependent variables are the things that the scientist focuses his or her observations on to see how they respond to the change made to the independent variable. If there is a direct link between the two types of variables (independent and dependent) then you may be uncovering a cause and effect relationship. The number of dependent variables in an experiment varies, but there can be more than one. Controlled variables are quantities that a scientist wants to remain constant, and she or he must observe them as carefully as the dependent variables. Some people refer to controlled variables as "constant variables. In the best experiments, the scientist must be able to measure the values for each variable". Weight or mass is an example of a variable that is very easy to measure. However, imagine trying to do an experiment where one of the variables is love. There is no such thing as a "love-meter." You might have a belief that someone is in love, but you cannot really be sure, and you would probably have friends that do not agree with you. So, love is not measurable in a scientific sense; therefore, it would be a poor variable to use in an experiment.

The Process of Measurement: The process starts with conceptualization followed by operalisation as detailed below

***Conceptualization*:** The process of conceptualization includes coming to some agreement about the meaning of the concept. In practice, you often move back and forth between loose ideas of what you are trying to study and searching for a word that best describes it. Sometimes you have to "make up" a name to encompass your concept. As you flush out the pieces or aspects of a concept, you begin to see the dimensions; the terms that define subgroups of a concept. With each dimension, you must decide on indicators that are signs of the presence or absence of that dimension. Dimensions are usually concepts themselves (Bhattacharya, Anol.2012).

***Operationalizing Choices*:** You must operationalize process of converting concepts into measurable terms. The process of creating a definition(s) for a concept that can be observed and measured. The development of specific research procedures that will result in empirical observations is defined as a combination of income and education and you want measure each. The development of questions or characteristics of data in qualitative work is that will indicate a concept.

***Precision in Variable Attribute Choices*:** Variable attributes need to be exhaustive and exclusive to represent full range of possible variation. Degree *of* Selection depends on your research interest. **The dependent variable** is the variable that the researcher measures; it is called a dependent variable because it depends upon the independent variable. The **independent variable** is the one that the researcher manipulates. For example, if you are studying the effects of a new educational program on student achievement, the program is the independent variable and your measures of achievement are the dependent ones.

Various Levels and Scales of Measurement: This is the most important part of any study. Therefore, for clear understanding of level and scale of measurement is essential to proceed rightly in research studies.

Level of Measurement

- ***Nominal Scale***: Categories, labels, data carry no numerical value
- ***Ordinal Scale***: Rank ordered data, but no information about distance between ranks
- ***Interval Scale***: Degree of distance between scores can be assessed with standard sized intervals
- ***Ratio Scale***: Same as interval scale with an absolute zero point.

Nominal - Categorical variables with no inherent order or ranking sequence such as names or classes (e.g., gender). Value may be a numerical, but without numerical value (e.g., I, II, III). The only operation that can be applied to Nominal variables is enumeration.

Nominal Measures

- Only offer a name or a label for a variable
- There is not ranking
- They are not numerically related
- Gender; Race

Ordinal - Variables with an inherent rank or order, e.g. mild, moderate, severe. Can be compared for equality, or greater or less, but not *how much* greater or less.

Ordinal Measures

- Variables with attributes that can be rank ordered
- Can say one response is *more or less* than another
- Distance between does not have meaning lower class, middle and upper class

Interval - Values of the variable are ordered as in Ordinal, and additionally, differences between values are meaningful, however, the scale is not absolutely anchored. Calendar dates and temperatures on the Fahrenheit scale are examples. Addition and subtraction, but not multiplication and division are meaningful operations.

Interval Measures

- Distance separating attributes has meaning and is standardized (equidistant)
- "0" value does not mean a variable is not present
- Score on an ACT test 50 vs. 100 does not mean person is twice as smart

Ratio - Variables with all properties of Interval plus an absolute, non-arbitrary zero point, e.g. age, weight, temperature (Kelvin). Addition, subtraction, multiplication, and division are all meaningful operations.

Ratio Measures

I. Attributes of a variable have a "true zero point" that means something

II. Waist measures and Biceps measures

III. Allows one to create ratios

Scales of Measurement

Measurement scales are used either to categorize or quantify variables. Description of Scales of Measurement with examples is presented in the following table-17.1.

Table 17.1: Description of Scales of Measurement with Examples

Scale	Description	Example	Distinction
Nominal	Categories with no numeric scales	Males / females introverts / extroverts	Impossible to define any quantitative values
Ordinal	Rank ordering numeric values limited	2, 3, 4-star restaurants Ranking TV programs by popularity	Intervals between items is unknown
Interval	Numeric properties are literal Assume equal intervals between values	Intelligence Aptitude test scores Temperature (Fahrenheit and Celsius)	No true zero point
Ratio	Zero indicates absence of variable measured	Reaction time Weight Age Frequency of behaviour	Can form ratios (someone weighs twice as much as another

Applying the Power of Statistics in Making Sense in Social Sciences Research

Qualitative Research

Qualitative research is a field of inquiry that crosscuts disciplines and subject matters. There are three major approaches to qualitative research: ethnography; phenomenology and grounded theory drawn from sociology. Typically, the research questions addressed by qualitative methods are discovery-oriented, descriptive and exploratory in nature. Qualitative researchers aim to gather an in-depth understanding of human behaviour and the reasons that govern human behaviour. Various aspects of behaviour could be based on deeply held values, personal perspectives, experiences and contextual circumstances. Qualitative research investigates the why and how of decision making, not just what, where, and when. Therefore, the need is for smaller but focused samples rather than large random samples. Qualitative analysis involves categorizing data into patterns as the primary basis for organizing and reporting results. Qualitative researchers typically rely on several methods for gathering information: (1) participation in the setting, (2) direct observation, (3) in depth interviews, (4) focus groups, and (5) analysis of documents and materials. Although it is common to draw a distinction between qualitative and quantitative aspects of scientific investigation, a mixed-methods approach (a combination of qualitative and quantitative techniques) is often used. Qualitative research is, in some cases, instrumental to developing an understanding of a phenomenon as a basis for quantitative research. In other cases, it can inform or enrich our understanding of quantitative results. Similarly, quantitative research may inform, or be drawn upon in the process of qualitative research. Qualitative analysis is one of the best methods of research for many fields of study. It provides depth and detail by analyzing things more than just numbers and sizes. It does this through examining attitudes, feelings, and behaviors. This type of analysis encourages people to expand their responses and also helps people to have a better and more detailed picture of the causes and conclusion of different actions and reactions in the world. Finally, qualitative analysis deals with the exclusion of prejudices and prejudgments, because every thought, action, and behavior is researched through the cause and the reality

In scientific research, **a variable is a measurable representation** of an abstract construct. As abstract entities, constructs are not directly measurable, and hence, we look for proxy measures called variables. For instance, a person's intelligence is often measured as his or her IQ (intelligence quotient) score, which is an index generated from an analytical and pattern-matching test administered to people

Types of Variables

- **Qualitative Variable**: Composed of categories which are not comparable in terms of magnitude
- **Quantitative Variable:** Can be ordered with respect to magnitude on some dimension
- **Continuous Variable:** A quantitative variable, which can be measured with an arbitrary degree of precision. Any two points on a scale of a continuous variable have an infinite number of values in between. It is generally measured.
- **Discrete Variable:** A quantitative variable where values can differ only by well-defined steps with no intermediate values possible. It is generally counted.

Qualitative Data Analysis

Qualitative data which is also known as descriptive data is a non-numerical data that captures concepts and opinions. Some examples of qualitative data include transcripts from interviews, audio/video recordings and notes from an observation. Qualitative modes of data analysis provide ways of discerning, examining, comparing and contrasting, and interpreting meaningful patterns or themes (Miles *et al.,* 2003). Meaningfulness is determined by the particular goals and objectives of the project at hand. The same data can be analyzed and synthesized from multiple angles depending on the particular research or evaluation questions being addressed. The varieties of approaches - including ethnography, narrative analysis, discourse analysis, and textual analysis correspond to different types of data, disciplinary traditions, objectives, and philosophical orientations (Patton, M.Q. 1990). However, all share several common characteristics that distinguish them from quantitative analytic approaches. Further; qualitative data analysis is simply the process of examining qualitative data to derive an explanation for a specific phenomenon. Qualitative data analysis gives us an understanding of our research objective by revealing patterns and themes in your data. Qualitative data analysis is a vital part of all qualitative research. Every research starts off with the collection of quality information. The collected information is then organized and analyzed to draw conclusion on the theme of the research. This process of organizing and analyzing the information collected during research is what is commonly known as data analysis in the world of research. Analyzing qualitative data can be very confusing due to its unstructured nature. However, data analysis in whatever form can be easily carried out using the right methodology. It is important for us to know that the validity of our research depends heavily on our data analysis. Now I am going to show you how to effectively carry out a qualitative data analysis. At this stage it is important for us to understand exactly what 'qualitative data' and 'qualitative data analysis' really mean.

The Purpose of Qualitative Data Analysis

i. Data organisation

ii. Data interpretation

iii. Pattern Identification

iv. Ties field data to research objective(s)

v. Forms the basis for informed and verifiable conclusion

Techniques to Analyze Qualitative Data

Qualitative analysis involves the why and how of decision making as opposed to the what, where, and when of quantitative analysis (Coffey, A., and Atkinson, P. 1996). Since qualitative analysis focuses on in-depth reasoning and quality of results, many researchers prefer it over the quantitative analysis that focuses on bigger sample sizes. Qualitative analysis needs small and focused samples instead of the large random samples that quantitative analysis uses. Qualitative analysis classifies data into patterns in order to arrange and conclude results. The data used can be in many forms such as texts, images, sounds, etc. The most important requirement needed is the presence of in-depth knowledge of the target subject. Qualitative analysis is applied in almost all possible fields of social sciences. In the field of psychology, qualitative analysis is used to understand diverse human behaviors their general trends and their causes and consequences. If you have taken the time to research on how to analyze qualitative data, it is very possible that you have come across several different steps and rules. It is also possible that you are wondering which of these steps are right for you. The truth is that any of those steps could be right for you. This is because qualitative analysis, though based on certain ground rules, does not follow a rigid process. This knowledge is vital for your data analysis because you do not want to find yourself in a corner as a result of following a rigid set of rules. However, there are ground rules for qualitative analysis. Understanding these ground rules begins with knowing the two main approaches to qualitative analysis.

1. ***Deductive Approach***: The deductive approach to qualitative data analysis involves analyzing data based on a structure predetermined by the researcher. In this case, you can use your research questions as a guide for grouping and analyzing your data. This is a quick and easy approach to qualitative data analysis and can be used when you as a researcher have an idea of likely responses from your sample population.
2. ***Inductive Approach:*** The inductive approach on the other hand, is not based on a structured or predetermined framework. This is a more thorough and time consuming approach to qualitative data analysis. This approach is often used when the researcher knows very little of their research phenomenon.

Steps to Effectively Analyzing Qualitative Data

Whether one is looking for how to analyze qualitative data from an interview or how to analyze qualitative data from a questionnaire, these simple steps in qualitative data analysis will ensure a robust data analysis.

Step 1: Transcribe all Data

After you have collected data from the field, it is largely unstructured and sometimes makes no sense. It is therefore, your duty as the researcher to make sense out of field data though transcription. The first step of analyzing you data is to transcribe all data. Transcription simply means converting all data into textual form. Technology has made it very easy for you to transcribe data. You can choose out of the many computer-assisted qualitative data analysis software (CAQDAS) to transcribe your data. Using tools, like ATLAS.ti, NVivo, and of course our favourite, EvaSys, you can transcribe data effectively and a faster rate than when done manually.

Step 2: Organize Your Data

After transcribing data, you will most likely be left with large amounts of information all over the place. A lot of new researchers get confused and frustrated at this point. However, you can get back on track by simply organizing your data. You must resist the temptation of working with unorganized data because it will only make your data analysis more difficult. One great way to organize your research data is by going back to your research objectives or questions and then organizing the collected data according to these objectives/ questions. You have to make sure to organize your data in a visually clear way. You can achieve this by using tables. Input your research objectives into the table and assign data according to each objective. You can also use any of the research software in Step 1 to simplify your data organization process.

Step 3: Code Your Data

Coding is the best way to compress your data into easily understandable concepts for a more efficient data analysis process. Coding in qualitative analysis simply involves categorizing your data into concepts, properties and patterns. Coding is a vital step in any qualitative data analysis and helps the researcher give meaning to data collected from the field. You can derive the codes for your analysis from the data you have collected, from relevant research findings or from your research objectives. Some popular coding terms include:

i. ***Descriptive coding***: Summarizing the central theme of your data

ii. ***In-Vivo Coding***: Using the language of your respondents to code

iii. ***Pattern Coding***: Finding patterns in your data and using them as the basis of your coding

After coding your data, one can then begin to build on the themes or patterns to gain deeper insight into the meaning of the data.

Step 4: Validate Data

Data validation is one of the pillars of successful research. Since data is at the heart of research, it becomes extremely vital to ensure that it is not flawed. You should note here that data validation is not just a step in qualitative data analysis; it is something you do all through your data analysis process. It has been listed as a step here to just highlight its importance. There are two sides to data validation. First is validity which is all about the accuracy of your design/methods and the second is reliability which is the extent to which your procedures produced consistent and dependable results.

Step 5: Conclusion of Data Analysis

Conclusion here simply means stating your findings and research outcomes based on the research objectives. While concluding your research, you have to find a valid link between the analyzed data and your research questions/objective. The next vital step in concluding your data analysis is presenting your data analysis as a final report. Your report has to state the processes and methods of your research, pros and cons of your research, and of course study limitations. In the final report, you should also state the implications of your findings and areas of future research.

The Process of Qualitative Analysis: Throughout the process of qualitative analysis, the analyst should be asking and reasking the following questions (Wolcott, H.F. 1994)

i. What patterns and common themes emerge in responses dealing with specific items? How do these patterns help to illuminate the broader study questions?

ii. Are there any deviations from these patterns? If yes, are there any factors that might explain these atypical responses?

iii. What interesting stories emerge from the responses? How can these stories help to illuminate the broader study questions?

iv. Do any of these patterns or findings suggest that additional data may need to be collected? Do any of the study questions need to be revised?

v. Do the patterns that emerge corroborate the findings of any corresponding qualitative analyses that have been conducted? If not, what might explain these discrepancies?

Contrast between Qualitative and Quantitative Data Analysis: An understanding of the difference between qualitative and quantitative data analysis is very important. The simple distinction between these two methods is that *qualitative data analysis* deals with the analysis of subjective and non-numerical data while quantitative data analysis focuses on analyzing data through a numerical or statistical means. *In Quantitative Analysis*, numbers are the material of analysis. By contrast, qualitative analysis deals in words and is guided by fewer universal rules and standardized procedures than statistical analysis. This relative lack of standardization is at once a source of versatility and the focus of considerable misunderstanding. That qualitative analysts will not specify uniform procedures to follow in all cases draws critical fire from researchers who question whether analysis can be truly rigorous in the absence of such universal criteria. These analysts may have helped to invite this criticism by failing to adequately articulate their standards for assessing qualitative analyses, or even denying that such standards are possible. Their stance has fed a fundamentally mistaken but relatively common idea of qualitative analysis as unsystematic, undisciplined, and "purely subjective."Although distinctly different from *quantitative statistical analysis* both in procedures and goals, good qualitative analysis is both systematic and intensely disciplined. Quantitative evaluation is more easily divided into discrete stages of instrument development, data collection, data processing, and data analysis. By contrast, in *qualitative evaluation*, data collection and data analysis are not temporally discrete stages as soon as the first pieces of data are collected; the evaluator begins the process of making sense of the information. Qualitative analysis is fundamentally an *iterative* set of processes. At the simplest level, qualitative analysis involves examining the assembled relevant data to determine how they answer the evaluation question(s) at hand. In analysis of survey results, for example, frequency distributions of responses to specific items on a questionnaire often structure the discussion and analysis of findings. By contrast, qualitative data most often occur in more embedded and less easily reducible or distillable forms than quantitative data. Because each qualitative study is unique, the analytical approach used will be unique. Because qualitative inquiry depends, at every stage, on the skills, training, insights, and capabilities of the researcher, qualitative analysis ultimately depends on the analytical intellect and style of the analyst. The human factor is the greatest strength and the fundamental weakness of qualitative inquiry and analysis. As should by now be obvious, it is truly a mistake to imagine that qualitative analysis is easy or can be done by untrained novices. *Applying guidelines requires judgment and creativity.*

Statistical Tests for Quantitative or Qualitative Data

Statistical methods analyze data on variables, which are characteristics that vary among subjects. Statistical methods depend on the type of variable.

i. Numerically measured variables, such as family income and number of children in a family, are quantitative. They are measured on an interval scale.

ii. Variables taking values in a set of categories are categorical or qualitative. Those measured with unordered categories, such as religious affiliation and blood group, have a nominal scale. Those measured with ordered categories, such as social class and political ideology, have an ordinal scale of measurement.

iii. Variables are also classified as discrete, having possible values that are a set of separate numbers or continuous, having a continuous, infinite set of possible values. Categorical variables, whether nominal or ordinal, are discrete. Quantitative variables can be of either type, but in practice are treated as continuous if they can take a large number of values.

Much social science research uses observational studies, which use available subjects to observe variables of interest. One should be cautious in attempting to conduct inferential analyses with data from such studies. ***Inferential statistical methods*** require probability samples, which incorporate randomization in some way. Random sampling allows control over the amount of sampling error, which describes how results can vary from sample to sample. Random samples are much more likely to be representative of the population than are non-probability sample such as volunteer samples

Statistical Tests Applicable for Quantitative and Qualitative Data

Key Differences in Approaches in Analyzing Quantitative Data and Qualitative Data: Recall the discussions on the differences between quantitative and qualitative data in the earlier paragraphs. To reiterate the following points are added.

Quantitative (numerical) data is any data that is in numerical form, such as statistics, percentages, etc. A researcher studying quantitative data asks a specific, narrow question and collects a sample of numerical data from participants to answer the question. The researcher analyzes the data with the help of statistics and hopes the numbers will yield an unbiased result that can be generalized to some larger population.

Qualitative (categorical) research, on the other hand, asks broad questions and collects word data from participants. The researcher looks for themes and describes the information in themes and patterns exclusive to that set of participants. Examples of qualitative variables are male/female, nationality, colour, etc.

How do you know what kind of test to use? Following paragraphs will clarify doubts as to the types of tests to be used either for qualitative or quantitative data collected by the researchers. I will give a brief overview of these tests here.

Quantitative Data Tests: Paired and unpaired t-tests and z-tests are just some of the statistical tests that can be used to test quantitative data. A z-test is any statistical test for which the distribution of the test statistic under the null hypothesis can be approximated by a normal distribution. Because of the central limit theorem, many test statistics are approximately normally distributed for large samples. For each significance level, the z-test has a single critical value. This fact makes it more convenient than the t-test, which has separate critical values for each sample size. Therefore, many statistical tests can be conveniently performed as approximate z-tests if the sample size is large or the population variance known.

Qualitative Data Tests: One of the most common statistical tests for qualitative data is the chi-square test both the goodness of fit test and test of independence. The chi-square test tests a null hypothesis stating that the frequency distribution of certain events observed in a sample is consistent with a particular theoretical distribution. The events considered must be mutually exclusive and have total probability. A common case for this test is where the events each cover an outcome of a categorical variable. A test of goodness of fit establishes whether or not an observed frequency distribution differs from a theoretical distribution, and a test of independence assesses whether paired observations on two variables, expressed in a contingency table, are independent of each other.

Types of Statistical Tests: There is a wide range of statistical tests. The decision of which statistical test to use depends on the research design, the distribution of the data, and the type of variable. In general, if the data is normally distributed, you will choose from parametric tests. If the data is non-normal, you will choose from the set of non-parametric tests. Below is a table listing just a few common statistical tests and their use in table-17.2.

Table 17.2: Types of Statistical Tests and their Appropriate Application

Type of Test	Use
Correlational	These tests look for an association between variables
Pearson correlation	Tests for the strength of the association between two continuous variables
Spearman correlation	Tests for the strength of the association between two ordinal variables (does not rely on the assumption of normally distributed data)
Chi-square	Tests for the strength of the association between two categorical variables
Comparison of Means:	**look for the difference between the means of variables**
Paired T-test	Tests for the difference between two related variables
Independent T-test	Tests for the difference between two independent variables
ANOVA	Tests the difference between group means after any other variance in the outcome variable is accounted for
Regression: assess if change in one variable predicts change in another variable	
Simple regression	Tests how change in the predictor variable predicts the level of change in the outcome variable
Multiple regression	Tests how change in the combination of two or more predictor variables predict the level of change in the outcome variable
Non-parametric: used when the data does not meet assumptions required for parametric tests	
Wilcoxon rank-sum test	Tests for the difference between two independent variables—takes into account magnitude and direction of difference
Wilcoxon sign-rank test	Tests for the difference between two related variables—takes into account the magnitude and direction of difference
Sign test	Tests if two related variables are different—ignores the magnitude of change, only takes into account direction

In most research studies, the analysis section follows these three phases of analysis

- Cleaning and organizing the data for analysis (Data Preparation)
- Describing the data (Descriptive Statistics)
- Testing Hypotheses and Models (Inferential Statistics)
- Data Preparation involves checking or logging the data in; checking the data for accuracy; entering the data into the computer; transforming the data; and developing and documenting a database structure that integrates the various measures.

Contrast between Descriptive and Inferential Statistics

Descriptive Statistics: A branch of mathematics dealing with summarization and description of collections of data sets, including the concepts of arithmetic mean, median, and mode. It is the discipline of quantitatively describing the main features of a collection of data, or the quantitative description itself. Descriptive statistics are distinguished from inferential statistics in that descriptive statistics aim to summarize a sample, rather than use the data to learn about the population that the sample of data is thought to represent. This generally means that descriptive statistics, unlike inferential statistics, are not developed on the basis of probability theory. Even when a data analysis draws its main Conclusion using inferential statistics, descriptive statistics are generally also presented. ***Descriptive statistics* is** dealing with summarization and description of collections of data sets, including the concepts of arithmetic mean, median, and mode as against **inferential statistics** that involves drawing Conclusion about a population based on sample data drawn from it.

Descriptive statistics are used to describe the basic features of the data in a study. They provide simple summaries about the sample and the measures. Together with simple graphics analysis, they form the basis of virtually every quantitative analysis of data. With descriptive statistics you are simply describing what is, what the data shows. *Examples*: Frequency distribution; Graphical representation of data; Measures of central tendency; Quartiles and percentiles; Measures of dispersions; Shapes of distributions; Determining symmetry by using the five number symmetry like a box plot of a set of data, gives a visual summary of five key numbers that are associated with the data involving: the minimum value, the lower quartile, the median, upper quartile and maximum value.

Description of Tables and Graphs

- ***A frequency distribution*** summarizes the counts for possible values or intervals of values. A relative frequency distribution reports this information using percentages or proportions.
- ***A bar graph*** uses bar over possible values to portray a frequency distribution for a categorical variable. For a quantitative variable, a similar graphic is called a histogram. It shows whether the distribution is approximately bell shaped, U shaped, skewed to the right (longer tail pointing to the right), or whatever.
- ***The stem-and-leaf plot*** is an alternative portrayal of data for a quantitative variable. It groups together observations having the same leading digit (stem), and shows also their final digit (leaf). For small samples, it displays the individual observations.
- The box plot portrays the quartiles, the extreme values, and outliers. A box plot and a stem and-leaf plot can also provide back-to-back comparisons of two groups of data.

Overview of Measures of Central Tendency

- Measures of central tendency describe the centre of the data, in terms of a typical observation.
- The mean is the sum of the observations divided by the sample size. It is the centre of gravity of the data.
- The median divides the ordered data set into two parts of equal numbers of observations. Half below and half above that point.
- The lower quarter of the observations fall below the lower quartile, and the upper quarter fall above the upper quartile. These are the 25th and 75th percentiles. The median is the 50th percentile. The quartiles and median split the data into four equal parts. They are less affected than the mean by outliers or extreme skew.
- The mode is the most commonly occurring value. It is valid for any type of data, though usually used with categorical data or discrete variables

Overview of Measures of Variability

- The range is the difference between the largest and smallest observations. The interquartile range is the range of the middle half of the data between the upper and lower quartiles. It is less affected by outliers.

- The variance averages the squared deviations about the mean. Its square root, the standard deviation, is easier to interpret, describing a typical distance from the mean.
- The empirical rule states that for a bell-shaped distribution, about 68% of the observations fall within one standard deviation of the mean, about 95% fall within two standard deviations, and nearly all, if not all, fall within three standard deviations.
- The coefficient of variation of a set of data is defined by the equation CV = (Standard Deviation/mean) %100.Coefficient of variation is a measure which is independent of the unit of measurement. Coefficient of variation is used to compare the variability of sets of data measured in different units. For example, we may wish to know, for a certain population, whether body masses, measured in kilograms, are more variable than heights, measured in centimetres.

Shape of Distribution

- Another important property of a set of data is the shape of its distribution. We can evaluate the shape of a distribution by considering two characteristics of data sets:
- Symmetry and kurtosis. Both symmetry and kurtosis, evaluate the manner in which the data are distributed around their mean.

Symmetry

A symmetrical distribution can be defined as one in which the upper half is a mirror image of the lower half of the distribution. If a vertical line is drawn through the mean of a distribution depicted by a histogram, the lower half could be "folded over" and would coincide with the upper half of the distribution. One way to identify symmetry involves a comparison of the mean and the median. If these two measures are equal, we may generally consider the distribution to be symmetrical. If the mean is greater than the median, the distribution may be described as positively or right skewed that is, has a long tail to the right. If the mean is less than the median, the distribution is considered to be negatively or left-skewed that is, has a long tail to the left. Now that you have looked at the distribution of your data and perhaps conducted some descriptive statistics to find out the mean, median, or mode, it is time to make some inferences about the data. As it is well known that inferential statistics are the set of statistical tests we use to make inferences about data. These statistical tests allow us to make inferences because they can tell us if the pattern we are observing is real or just due to chance.

Inferential Statistics

Inferential Statistics investigate questions, models and hypotheses. In many cases, the conclusion from inferential statistics extends beyond the immediate data alone. For instance, we use inferential statistics to try to infer from the sample data what the population thinks. We use inferential statistics to make judgments of the probability that an observed difference between groups is a dependable one or one that might have happened by chance in this study. Thus, we use inferential statistics to make inferences from our data to more general conditions and we use descriptive statistics simply to describe what is going on in our data. Statistical inference makes propositions about populations, using data drawn from the population of interest via some form of random sampling. More generally, data about a random process is obtained from its observed behavior during a finite period of time. Given a parameter or hypothesis about which one wishes to make inference, statistical inference most often uses a statistical model of the random process that is supposed to generate the data and a particular realization of the random process. The conclusion of a statistical inference is a statistical proposition. Some common forms of statistical proposition are:

- An estimate; i.e., a particular value that best approximates some parameter of interest
- A confidence interval (or set estimate); i.e., an interval constructed using a data set drawn from a population so that, under repeated sampling of such data sets, such intervals would contain the true parameter value with the probability at the stated confidence level
- A credible interval; i.e., a set of values containing, for example, 95% of posterior belief
- Rejection of a hypothesis
- Clustering or classification of data points into groups

Hypothesis Tests or Confidence Intervals

The Difference between Hypothesis Testing and Confidence **Intervals**

When we conduct a hypothesis test, we assume the true parameters of interest is known. When we use confidence intervals, we are estimating the parameters of interest. Confidence intervals are closely related to statistical significance testing. As for an example one wants to test estimated parameter against the alternative and a test can be performed.

Hypothesis test: A test that defines a procedure that controls the probability of incorrectly deciding that a default position (null hypothesis) is incorrect based on how likely it would be for a set of observations to occur if the null hypothesis were true.

Confidence interval: A type of interval estimate of a population parameter used to indicate the reliability of an estimate.

Inferential Analysis

We have just reviewed descriptive analysis, which includes descriptive statistics such as range, minimum, maximum, and frequency. It also includes measures of central tendency such as mean, median, mode, and standard deviation. Once you have appropriately described your data, then you can move on to making inferences based on your data. Inferential analysis uses statistical tests to see whether a pattern we observe is due to chance or due to the program or intervention effects. Research often uses inferential analysis to determine if there is a relationship between an intervention and an outcome as well as the strength of that relationship. This section provides an overview of things to consider before starting inferential analysis, examples of common statistical tests, and the meaning of statistical significance.

The first step in inferential analysis: One of the first steps in inferential analysis is to answer the question what does the distribution of data look like? The type of test you choose will be guided by the distribution of the data. Distributions fall into two categories, normal and non-normal. You should always check the distribution of your data before beginning inferential analysis.

Normal Distribution: This type of distribution looks like the Bell Curve. The graph below is an example of what a *normal distribution* looks like (Figure 17.1). If your distribution looks like the one in the image below or close to it, your distribution is normal. This type of distribution shows us that the majority of the data is clustered around one number or value. Usually if the data is normal, then we choose statistical tests called parametric tests.

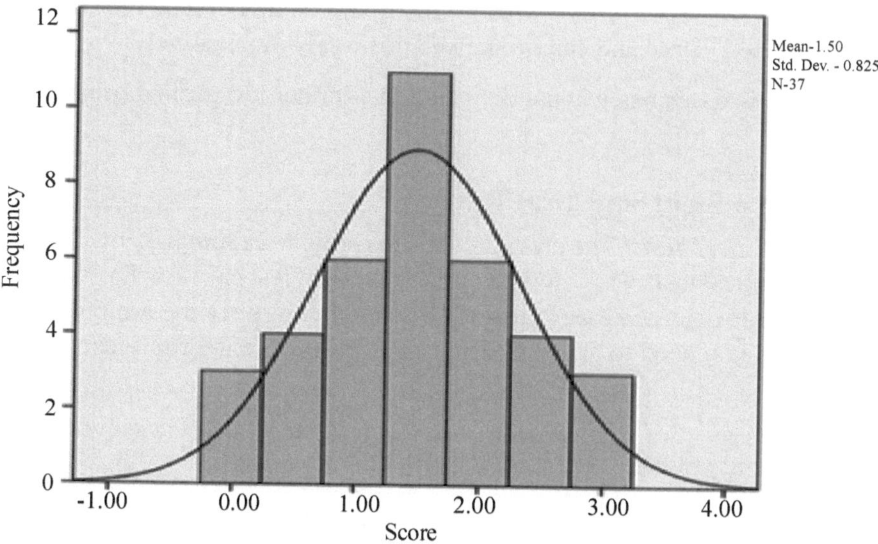

Fig. 17.1: Normal Distribution

Non-Normal Distributions: A small sample size or unusual sets of responses are common reasons that data may not be normally distributed (Fig. 17.2). Generally if the data is non-normal, we choose from a set of statistical tests called non-parametric tests.

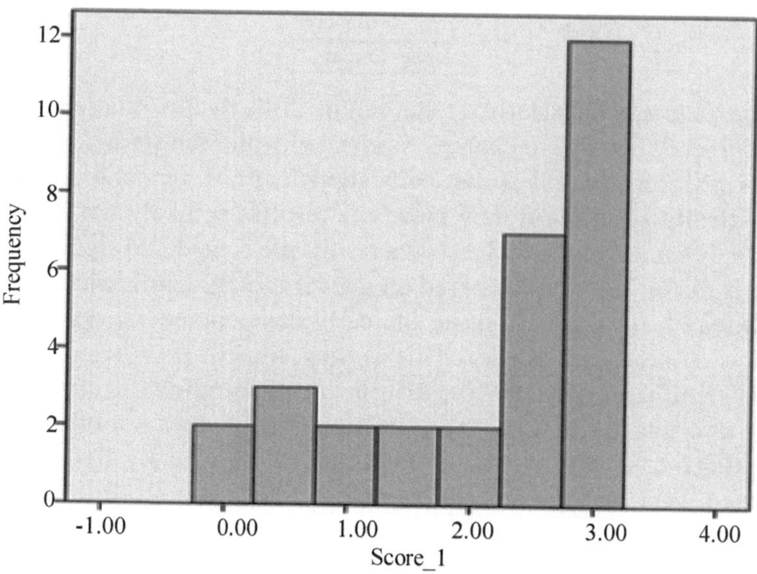

Fig. 17.2: Non-Normal Distribution

Skewed: ***The two types of non-normal*** distributions do not take the shape of the familiar Bell Curve and can be skewed positively or negatively.

Kurtosis: This describes when a distribution is either too peaked (pointy) or too flat.

Deciding on a Right Statistical Test

***Types of Statistical Tests*:** The previous sections provided a summary of different kinds of statistical tests. How do you choose the right test based on the research design, variable type, and distribution? The chart below provides a summary of the questions you need to answer before you can choose the right test.

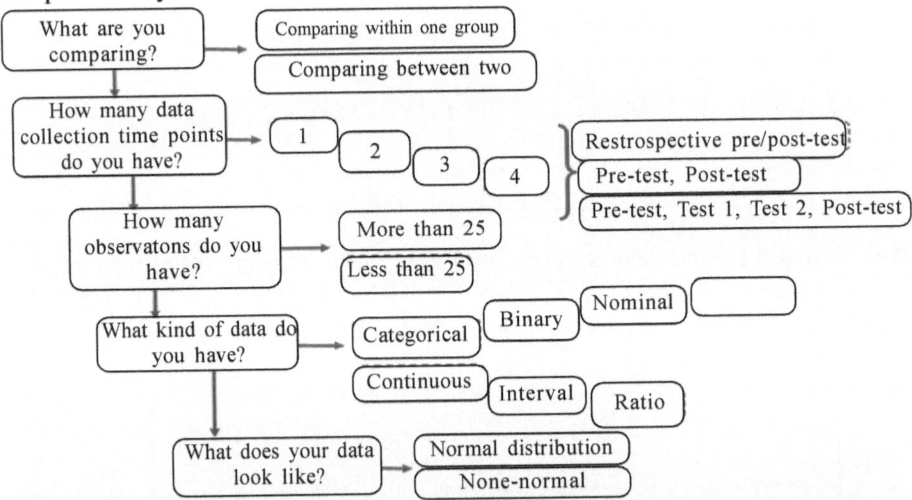

Statistical Significance: This statistics can tell us if the results/relationships we observe are real or just due to chance. Associated with each statistic is a p-value that tells us if something is **statistically significant**. If someone says the test was **statistically significant** they mean the results are likely not due to random chance. For many statistical tests the results are considered significant if the p-value is .05 or less. This is based on a level of 95% confidence. This cut-off or criteria of .05 was set during the early development of statistical methods and is somewhat arbitrary. This is why sometimes a result may be **statistically significant** without necessarily being **meaningful**. It is important to understand also that a result of statistical significance does not mean the effect is **meaningful.** Sometimes small effects can be statistically significant simply because a very large number of samples in the research. In order to account for this, the effect size is often calculated in research. The effect size is a measure of magnitude or strength versus the p-value of statistical tests which tell us if a relationship is due to chance or not and it lets researchers know if the results are meaningful or just due to the large number of people who participated in the research.

Common Rating Scales used in Social Science Research with Applicable Appropriate Statistical Methods

The scale or index construction in social science research is a complex process involving several key decisions. Some of these decisions are:

i. Should you use a scale, index, or typology?

ii. How do you plan to analyze the data?

iii. What is your desired level of measurement (nominal, ordinal, interval, or ratio) or rating scale? How many scale attributes should you use (e.g., 1 to 10; 1 to 7; 3 to +3)?

iv. Should you use an odd or even number of attributes (i.e., do you wish to have neutral or mid-point value)?

v. How do you wish to label the scale attributes (especially for semantic differential scales)?

vi. Finally, what procedure would you use to generate the scale items (e.g., Thurstone, Likert, or Guttman method) or index components?

This chapter examines the process and outcomes of scale development and appropriate statistical techniques used for analysis. The process of creating the indicators is called scaling. More formally, scaling is a branch of measurement that involves the construction of measures by associating qualitative judgments about unobservable constructs with quantitative, measurable metric units. It is well known that "Scaling is the assignment of objects to numbers according to a rule." This process of measuring abstract concepts in concrete terms remains one of the most difficult tasks in empirical social science research. The outcome of a scaling process is a scale, which is an empirical structure for measuring items or indicators of a given construct. Understand that "scales", as discussed in this section, are a little different from "rating scales" discussed in the previous section. A rating scale is used to capture the respondents' reactions to a given item, for instance, such as a nominal scaled item captures a yes/no reaction and an interval scaled item captures a value between "strongly disagree" to "strongly agree." Attaching a rating scale to a statement or instrument is not scaling. Rather, scaling is the formal process of developing scale items, before rating scales can be attached to those items.

Scales can be un-dimensional or multidimensional, based on whether the underlying construct is un-dimensional (e.g., weight, wind speed, firm size) or multidimensional (e.g., academic aptitude, intelligence). Multi-dimensional scales, on the other hand, employ different items or tests to measure each dimension of the construct separately, and then combine the scores on each dimension to create an overall measure of the multidimensional construct. For instance, academic aptitude can be measured using two separate tests of students' mathematical and verbal ability, and then combining these scores to create an overall measure for academic aptitude. Since most scales employed in social science research are un-dimensional, we will next examine three approaches for creating un-dimensional scales.

Un-dimensional scaling methods: The three most popular un-dimensional scaling methods are: (1) Thurstone's equal-appearing scaling, (2) Likert's summative scaling, and (3) Guttman's cumulative scaling. The three approaches are similar in many respects, with the key differences being the rating of the scale items by judges and the statistical methods used to select the final items. Each of these methods is discussed next with appropriate statistical measure.

Thurstone's equal-appearing scaling method: Louis Thurstone, one of the earliest and most famous scaling theorists published a method of equal-appearing intervals in 1925. This method starts with a clear conceptual definition of the construct of interest. Based on this definition, potential scale items are generated to measure this construct. These items are generated by experts who know something about the construct being measured. The initial pool of candidate items (ideally 80 to 100 items) should be worded in a similar manner, for instance, by framing them as statements to which respondents may agree or disagree (and not as questions or other things). Next, a panel of judges is recruited to select specific items from this candidate pool to represent the construct of interest. The selection process is done by having each judge independently rate each item on a scale from 1 to 11 based on how closely, in their opinion, that item reflects the intended construct (1 represents extremely unfavourable and 11 represents extremely favourable).

Statistical Methods: For each item, compute ***the median and inter-quartile range*** (the difference between the 75th and the 25th percentile – a measure of dispersion), which are plotted on a histogram. The final scale items are selected as statements that are at equal intervals across a range of medians. This can be done by grouping items with a common median, and then selecting the item with the smallest inter-quartile range within each median group. *The median value of each scale item represents the weight to be used for aggregating the items into a composite scale score representing the construct of interest.*

Likert's summative scaling method

The Likert method, an un-dimensional scaling method is possibly the most popular of the three scaling approaches described in this chapter. As with Thurstone's method, the Likert method also starts with a clear definition of the construct of interest and using a set of experts to generate about 80 to 100 potential scale items. These items are then rated by judges on a 1 to 5 (or 1 to 7) rating scale as follows: 1 for strongly disagree with the concept, 2 for somewhat disagree with the concept, 3 for undecided, 4 for somewhat agree with the concept and 5 for strongly agree with the concept. Following this rating, specific items can be selected for the final scale can be selected in one of several ways:

Statistical Methods: (1) By computing bi-variate correlations between judges rating of each item and the total item (created by summing all individual items for each respondent), and throwing out items with low (e.g., less than 0.60) item-to-total correlations, or

(2) By averaging the rating for each item for the top quartile and the bottom quartile of judges, doing a t-test for the difference in means, and selecting items that have high t-values (i.e., those that discriminates best between the top and bottom quartile responses). In the end, researcher's judgment may be used to obtain a relatively small (say 10 to 15) set of items that have high item-to-total correlations and high discrimination (i.e., high t-values). The Likert method assumes equal weights for all items, and hence, respondent's responses to each item can be summed to create a composite score for that respondent. Hence, this method is called a summated scale. Note that any item with reversed meaning from the original direction of the construct must be reverse coded (i.e., 1 becomes a 5, 2 becomes a 4, and so forth) before summating.

Guttman's cumulative scaling method

Designed by Guttman, the cumulative scaling method is based on Emory Bogardus' social distance technique, which assumes that people's willingness to participate in social relations with other people vary in degrees of intensity, and measures that intensity using a list of items arranged from "least intense" to "most intense". The idea is that people who agree with one item on this list also agree with all previous items. In practice, we seldom find a set of items that matches this cumulative pattern perfectly.

Statistical Methods

A **Scalo gram** analysis is used to examine how closely a set of items corresponds to the idea of cumulativeness. Like previous scaling methods, the Guttman method also starts with a clear definition of the construct of interest, and then using experts to develop a large set of candidate items. A group of judges then rate each candidate item as "yes" if they view the item as being favourable to the construct and "no" if they see the item as unfavourable. Next, a matrix or table is created showing the judges' responses to all candidate items. This matrix is sorted in decreasing order from judges with more "yes" at the top to those with fewer "yes" at the bottom. Judges with the same number of "yes", the statements can be sorted from left to right based on most number of agreements to least. To determine a set of items that best approximates the cumulativeness property, *a data analysis technique called Scalo gram analysis can be used* (or this can be done visually if the number of items is small). The statistical technique also estimates a score for each item that can be used to compute a respondent's overall score on the entire set of items.

Nominal scales also called categorical scales, measure categorical data. These scales are used for variables or indicators that have mutually exclusive attributes. Examples include gender (two values: male or female), industry type (manufacturing, financial, agriculture, etc.), and religious affiliation (Christian, Muslim, Jew, etc.). Even if we assign unique numbers to each value, for instance 1 for male and 2 for female, the numbers do not really mean anything (i.e., 1 is not less than or half of 2) and could have been easily been represented non-numerically, such as M for male and F for female. Nominal scales merely offer names or labels for different attribute values.

Statistical Methods: The appropriate measure of central tendency of a nominal scale is **mode**, and neither the mean nor the median can be defined. Permissible statistics are chi-square and frequency distribution, and only a one-to-one (equality) transformation is allowed (e.g., 1=Male, 2=Female).

Ordinal scales are those that measure rank-ordered data, such as the ranking of students in a class as first, second, third, and so forth, based on their grade point average or test scores. However, the actual or relative values of attributes or difference in attribute values cannot be assessed. For instance, ranking of students in class says nothing about the actual GPA or test scores of the students, or how they well performed relative to one another. Ordinal scales can also use attribute labels (anchors) such as "bad", "medium", and "good", or "strongly dissatisfied", "somewhat dissatisfied", "neutral", or "somewhat satisfied", and "strongly satisfied.

Statistical Methods: *The central tendency measure of an ordinal scale can be its **median or mode**,* and means are un-interpretable. Hence, statistical analyses may involve percentiles and non-parametric analysis, but more sophisticated techniques such as correlation, regression, and analysis of variance, are not appropriate. Monotonically increasing transformation (which retains the ranking) is allowed.

Interval scales are those where the values measured are not only rank-ordered but are also equidistant from adjacent attributes. For example, the temperature scale (in Fahrenheit or Celsius), where the difference between 30 and 40-degree Fahrenheit, is the same as that between 80 and 90-degree Fahrenheit.

Interval scale allows us to examine "how much more" is one attribute when compared to another, which is not possible with nominal or ordinal scales.

Statistical Methods: Allowed central tendency measures include mean, median, or mode, and measures of dispersion, such as range and standard deviation. Permissible statistical analyses include all of those allowed for nominal and ordinal scales, plus correlation, regression, analysis of variance, and so on. Allowed scale transformation are positive linear.

Ratio scales are those that have all the qualities of nominal, ordinal, and interval scales, and in addition, also has a "true zero" point (where the value zero implies lack or non-availability of the underlying construct). Most measurement in the natural sciences and engineering, such as mass, incline of a plane, and electric charge, employ ratio scales, as are some social science variables such as age, tenure in an organization, and firm size (measured as employee count or gross revenues). These scales are called "ratio" scales because the ratios of two points on these measures are meaningful and interpretable.

Statistical Methods: All measures of central tendencies, including geometric and harmonic means, are allowed for ratio scales, as are ratio measures, such as studentized range or coefficient of variation. All statistical methods are allowed. Sophisticated transformation such as positive similar (e.g., multiplicative or logarithmic) are also allowed.

Table 17.3: Statistical Properties of Rating Scales

Scale	Central Tendency	Statistics	Transformations
Nominal	Mode	Chi-square	One-to-one (equality)
Ordinal	Median	Percentile, non-parametric statistics	Monotonic increasing (order)
Interval	Arithmetic mean, range, standard deviation	Correlation, regression, analysis of variance	Interval Arithmetic mean, range, standard deviation
Ratio	Geometric mean, Harmonic mean	Coefficient of variation	Positive similarities (multiplicative, logarithmic)

Indexes and Typologies: An index is a composite score derived from aggregating measures of multiple constructs called components using a set of rules and formulas. It is different from scales in that scales also aggregate measures, but these measures measure different dimensions or the same dimension of a single construct. A well-known example of an index is the *consumer price index (CPI)*, which is computed every month by the Bureau of Labour Statistics of the Department of Labour. The CPI is a measure of how much consumers have to pay for goods and services in general and is divided into eight major categories (food and beverages, housing, apparel, transportation, healthcare, recreation, education and communication, and "other goods and services"), which are further subdivided into more than 200 smaller items. Each month, government employees call all over the country to get the current prices of many items. *Another example of index is socio-economic status (SES), also called the Duncan socioeconomic index (SEI).* This index is a combination of three constructs: *income, education, and occupation.* Income is measured in dollars, education in years or degrees achieved, and occupation is classified into categories or levels by status. These very different measures are combined to create an overall SES index score, using a weighted combination of "occupational education" (percentage of people in that occupation who had one or more year of college education) and "occupational income" (percentage of people in that occupation who earned more than a specific annual income). However, SES index measurement has generated a lot of controversy and disagreement among researchers.

The process of creating an index is similar to that of a scale. First, conceptualize (define) the index and its constituent components. Though this appears simple, there may be a lot of disagreement among judges on what components (constructs) should be included or excluded from an index.

Typologies: Scales and indexes generate ordinal measures of un-dimensional constructs. However, researchers sometimes wish to summarize measures of two or more constructs to create a set of categories or types called a typology.

Computers and Statistical Analysis

The recent widespread use of computers has had a tremendous impact on statistical analysis. Computers can perform more calculations faster and far more accurately than can human technicians. The use of computers makes it possible for investigators to devote more time to the improvement of the quality of raw data and the interpretation of the results. The current prevalence of microcomputers and the abundance of statistical software packages have further revolutionized statistical computing. The researcher in search of a statistical software package will find the book by Woodward et al. (1987) extremely helpful. This book describes approximately 140 packages. Among the most prominent ones are: Statistical Package for the Social Sciences (SPSS), S-plus, MINITAB, SAS and GENSTAT. The spreadsheet, Excel, also has facilities for statistical analysis

Statistical Software for Qualitative and Quantitative Analysis

Another good place to look for help in starting statistical analysis is a local college. Many universities and colleges offer introductory statistics courses. There are also many books on quantitative and qualitative analysis that can help you begin your data analysis. An excellent resource for beginning SPSS users is *Discovering Statistics Using SPSS by Andy Field.*

Statistical Software for Quantitative Data Analysis

SPSS: A general-purpose statistical package widely used in academic research for editing, analysing and presenting numerical data. It is compatible with all file formats that are commonly used for structured data such as Excel, plain text files and relational (SQL) databases.

Stata: A powerful and flexible general-purpose statistical software package used in research, among others in the fields of economics, sociology, political science. Its capabilities include data management, statistical analysis, graphics, simulations, regression, and custom programming.

R: A free software environment for statistical computing and graphics. It compiles and runs on a wide variety of UNIX platforms, Windows and MacOS. R provides a wide variety of statistical (linear and nonlinear modeling, classical statistical tests, time-series analysis, classification, clustering, etc.) and graphical techniques, and is highly extensible.

R is freely available online.

Statistical Software for Qualitative Data Analysis

Software packages comprised of tools designed to facilitate a qualitative approach to qualitative data, which include texts, graphics, audio or video. These packages sometimes referred as CAQDAS - (Computer Assisted/Aided Qualitative Data Analysis) may also enable the incorporation of quantitative (numeric) data and/or include tools for taking quantitative approaches to qualitative data.

NVivo: A qualitative data analysis (QDA) computer software package produced by QSR International. It has been designed for qualitative researchers working with very rich text-based and/or multimedia information, where deep levels of analysis on small or large volumes of data are required.

MAXQDA: An alternative to Nvivo and handles a similar range of data types allowing organisation, colour coding and retrieval of data. Text, audio or video may equally be dealt with by this software package. A range of data visualisation tools are also included.

Atlas.ti: Software for the qualitative analysis of large bodies of textual, graphical, audio and video data. It offers a variety of tools for accomplishing the tasks associated with any systematic approach to "soft" data, i.e. material which cannot be analysed by formal, statistical approaches in meaningful ways.

Data Visualization Tools

ArcGIS: A geographic information system (GIS) that helps to explore highly accurate geospatial data; you can create maps, analyze data for land use studies and other reports, and prepare data for use in an application or database.

Blender: This free and open source 3D creation suite supports the entirety of the 3D pipeline modeling, rigging, animation, simulation, rendering, compositing and motion tracking, in the context of research data in particular.

Datawrapper: An online data-visualization tool for making interactive charts which are responsive and embeddable in a website.

QGIS: A cross-platform, free and open-source desktop geographic information system (GIS) application.

R and Shiny: *R* is a tool used for data analysis and visualisation.

Using the free *Shiny* package, these analyses and visualisations can be published as interactive webpages just using R.

Explorer: A suite of online tools and data that allow users to visually explore hundreds of thousands of data indicators across demography, economy, health, religion, crime and more. Users can visualize and interact with data, create reports and downloads for offline processing.

Demographic Profiles: A new tool designed to provide users with an overview of the most popular demographic and socio-economic topics for a given geographical and/or administrative area. It helps to explore census data, finding the right facts, to analysesocio-economic data and discover trends, to visualise the data and groups with charts by topic.

Tableau Public: An easy to use, free and powerful tool for creating interactive dashboards and data visualisations that can be shared publically and embedded in your personal site.

Conclusion

Statistics are used to inform policies that has impact on millions of people. Statistical methods are applied in all fields that involve decision-making, using a body of data to make informed guesses despite lack of definite knowledge. Statistics have become a key feature of social science, and are applied across a range of disciplines including economics, psychology, political science, sociology and anthropology. Using sources such as surveys, censuses and administrative records, statisticians collect and analyse data to give us large-scale and small-scale impressions of our complex society building a clearer picture of where we are and where we are going. Elaborated the process of defining the research problem; develop and implement a sampling plan; conceptualize, operationalize and test the measures. Qualitative analysis is one of the best methods of research for many fields of study. It provides depth and detail by analyzing things more than just numbers and sizes. It does this through examining attitudes, feelings, and behaviors. This type of analysis encourages people to expand their responses and also helps people to have a better and more detailed picture of the causes and Conclusion of different actions and reactions in the world. In qualitative analysis deals with thought, action, and behavior is researched through the cause and the reality. Finally, the challenges of data analytics are very specific in the field of social sciences and must be addressed with the appropriate methodology and technology. Because each qualitative study is unique, the analytical approach used will be unique. Because qualitative inquiry depends, at every stage, on the skills, training, insights, and capabilities of the researcher. The human factor is the greatest strength and the fundamental weakness of qualitative inquiry and analysis. How to choose the right type of test is clearly tabulated for the researchers. Review on the Role of Economics in Fisheries Research is dealt with elaborately. This chapter is expected to provide working knowledge to the researchers/professionals engaged in social science particularly qualitative data analysis.

Acknowledgements: The author is grateful to the students for all out support in preparing the manuscript.

References

Bavinck, M., S. Jentoft, and J. Scholtens. (2018). Fisheries as social struggle: a reinvigorated social science research agenda. *Marine Policy* 94: 46–52

Bennett, N.J., R. Roth, S.C. Klain, K. Chan, P. Christie, D.A. Clark, G. Cullman, D. Curran, T.J. Durbin, G. Epstein, A. Greenberg, M.P. Nelson, J. Sandlos, R. Stedmann, T.L. Teel, R. Thomas, D. Veríssimo, and C. Wyborn (2017). Conservation social science: Understanding and integrating human dimensions to improve conservation. *Biological Conservation* 205: 93–108

Berkowitz, S. (1996). Using Qualitative and Mixed Method Approaches. Chapter 4 in Needs Assessment: A Creative and Practical Guide for Social Scientists, R. Reviere, S. Berkowitz, C.C. Carter, and C. Graves-Ferguson, Eds. Washington, DC: Taylor & Francis.

Bhattacherjee Anol (2012). Social Science Research: Principles, Methods, and Practices, University of South Florida, USA.

Coffey, A., and Atkinson, P. (1996). Making Sense of Qualitative Data: Complementary Research Strategies. Thousand Oaks, CA: Sage.

Food and Agriculture Organisation of the United Nations. FAO (2018). The State of World Fisheries and Aquaculture 2018. FAO, Rome. http://www.fao.org/3/i9540en/I9540EN.pdf

Giorgi, A and Giorgi, B (2003). Phenomenology. In J A Smith (ed.) Qualitative Psychology: A Practical Guide to Research Methods. London: Sage Publications.

Howe, K., and Eisenhart, M. (1990). Standards for Qualitative and Quantitative Research: A Prolegomenon. Educational Researcher, 19(4):2-9.

Johnsen, J.P., and B. Hersoug. (2014). Local empowerment through the creation of coastal space? *Ecology and Society* 19 (2): 60.

Johnsen, J.P., and S. Jentoft. (2018). Transferable quotas in Norwegian fisheries. In Fisheries, quota management and quota transfer: Rationalization through bio-economics, ed. G.M. Winder. Cham: Springer. Knott, C., and B. Neis (2017). Privatization, financialization and ocean grabbing in New Brunswick herring fisheries and salmon aquaculture. *Marine Policy 80: 10–18*

Knol, M. (2011). Mapping ocean governance: from ecological values to policy instrumentation. *Journal of Environmental Planning and Management* 54 (7): 979–995.

Knott, C., and B. Neis. (2017). Privatization, financialization and ocean grabbing in New Brunswick herring fisheries and salmon aquaculture. Marine Policy 80: 10–18

Lascoumes, P., and P. Le Galès. (2007). Introduction: understanding public policy through its instruments: from the nature of instruments to the sociology of public policy instrumentation. Governance: *An International Journal of Policy, Administration, and Institutions* 20 (1): 1–21

Miles, M.B, and Huberman, A.M. (1994). Qualitative DataAnalysis, 2nd Ed., p. 10-12. Newbury Park, CA: Sage

OECD. (2016). The ocean economy in 2030. Paris: Organisation for Economic Co-operation and Development

Patton, M.Q. (1990). Qualitative Evaluation and Research Methods, 2nd Ed. Newbury Park: CA, Sage.

Ricker, W.E. (1954). Stock and recruitment. *Journal of the Fisheries Board of Canada* 11(5): 559–623

Roy, A. K. and Nibedita Jena, (2008). Econometric Approach for Estimation of Technical Efficiency of Aquaculture Farms. *In*: Applied Bioinformatics, Statistics and Economics in Fisheries Research. (Eds. Roy, A. K and N. Sarangi), New India Publishing Agency, New Delhi, and PP: 501-518.

Roy, A.K. (2010). A Study on Statistical Methods for Impact Assessment of Freshwater Aquaculture with Particular Reference to Kolleru Lake, Summer School Proc., WBUF&AS, Kolkata.

Roy, A.K. (2010). Quantitative methods for Social Science Research. Issues and Tools for social sciences research. (Eds.Kathia et al.), CIFRI, Barrackpore.pp:391-408

Roy, A.K. (2018). Partial Budgeting Analysis of an Intervention of 'Land Shaping Technology' at South 24 Parganas, West Bengal, India. *Open Acc J Envi Soi Sci.*, 1(4):94-102.

Schaefer, M.B. (1957). Some considerations of population dynamics and economics in relation to the management of the commercial marine fisheries. *Journal of the Fisheries Board of Canada* 14 (5): 669–681

Silver, J.J., N.J. Gray, L.M. Campbell, L.W. Fairbanks, and R.L. Gruby (2015). Blue economy and competing discourses in international oceans governance. *The Journal of Environment & Development* 24 (2):135–160

Soma, K., S.W.K. van den Burg, E.W.J. Hoefnagel, M. Stuiver, and C.M. van der Heide (2018). Social innovation – a future pathway for blue growth? *Marine Policy* 87: 363–370

Song, A.M., J.P. Johnsen, and T.H. Morrison. (2018). Reconstructing governability: how fisheries are made governable. *Fish and Fisheries* 19 (2): 377–389.

Urquhart, J., T.G. Acott, D. Symes, and M. Zhao (2014). Social issues in sustainable fisheries management. Berlin: Springer.

Wolcott, H.F. (1994). Transforming Qualitative Data: Description, Analysis and Interpretation, Thousand Oaks: CA, Sage.

18

ERP Software Application in Aquaculture Industry – Need and Status

Subhabaha Pal and Satyabrata Pal

Introduction

Aquaculture, also known as aqua-farming, embodies the procedure of the farming of fish, crustaceans, molluscs, aquatic plants, algae, and other organisms. Aquaculture involves cultivation of freshwater and saltwater fish-populations under controlled conditions (Wikipedia, 2019). Aquaculture industry has undergone a lot of change in the last two decades. The application of the sophisticated technologies in the aquaculture industry has greatly altered the face of the industry in special relation to the Western and developed countries. New tools related to the measurement of the different environmental parameters needed to be controlled in different layers of the production process like Hatchery, Nursery and Grow-out have come up. Lots of data generated from the above stages relate to include the parameters, like, cost of production, health condition of produced fish, etc., which ultimately leads to the generation of the appropriate pricing of produce. With the advent of globalization, all aquaculture product companies need to look for exporting their respective products along with meeting the full satisfaction related to the demand from the domestic markets with an ultimate objective to increasing the profit. Export of the fisheries products in the western countries needs proper certification on the quality of the products duly by the appropriate authorities. The certification authorities need to check the detailed production process related to the fisheries- products in order to be sure about the quality imbibed in the same. The development of integrated software maintaining the details of the production process makes the task easy for both the production company and the certification authority, thus satisfying the needs of the end-customers in the western markets. Useful ERP software needs to maintain all relevant information on the production process commencing from the very early stage of the production till the act of selling the products in the market.

Traceability is an aspect which is highly needed in the aquaculture production process in order to maintain the quality of aqua-farming products. The aspect of Traceability can provide food safety and security in relation to the aquaculture production. The Retailers ask for more detailed information on the products and such product information are required for the certification purpose too. Building up of the aspect of traceability in the production process assists in securing the production control and optimization, which ultimately reflect in reaping the benefits related to attainment of, so to say, the competitive advantage. Briefly speaking, Traceability brings the competitive advantage and increases the confidence on the product by the consumers. The benefits derived from absorbing the aqua-culture Traceability- relationship include market benefits, assuring quality and safety management, and reduced cost of production due to optimization and product recall. Any company intending to incorporate the aspect of Traceability in its overall production process needs to perform the following tasks:

i. The company should track information from each stage starting from hatchery up to harvest

ii. The company should keep records of the followings: all feeds, medicaments and other items given to fish

iii. The company should make sure that the information collected should not be lost during the operations, like, transferring, grading or mixing the fishes

iv. The company should be able to respond to customer questions and to issue product certificates

v. The company should perform the activities related to evaluating suppliers, feed types, and feeding and management practices

It is difficult to sustain a good traceability monitoring without having built a system. Huge amount of varied data gets generated at each stage of production process which may create confusion and hence requires creation of a procedure which entails a good and accurate coordination among processes related to storing and retrieval. Thus if the above functions are not done properly, there remains a high chance of loss of some tracking information. A software system should support the full traceability of the produced fish and the production processes, which falls under the main requirement of the certification bodies. The system should gather relevant (and important) information during the life time of the fish in order to provide those at the time of certification. It should also be able to document all complaints and replies also.

Use of a software system provides the following benefits (AquaManager, 2019):

i. Lower production costs

ii. Improved profitability

iii. Improved operational efficiency

iv. Management support for decision-making

v. Efficient and effective management of equipment and human resources

vi. Improved product quality through the identification of problems and the effective monitoring of quality characteristics

vii. Optimization of stock levels and purchasing policies

Aquaculture Production Process and Software Requirement

Each aquaculture/aqua farming product needs to go through generally three stages at the time of production (AquaManager, 2019).

1. Hatcheries - A fish hatchery is a place for artificial breeding, hatching, and rearing through the early life stages of aquatic farm products (Wikipedia, 2019). Hatcheries produce larval and juvenile fish, shellfish, and crustaceans, primarily to support the aquaculture industry where they are transferred to the on-growing systems, such as fish farms, to reach harvest size. Some aquatic species that are commonly raised in hatcheries include, economically important finfishes, catfishes, shrimp, Indian prawns, salmon, tilapia, Pacific oysters, scallops, etc. The world total aquaculture production in 2015 was 106 million tonnes in live weight, with an estimated farm gate value of US$163 billion. This total is comprised of farmed aquatic animals (76.6 million tonnes, US$157.9 billion), aquatic plants (29.4 million tonnes; US$4.8 billion) and non-food products (41.1 thousand tonnes; US$208.2 million) (FAO, 2017). The China significantly dominating the global aquaculture market; however, the value of aquaculture hatchery and nursery production has yet to be estimated. A fish hatchery is a complex system. It includes a long range of activities ranging from the selection, collection and manipulation of brood stock and the production of live feed up to the production and harvest of larvae in the shortest time possible.

Any ERP system specific for aquaculture domain must provide an integrated hatchery management module that should help in optimal use of the hatcheries, both from the economical and production points of views. It should register, control and optimise all the major parameters that affect production efficiency and fish quality. All the information related to rearing environment (temperature, oxygen, salinity, etc.), the live feed production and the fish behaviour should be captured into the system.

2. Nursery - The purpose of the nursery phase is to grow the fish up to a size that is more suitable for stocking into large production units (AquaManager, 2019). The nursery phase generally determines the variation in size that the farm will have when the fish reach market size. If not managed properly, the fish farm manager will be facing difficulty in grading and harvesting the final product. It is therefore, very important to ensure that fish are not overcrowded, and are fed an appropriately-sized diet, while maintaining excellent water quality. The software must provide this kind of control, enabling the farm to optimize all the variables driving the success of the nursery process. In addition, the software should import data from the hatchery stage and it should export all the data to next stage to have a swift transition from one phase to the other.

3. Grow Out- The grow-out stage is the final and most expensive part of the fish farming process (AquaManager, 2019). To obtain optimal growth, reduce fish farming costs and improve efficiency, the farm needs to be able to continuously evaluate performance, identify problems immediately as soon as they arise, check every day if the management of the fish is done in the right way, evaluate feed suppliers and feeding policies, optimize feeding tables and growth models, know precisely the cost of the fish, make the best use of the available resources. The software needs to perform all these tasks for effective continuation of the production process.

Any ERP software for aquaculture should possess all capabilities to handle data emanated at these three stages. The AquaManager is fish-farming ERP software which works specifically in this direction.

Aqua Manager

The AquaManager software covers the entire fish production process over all the above three stages and it has different modules to cover these three stages. The interaction of the three modules, namely, Hatchery, Nursery and Grow Out, increase the traceability of the fish production up to a maximum level. The AquaManager allows the farm to understand the trio functions, namely, what is happening, why is happening and how is happening during the fish production process. The AqupaManager software provides tools (AquaManager, 2019) that –

a) Help the users to continuously evaluate performance

b) Help the users to visualize, to analyse the data, and to allow the users to prevent problems before the escalation stage. Such functions are very difficult to do consistently through a manual process.

c) Help the users to optimize the feeding strategy based on the specific conditions of the farm.

d) Allow the farm to understand what is happening and when to take corrective actions. Without a good system, this information comes a bit late.

A. Hatchery

The AquaManager software starts with providing solution for hatchery stage and module is named as 'Hatchery'. In the 'Hatchery' module, the following functions are generally performed

1. ***Broad Stock Management***: The module tracks all information related to the stock at population or at individual level. It helps in evaluating performance, improving the breeding process and ensuring traceability from egg to harvest. The AquaManager Hatchery module provides the tools required for efficient management of brood-stock fish which is one of the most valuable assets of the hatchery. The 'Hatchery' module facilitates the following activities:

   a. Tracking fish individually by gender and/or tag ID

   b. Registering transactions both at tank or fish level

   c. User-defined brood-stock fish attributes

   d. User-defined attributes in each and every brood-stock transaction

   e. Registration and analysis of spawning at tank or fish level

   f. Immediate access to the history of the brood-stock tank or each individual fish

   g. Management of incubation, fertilization and survival rates

   h. Graphical layout of brood-stock and incubator tanks

   i. Record user-defined aquaculture data on brood-stock and incubation process

2. ***Live Feed Management***: The module can efficiently control the production of live feed, gain access to real-time information, improve productivity, eliminate quality problems, unlock capacity and reduce inventory and labour costs. Some of the facilities available in the system are the following:

   a. Management of algae, rotifer and artemia production

   b. User defined measurements

   c. Supports batch code registration of every item used in production of live feed (additives, enrichments, feed, etc.)

  d. Full traceability to item batch code level
  e. Smart screens for fast data entry

3. ***Larvae Rearing and Weaning***: The module facilitates improvement of production control, reduction of inventories and increase of quality. The AquaManager provides all facilities that one farm needs to control and optimize the management of its hatchery. All the information related to the rearing environment, the feed consumption, the transactions and the fish behaviour can be easily registered into the system. The application is fully configurable, giving the users great flexibility in scheduling and management of its fish production. It is integrated aquaculture software that ensures complete traceability of the produced larvae, allowing the users to trace products and their associated attributes backwards and forwards through the production chain from brood-stock to live feed production, to rearing and weaning.

4. ***Production Planning***: The AquaManager provides advanced planning and scheduling tools to help the users produce the highest possible output with the lowest possible cost, while ensuring high quality and production performance. The module reduces production plan development time. It also improves transparency and accuracy both of the planning process as well as the resulting financial impact. It supports decision making at any level. In the module, the users can create, run and evaluate production plans at any level, i.e., group, region or site. The plans take into account production targets as well as capacity restrictions, survival rates, culturing policies, etc. – all these information are user-defined and fully configurable. The AquaManager gives the competitive edge through helping in increasing efficiency by making best use of hatchery resources, avoiding mistakes, optimizing inventory levels and increasing profitability.

5. ***Cost Analysis***: The software helps tracking the cost of all items, from live feed to larvae, either finished or in production, taking into account the direct and indirect costs. The AquaManager helps the users to improve transparency and accuracy both of the costing process as well as of the financial results. It allows the users to identify trends, problems and causes and supports decision making at any level.

6. ***Hatchery Inventory Management***: The software maximizes the return on inventory investment. Most of the items that are used in a hatchery are expensive and the software helps in avoiding over-purchasing or under-purchasing and optimizing inventory.

The AquaManager software helps to figure out how much inventory the farm needs, gain real-time visibility across different business and optimize stock levels. The software facilitates:

Powerful batch code tracking utilities

Full traceability on all items

Support for evaluation of suppliers and products

Unlimited number of inventory locations per region

All kinds of transactions are supported: purchase, transfer, count, and disposal

Integration with the user's current ERP system of financial management application is possible

Automatic notification when items go below attention or re-order level

7. ***Analysis and Reporting***: The software eases the analysis of hatchery performance. With all data in one place, the AquaManager makes it easy to access and analyse data, take decisions at any level and move the management of hatchery to the next level.

   a. Many predefined report templates are provided.

   b. All the reports that can be fully customized by the end-user.

   c. Reports can be either printed or exported into standard formats.

   d. All the analyses are drill-down and multi-dimensional.

B. Nursery

It is very important to ensure that fish are not overcrowded, and are fed an appropriately-sized diet, while maintaining excellent water quality. AquaManager nursery is fish-farm software that provides the users this kind of control, enabling them to optimize all the variables driving the success of the nursery process. In addition, as AquaManager Nursery can import data from the hatchery module it also exports all the data to aquaManager Grow out module to have a swift transition from one phase to the other.

The Nursery module performs the following functions –

1. **Efficient Production Control**–The software module facilitates –

a. Improved management of the nursery site, day-to-day control of all production parameters, and timely identification of production problems or trends.

b. Quick and easy registration of all daily activities through user friendly screens or mobile devices.

c. Real-time, accurate view on performance that helps the users to make the right decisions and react quickly on any deviations.

d. Full traceability on the history of every fish population without any restriction on the number of movements (grading, transfer, etc.).

e. Immediate estimation of the expected fish number at the end of the production and the expected growth per fish size

f. Optimization of feeding strategies, feed prognosis.

g. Full overview of the production, at any level (site, region or group) in tabular or graphical formats.

2. **Successful Inventory Management**– This software module facilitates:

a. Optimize stock levels and purchasing policies, accurately track inventory, improve visibility, and gain full traceability on each item.

b. Integrated work flow for purchase orders and inventory transactions.

c. Complete traceability at the batch code level on all items used in the fish production (feed, medicaments, etc.).

d. Unlimited number of inventory locations per region.

e. Support of all types of transactions - purchase, transfer, count, and disposal.

f. Integration with the users' ERP system of financial management application is possible.

3. **Cost Analysis**–The software facilitates tracking of the cost of each batch or unit, either finished or in production, taking into account direct and indirect costs. The AuaManager supports real-time cost control through all stages of production, thereby helping the users make the best decisions to minimize production costs, improve profitability and make informed decisions on pricing policies and profit margins.

4. **Analysis & Reporting**–The fish farm software provides powerful reporting facilities, personalized dashboards, flexible reporting tools, and powerful drill-down capabilities. The software facilitates –

   a. Transformation of data into knowledge

   b. Analysis of performance and taking decisions at any level.

   c. Generation of professional business reports and attractive charts.

d. Predefined report templates

e. Full customization of all reports by the end-user

f. Printing or export of Reports into standard formats (Excel, pdf, rtf, html, etc)

C. Grow Out

Any fish-farming software should provide real time production control. The users should understand what is happening, as it happens, identify problems or trends, react quickly, and be proactive. The grow-out stage is the final and most expensive part of the fish farming process. It is extremely sensitive to feed, feed conversion, nutrient retention, health and bio-security with feed and health costs representing about 70 % of operation expenditure. To produce optimal growth, reduce fish farming costs and improve efficiency, the firm needs to be able to continuously evaluate performance, identify problems immediately as soon as they arise, check every day if the management of the fish is done in the right way, evaluate feed suppliers and feeding policies, optimize feeding tables and growth models, know precisely the cost of the fish, make best use of the available resources and much more. AquaManager is a powerful fish farming software that provides this kind of control. It makes sure that all critical information is collected and registered. It also helps to use this information to dramatically improve performance. It also offers to the management the support they need for efficient decision making and helps to improve profitability through timely identification of problems or trends, cost control and better planning. The major advantages of the 'Grow Out' module are:

1. **Efficient Production Management In An Easy Way** – The users can make better use of the feed, lower fish farming costs, improve efficiency and ensure production can meet customer demand. The software facilitates the following activities:

a. Monitor, control and improve the use of the feed, based on fish behaviour and real time data

b. Quick and easy registration of all daily activities through user friendly screens or mobile devices.

c. Continuous, day-to-day performance evaluation and improvement.

d. Improvement of fish growth models based on actual data.

e. Identification of problems immediately as soon as they arise.

f. Optimization of fish feeding strategies.

g. Accurate estimation of the required feed for the next period.

h. Full overview of the fish farm, region or group status in tabular or graphical formats.

2. **Knowing the actual cost of Fish**– The users can control all the cost drivers; know precisely the cost of each individual batch, cage or tank, make informed decisions on pricing policies and profit margins. An aquaculture company cannot get the fish farming costs from the conventional profit and loss statement. It needs to track costs at a much more detailed level and control the costs at the level of each species, batch or unit. Without product-specific information, it cannot tell which products are doing well and which need special attention or corrective actions. The AquaManager supports real-time cost control through all stages of fish production, thereby helping the farm make the best decisions to minimize production costs, improve profitability and make informed decisions on pricing policies and profit margins. Aquaculture management software helps to control fish farming cost. The features of the 'Grow Out' module are the following:

a. Both direct costs (feed, fry, medicine, etc) and production overheads can be taken into account

b. Cost can be calculated at any level: group, region, fish farm, batch or each individual unit

c. In case of transfer transactions, all data are distributed to the destination units

d. Detailed cost analysis both per cost type (e.g. feed, fry, overheads) as well as per fish quantity (remaining, mortalities, harvests, adjustments)

e. ERP or accounting system integration

3. **Successful Inventory Management** – The software facilitates optimization of stock levels and purchasing policies, knowledge of the quantity on hand at any time, controlling of the stock per location and tracking all the related transactions. The features of this software includes:

a. Integrated workflow for purchase orders, inventory transactions, costing

b. Unlimited number of inventory locations per region

c. All types of transactions are supported: purchase, transfer, count, and disposal

d. Integration with the farm's ERP system of financial management application is also possible - automatic calculation of average purchase prices, stock monitoring.

4. **Analysis & Reporting–** The software facilitates transformation of data to knowledge and taking appropriate decisions based on real-time analysed data. The AquaManager is not only a data registration tool but it is also an intelligent fish farming software that transforms data into knowledge. It helps the users quickly and easily analyze the performance at any level and identifies problems and areas for improvement. It helps in generating professional business reports and attractive charts.

Data Loggers

The AquaManager has the capability for real-time monitoring of aquaculture data (water quality data, environmental parameters, weather information, etc.) that is accurate and reliable. It helps in optimizing fish management, improvement of production efficiency and makes sure that the production runs smoothly, and responds quickly to emergency situations. The AquaManager Data Loggers module can be used for monitoring following environmental parameters in aquaculture farms:

a) Wind speed
b) Wind direction
c) Temperature
d) Atmospheric pressure
e) Solar / UV radiation
f) Rain height
g) Measured Parameters
h) Water Quality Parameters
i) Dissolved oxygen
j) Temperature
k) Salinity
l) Conductivity
m) pH
n) ORP
o) Turbidity

The AquaManager software can be easily synced with aquaculture monitoring devices to measure water quality parameters and environmental data.

The major benefits of Data Loggers are:

a) Easy to use aquaculture monitoring devices, with minimum requirements for electric supply and communication infrastructure.

b) Flexibility and modularity. The solution can be engineered to users' needs.

c) Low initial and maintenance cost.

d) Straight forward installation and maintenance.

Conclusion

The paper discusses how an ERP software can assist in the aquaculture business and it discusses the use of a software, entitled, AquaManager in this perspective. The AquaManager is a unique fish farming software that provides visibility and control over the full range of variables driving the success of aqua-farm production process: cost, quality, volume, scheduling, and profitability—as they occur. The software facilitates the following:

a) Reduction of costs by enabling production managers (in fish farm or hatchery) to view, analyse and control costs during the production. In the aquaculture industry almost 80% of the production costs are in the hands of the farm manager.

b) Improvement of product quality by analysing all the factors contributing to fish health and growth and identification of problems as they arise

c) Improvement of scheduling by providing immediate feedback on plan performance

d) Increase of efficiency and utilization throughout the fish farm, and beyond.

The AquaManager is a comprehensive, integrated software solution for improved efficiency in aquaculture industries. It is a complete fish farming software that supports all stages of fish production, from hatchery to harvest.

References

'Aquaculture,' URL - https://en.wikipedia.org/wiki/Aquaculture . (Last updated on 07.01.2019).

'AquaManager-Tracking the traceability in the aquaculture sector and value chain', URL - http://aqua-int.com/wp-content/uploads/2014/07/Victor-Duval-AquaManager.pdf (Last updated on 07.01.2019).

FAO (2017). GLOBEFISH - Analysis and information on world fish trade. The 11th Global Aquaculture Summit 2017, organized by the China Aquatic Products Processing and Marketing Alliance (CAPPMA) and the U.S. Soybean Export Council (USSEC) on 26/06/2017 in .

'Fish Hatchery', URL - https://en.wikipedia.org/wiki/Fish_hatchery. (Last updated on 08.01.2019).

'Nursery', URL-https://www.aqua-manager.com/fish-farm-software-nursery. (Last updated on 08.01.2019).

'Grow Out', URL-https://www.aqua-manager.com/fish-farming-software-grow-out. (Last – updated on 08.01.2019).

19

Fisheries Resource Assessment and Management: Concepts and Techniques

K.V. Radhakrishnan

Introduction

According to the Food and Agriculture Organization (FAO), fish and fishery products accounts for 17% of the global human consumption of animal protein. Per capita food fish supply has grown to 18.4kg per person, an 8% rise since 2006. Approximately 54% of the seafood protein supply for human consumption comes from wild catch; unfortunately about 30% of the world's wild-capture fisheries are overexploited and many fisheries are in bad shape and getting worse. Solving this problem will require innovative monitoring and management tools, but we can provide tremendous benefits if we act now to reverse our course. Taking a serious, in-depth look at fisheries could help feed the world's growing population, and conserve marine species. Humanity is reaching farther and deeper into the oceans and, until recently, there were only anecdotes and guesses about the complete effects of our actions. Recent reports show that some fisheries are being managed sustainably and good progress is being made in reducing exploitation rates and restoring overexploited fish stocks and marine ecosystems.

Sustainable fisheries management is an integrated process that seeks to attain an optimal state that balances ecological, economic, social and cultural objectives for fisheries. It involves information gathering, analysis, planning, consultation, decision-making, allocation of resources and formulation and implementation, with enforcement as necessary, which govern fisheries activities to ensure the above mentioned fisheries objectives. A thorough knowledge on the fish resource assessment is of utmost importance for a fishery manager to effectively make decisions and implement it. In this chapter, an attempt is made to briefly explain the basic concepts of fish resource under management, its assessment and techniques to be applied.

Fish resources and Assessment

Among all the exploited natural animal resources, fisheries resources are the largest. Magnitude, dynamics and balancing of fish stocks possess great challenges to their assessment and management. Many species have wide distribution, multiple stocks, and temporal variation in abundance, above all, the resources cannot be seen visually. However, accurate information on fish stocks is essential for their sustainable exploitation and conservation. Objective of fish stock assessment is to predict changes in yield of stock as function of both fishery dependent (fishing effort) as well as fishery independent factors so that optimum levels of effort and yield could be determined. In other words, a stock assessment provides information on optimum exploitation of fish resources which will give maximum yields from the particular fishery in the long run. Generally two important models are applied for fish resource assessment: Holistic models: Applicable when data is limited. Take stock as a homogeneous biomass and don't take in to account the length or age structure of the stock. E.g.: Swept area method, Surplus production models. Second is analytical models which are applicable when quality and quantity of data are available. They give more reliable predictions on stock. E.g.: Thompson and Bell Model, Beverton and Holt model. A diagrammatic representation on significance of stock assessment in formulation of policies and implementation of fishing regulations are given in figure below.

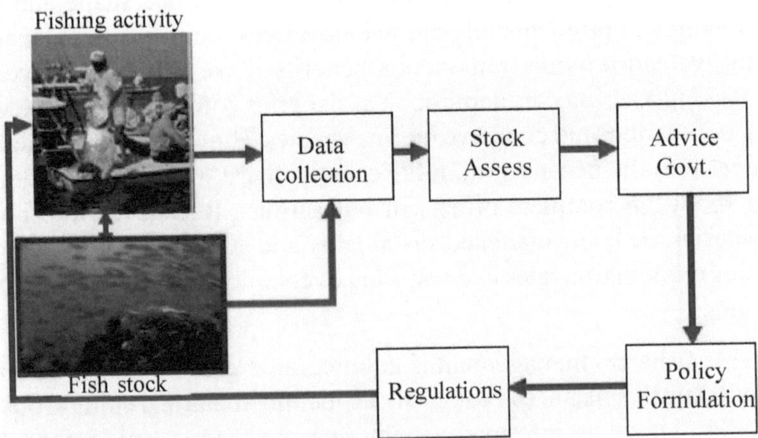

Fig. 19.1: Diagram Showing Significance of Stock Assessment in Policies Formulation on Fishing Regulation and its Implementation

Sustainable yield in fisheries

Fisheries scientists dedicate their time to studying how fishing can be balanced so as to allow aquatic species to maintain their resources at sustainable level, in a dynamically fluctuating and changing environment. Decades of research and managers' experiences, trying to apply scientific advice on the ground, have shaped current practices in monitoring and managing fisheries sustainably. Two key factors need to be balanced in order for fishing to be sustainable: 1) the weight, or biomass, of the targeted population (B or more specifically SSB). 2) The fishing pressure (F). Other factors also play a role, including the abundance of predators, availability of food, environmental variability, disruption of climate cycles, etc. The effectiveness of the management system in constraining fishing pressure within sustainable limits is also critical.

Maximum Sustainable Yield (MSY) – a 'sweet spot' for sustainability

When a 'virgin' or previously unfished population is first harvested, its biomass will initially decline as a result of fishing. But there is a point where a roughly constant harvest can be maintained indefinitely without causing decline in the population, and where the productivity of the population is at its maximum. This maximum catch that can be harvested sustainably is called the Maximum Sustainable Yield (MSY). The stock biomass required to support MSY is known as BMSY. At BMSY, fishing is considered to be sustainable indefinitely, with a fishing pressure of FMSY. Fishing more than this, leads to sub-optimal harvest, where increasing fishing pressure yields smaller catches. Overfishing can, in the long run, become unsustainable when the population declines to the point that there aren't enough large, mature fish (the 'spawners', making up the spawning stock biomass, SSB) to reproduce, or young fish to grow, and replenish the fish that are harvested. It is important to know how many larger mature animals there are because they produce more and healthier offspring, and it's also important to know how many juveniles there are that can grow into next year's available catch. This stage, when new generations are insufficient to replace the old, is known as the Point of Recruitment Impairment (PRI).

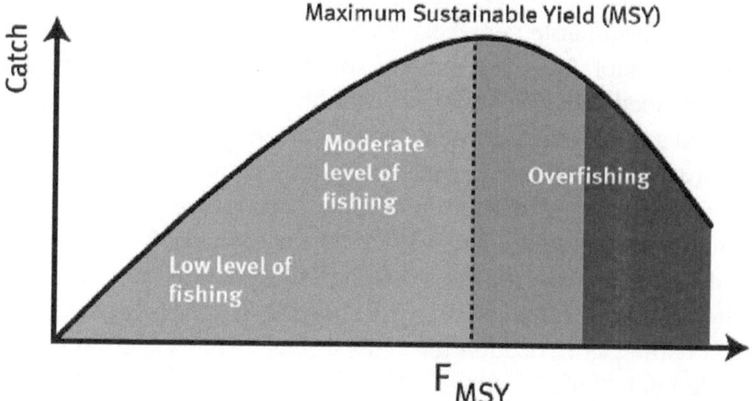

Fig. 19.2: Relationship of fishing effort, catch and MSY

Each aquatic population has its own specific BMSY, FMSY, and MSY. In addition, as other factors such as climate, food and predators vary, so does the maximum biomass that the environment can support or the amount of fishing that the population can compensate for. Therefore MSY is not a static number: the size of populations and the amount of fishing they can sustain changes dynamically.

The maximum sustainable yield (MSY) can be estimated from the following input data:

f(i) = effort in year i, i = 1,2,...,n

Y/f = yield (catch in weight) per unit of effort in year i.

Y/f may be derived from the yield, Y(i), of year i for the entire fishery and the corresponding effort, f(i), by Y/f = Y(i)/f(i), i = 1,2,...,n..........

The simplest way of expressing yield per unit of effort, Y/f, as a function of the effort, f, is the linear model suggested by Schaefer (1954):

Y(i)/f(i) = a + b*f(i)

In this method,

MSY = $-a^2/4b$,

Optimum effort (fMSY) =-a/2b,

Yield for a given effort (Y/f) = $af-bf^2$

Where,

a = intercept; b the regression coefficient.

Maximum Economic Yield (MEY)

MEY is that yield level, which coincides with the level of harvest or effort that maximized the sustainable net profit from fishing. A MEY harvest is desirable because it is the catch level that enables society to do the best it can with what nature has provided. MEY is a long-run equilibrium concept which refers to the level of output and the corresponding level of effort that maximize the expected economic profits in a fishery. In most cases, this scenario results in yields and effort levels that are less than at maximum sustainable yield (MSY) and in stock biomass levels greater than at MSY. The net income from fishing and the subsequent use of income for the livelihood of fishers is of vital importance. Besides the cost and returns in fishing plays a significant role as incentives for engaging in fishing as an occupation. This thought gave way for the economics to be included in fisheries management.

The steps involved in the calculation of MEY are:

$p = a - by$

Where,

p is the price per unit weight of fish

y is the annual yield.

The average price per unit weight of fish (p) is generally a monotonically decreasing function of annual yield (y)

The profit will be obtained as a difference between total revenue (TR) and total cost (TC):

$$\Pi = TR - TC$$
$$= (p-c)y$$

Where

'c' is the cost of harvesting one unit weight of fish. From this, a cost function will be fit from the data collected

MEY = (a – c) / 2b,

fmey = [a +/- (a2 – 4 b MEY)]1/2 / 2b

Where,

a = intercept; b, c =regression coefficients.

Adjusting fishery for MSY and MEY

Reduction of number of fishing vessels will decrease the total capture efforts and level of fishing intensity and thus reduce total catches of the fishes. Reduction will reduce the total fishing efforts and thus raise the economic efficiency of the remaining fishing fleet. Theoretically, downsizing will make MR=MC where MEY (Maximum Economic Yield) is realized. This is illustrated in the Figure below.

TR stands for total fishing revenues and TC stands for total fishing cost. Suppose the current fishing efforts are at the point of Eo. Since Eo is beyond Em at which fishing effort is optimum level realizing MSY, current fishing is at the level of overfishing. It means from the biological point of view that fishing efforts should be reduced from Eo to Em. From the economic point of view, at the point of Eo, the amount of total fishing revenue is same as that of total fishing cost because MR=AC (or TR=TC). In other words, at the point Eo, economic overfishing is occurring as much as Ee-Eo.

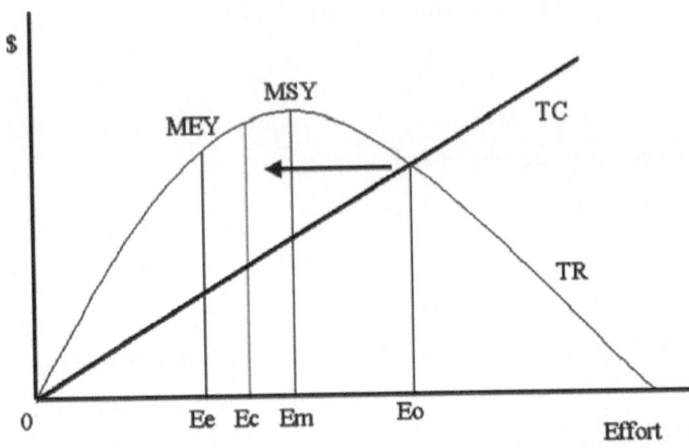

Fig. 19.3: Diagram showing TR, TC, MSY and MEY with respect to Fishing Effort

If the fishing is at the point of overfishing, the total amount of fishing efforts should be reduced to prevent over-catching. Level of reduction will be decided between Eo and Ee. For a biologically sustainable fishing, the fishing efforts should be reduced to the level of Em. In terms of economics, the level of fishing efforts should be reduced to the point of Em for maximized economic gains. However, in practical terms the level of reduction will be somewhere between Em and Ee, where conservation of fisheries resources can be maintained.

Alternative reference points

As it's impossible to count and measure all fish in the sea, biomass and fishing pressure are never measured directly. They are inferred through models, knowledge of species biology, and information collected every year on fisheries activities and scientific surveys.

So, how can we manage fisheries towards a goal that is ever changing and so elusive to measure?

Rather than expecting stocks to be exactly at BMSY at all times which would be incredibly challenging to monitor in real time, given the variability and natural fluctuations of wild populations, scientists delimit a range of values that are considered a 'safe space' within which the stock is not at risk. This space is bound by thresholds that, when crossed, signal to managers that they need to intervene. Good monitoring and management practices ensure that if, for any reason, stock biomass drops below a reference level, a remedial action is put in place, such as a reduction in fishing time or in catch limits, allowing stocks to recover.

Goals and objectives of Fisheries management

The primary goal of fisheries management is the long-term sustainable use of the fisheries resources. There are biological; ecological; economic and social, where social includes political and cultural goals. The main objectives could be summarized as under:

- to maintain the target species at or above the levels necessary to ensure their continued productivity (biological);
- to minimize the impacts of fishing on the physical environment and on non-target (by catch), associated and dependent species (ecological);
- to maximise the net incomes of the participating fishers (economic); and
- to maximize employment opportunities for those dependent on the fishery for their livelihoods (social).

Identifying specific goals is important in clarifying how the fish resources are to be used to benefit society, and they should be agreed upon and recorded, both at the policy level and for each fishery. In line with the objectives, the following considerations are important for a fishery manager for effective implementation of his measures.

1. Biological Considerations

Aquatic living resources are capable of on-going renewal through the processes of growth in size and mass of individuals and additions to the population or community through reproduction (recruitment). In a population at equilibrium, the additive processes of growth and reproduction on average equal the loss process of total mortality. A primary task of fisheries management is to ensure that fishing mortality does not exceed the amount which the population can withstand, in addition to natural mortality, without undue harm or damage to the sustainability and productivity of the population. This requires not only that the total population is maintained above a certain abundance or biomass, but also that age structure of population is maintained in a state in which it is able to maintain level of reproduction, and hence recruitment, necessary to replenish the losses through mortality. Further, fishing over a long period on selected portions of a stock can reduce the frequency of the particular genetic characteristics giving rise to that feature or behaviour. This has the effect of reducing the overall genetic diversity of the stock. With reduced genetic diversity, production potential of population can be adversely affected and it may also become less resilient to environmental variability and change. Fisheries management needs to be aware of this danger and avoid maintaining such selective pressures over a prolonged period.

2. Ecological and Environmental Considerations

The physical component of the ecosystem, the water itself, the substrate, inflows of freshwater or nutrients and other non-biological processes may also be very important. Different substrates may be essential for the production of food organisms, for shelter, or as spawning or nursery grounds. The environment of fish is very rarely static and the aquatic environment can vary substantially over time, from hourly variability, such as the tides, to seasonal variability in, for example, water temperature and currents, to decadal variability as in the occurrence of El Niño events and regime shifts. These changes frequently affect the population dynamics of fish populations, resulting in variability in growth rates, recruitment, natural mortality rates or any combination of these. Changes in any of biological, chemical, geological or physical components of the ecosystem can have impacts on the resource population and community. Fisheries management should both take into account their impacts on the resource and, in consultation with other relevant agencies and parties, take steps to minimize their impacts on the fishery ecosystem.

3. Technological Considerations

In most fisheries, the only mechanism the fishery manager has to ensure sustainable utilization of the resources is by regulating the quantity of fish caught, when and where they are caught and the size at which they are caught. This can be done through directly regulating the catch taken, by regulating the amount of effort allowed in the fishery, by specifying closed seasons and closed areas and by regulating the type of gear and fishing methods used. However, there are constraints on how precise the manager can be in setting such regulations. Catch controls are often difficult to monitor and therefore to implement. It is difficult to estimate fishing effort precisely, and normally improving technology and developing skills result in on-going increases in the efficiency of fishing operations, leading to continuing increases in effective effort, unless steps are actively taken to counter these improvements or their consequences. Fishing gear is rarely strongly selective and bycatch of non-target species or unwanted sizes of target species is frequent. A fundamental problem in many fisheries is the existence of too much effort.

4. Social and Cultural Considerations

Social changes take place continuously and on different scales, affected by changes in weather, employment, political circumstances, supply of and demand for fisheries products and other factors. Such changes can affect the appropriateness and effectiveness of management strategies, and therefore need to be considered and accommodated by them. However, as in other considerations, it can be difficult to identify and quantify the key social and cultural factors influencing fisheries management, generating additional uncertainties for manager. A major social constraint in fisheries management is that fishing families and communities may not be willing to move into other occupations or away from their normal homes when there is surplus capacity in a fishery, even when their quality of life may be suffering as a result of depleted fish resources. The problem is much worse when there are no other opportunities outside of fisheries in which they could earn a basic living. Under such circumstances, the political decision to reduce capacity in the fishery is an extremely unattractive option. A key requirement for ensuring that social and cultural considerations are properly considered is to involve the interested parties in fisheries management, keeping them well-informed on the management aspects of the fishery and providing them with the opportunity to express their needs and concerns.

5. Economic Considerations

In a stable fishery, market largely determines the economic efficiency. However, such stability can't be expected for long term in an exploited fishery. Uncertainties of unpredictable variability in resources, impacts of other fisheries on the target resources, subsidies, trade regulations, fiscal regulations and variability in markets and demand all lead to sub-optimum economic performance. It is important for the management authority to consider the broad economic context of a fishery, including relevant macroeconomic factors. Since most of the fisheries are open access type, people will continue to enter the fishery until the benefits from fishing are so low as to be unattractive to prospective new entrants. This in many cases lead to very poor economic efficiency and, unless strong and effective management measures are in place and enforced, to over-exploitation of resources.

6. Considerations Imposed By Other Parties

Other parties which use the fishing grounds can include tourism, conservation, oil and gas extraction, offshore mining and shipping, aquaculture and mariculture, coastal zone development for housing, business or industry, and agriculture. All of these can impose significant constraints on fishing activities and may be impacted by fishing activities. The manager therefore needs to be aware of such activities and of real or potential impacts in both directions. When developing management strategies and formulating management measures, potential conflicts with other users need to be identified and addressed, and the potential impacts of other users on the efficacy of the management strategy and measures need to be considered. The strategy must be adapted so as to account for and be robust to these impacts.

Limitations and Conflicting Interests of Fishery Management Goals

It is difficult to minimize impacts of the fishery on the ecosystem and simultaneously to maximize net incomes. Similarly, it is very probable that management strategies that aim to maximise net incomes will not also maximise employment opportunities. Some compromise between these goals has to be achieved before an effective management strategy can be devised.

Operational Objectives of Fisheries Management

The trade-offs between the biological, ecological, economic and social goals must have been agreed upon and the conflicts and contradictions resolved. All the goals could be then simultaneously achieved. To maintain the stock at all times above 50% of its mean unexploited level (biological); to maintain all non-target, associated and dependent species above 50% of their mean biomass levels in the absence of fishing activities (ecological).

Fisheries Co-Management

Fisheries co-management is a flexible and cooperative management of the aquatic resources by the user groups and the government. The responsibility of the resource is shared between the user groups and the government, both the community and the government are involved during the decision making, implementation and enforcement processes. Depending on the level of participation between the government and the community, five different types of co-management have been identified.

1. Instructive management is top down management from the government. The government instructs the fishermen as to what laws and policies they are required to follow. Information is only shared with the community towards the end of the planning process.

2. In advisory management, users decide what decisions should be made and advise the government as such, the government than endorses the decision.

3. In informative management, the user group makes all of the decisions, and informs the government once they have decided.

Co-management is developed as an attempt to improve the success of fisheries as many of the aquatic resources being risked or were already depleted. The aim was to involve the community in the decision making process so that there may be an increased adherence to the regulations set by the government. The top down approach of fisheries management has frequently failed as it often goes against the community and their internal structure. Furthermore, the reason for the new governmental regulations and laws such as a decrease in total allowable catch (TAC) or catch per unit effort (CPUE) were often poorly understood by the community due to lack of education. Co-management makes the community more responsible to not to disregard the laws and regulations set by the Government for improving the stock and fisheries.

Co-Management and Community Management

Co-management is different from community management as the government plays an important part in the decision making process. In community management, the laws are generally not enforced as governmental laws, but as a community guide and framework. This can make prosecution difficult. In co-management, the user group and the government develop laws and regulations together and work towards implementing them as a unit. If community based management forms part of national legislation, or developmental plans than it is classified as co-management.

Advantages vs. limitations of Co-management

The fishermen community, specifically the users of the resource often possess knowledge that can aid the government in the decision making process, therefore co-management is the combination of scientific and traditional knowledge, this process ensures the best possible outcome. If the community is involved with and agrees on the new laws and decisions they are more likely to comply and may even aid in ensuring the new laws are enforced and maintained. In dynamic environments, conditions can vary at a rapid rate. However, Co-management is known to be time consuming, as one must spend time collecting surveys and gaining trust within the community. Educating the community is often necessary so that more informed decisions can be made. Strong lines of communication between the government and the community is essential. Many of the more isolated communities have different languages. This may slow down effective communication and making use of vital information and benefits from Government. Lack of funding, data and resources are also main contributing factors to un-successful co-management. Co-management requires constant communication and effort, and therefore long term sustainability can be difficult. Third party involvement such as non-governmental organizations (NGO's), or student groups, often forms an essential part of a successful co-management.

Ecosystem Based Fisheries Management

Management strategies have increasingly turned towards the ecosystem approach to fisheries management (EAFM, or 'ecosystem-based fisheries management', EBFM) as an alternative to species-based management in order to account for the broad range of interdependent relationships that occur within ecosystems. Ecosystem-based fishery management (EBFM), also referred to as an ecosystem approach to fisheries (EAF), has been proposed as a more effective and holistic approach for managing world fisheries. The aim is to sustain healthy marine ecosystems and the fisheries they support by addressing some of the unintended consequences of fishing, such as habitat destruction, incidental mortality of non target species, and changes in the structure and function of ecosystems.

Fishery legislations

A wide range of bodies working at National and International levels to ensure the effective fisheries management through their regulatory measures. They are briefly summarized as below.

1. National

The scope of the national legislation varies substantially between countries. However, typically the primary legislation is broad, prescribing the principles and policy relating to fisheries. Primary legislation would usually be described in the form of Fisheries Act or equivalent legislation. Control measures such as the amount of effort allowed in a given fishery, or the annual TAC should not be included in the primary legislation. Subsidiary legislation, produced by the delegated regulatory authority is often referred to as regulations, orders, proclamations etc. They would include specifying the control measures which require frequent, typically annual, revision such as licenses, gear restrictions, closed areas and seasons and input and output controls.

2. International

The modern fisheries manager is required to be familiar also with the bewildering diversity of international legislation and voluntary instruments dealing directly with or impinging on fisheries. Chief amongst the international instruments is the United Nations Convention on the Law of the Sea of 10 December 1982 (LOS Convention), which entered into force in 1994. This convention sets the legal context for all subsequent international arrangements and agreements relating to the use of the oceans and seas. Arising directly from the LOS Convention and designed to strengthen its provisions relating to high seas fisheries and transboundary stocks, are the UN Fish Stocks Agreement and the FAO Compliance Agreement. Convention for International Trade in Endangered Species of Fauna and Flora (CITES) has had great impact on marine fisheries management by curbing some marine species subjected to international trading. Most countries involved in fisheries are or will become members of one or more regional bodies involved in utilization, management and conservation of marine living resources. These include bodies such as the various tuna commissions (e.g. the International Convention on the Conservation of Atlantic Tuna (ICCAT) and the Convention on Indian Ocean Tuna (IOTC)), the Convention on the Conservation of Antarctic Living Marine Resources (CCAMLR), various FAO regional fishery bodies such as the Fishery Committee for the Eastern Central Atlantic (CECAF) and the Asia-Pacific Fishery Commission (APFIC), and many others. The manager must be aware of those, in which his or her country is involved, and the implications and obligations of membership.

Criteria for Evaluating Effective Fisheries Resource Management

Ultimately, fishery management techniques must be evaluated against the objectives of management, and comparison of benefits (short- and long term) associated with the application of a management technique. Generally, the techniques are evaluated against the following criteria.

1) ***Competence in Controlling fishing mortality*:** The linkage between the resource sector and socioeconomic sector of the fishery system is fishing mortality. This is the rate (proportion of the population per unit time) at which fish are improved as a result of fishing. Thus man's ability to accomplish objectives related to the condition of the resource depends on controlling fishing mortality. The secondary impact of fishing mortality on growth, natural mortality, and recruitment as a result of its influence on the size and structure of the resource, is of concern in the evaluation of fishery management decisions.

2) ***Biological impacts of Fishery*:** Fisheries management actions may have more specific biological impacts than control of fishing mortality; some fishery management techniques are effective in restricting the impact of fishing on specific components of the resource. For example the techniques may be intended to protect small fish, spawning fish, or fish of a particular species.

3) ***Maintaining natural functioning of the fishery system*:** Fisheries management serves as an artificial form of control of system dynamics such as the natural biological processes which act to compensate for or buffer against perturbations and the socioeconomic interrelationships between fishing activity (commercial or recreational) and the activities of related industries and markets.

4) ***Encompass cost of enforcement and administration:*** The net benefit of fishery management must include the cost of administering and enforcing the management programs. Some fisheries management techniques are inherently more difficult and costly to administer and enforce than others.

5) ***Impact on non-target components of the fishery system:*** The impact of application of fishery management plan on non-target components of the resource should also be considered. In the case of species specific biological objectives, the effect of management actions on the capacity of the harvesting, processing or related sectors to redirect attention to other components of the resource is particularly important.

Conclusion

Fisheries are an important source of food, employment, economic activity and recreation for the people around the globe. However, the increased pressure on marine capture fisheries – from growing populations, rising demand for seafood, and a rapid increase in fisheries exploitation – has caused a decline in the productivity of many fisheries. Effective management of fisheries resources is a hot topic in most fisheries discussions not just from an environmental perspective but also because of its impact on the economy and the livelihood of our population. Sustainably managing the world's shared fish stocks, and securing the well-being of the habitats and ecosystems on which they depend, requires effective governance at every level. Policies must be informed by the best available science and backed by commitments from governments to enforce compliance.

References

Charles, A. T. 2001. *Sustainable Fishery Systems*. Blackwell Science, London. 384 pp.

Cochrane, K.L. 2000. Reconciling sustainability, economic efficiency and equity in fisheries: the one that got away? *Fish and Fisheries*, 1: 3-21.

FAO. 1995. Code of Conduct for Responsible Fisheries. FAO, Rome. 41pp.

FAO. 1997. FAO Technical Guidelines for Responsible Fisheries No. 4: Fisheries Management. FAO, Rome. 82pp.

FAO. 2000. *The State of World Fisheries and Aquaculture*. 2000. FAO, Rome.

Gordon, H. Scott, 1954. "The Economic Theory of a Common Property Resource". Journal of Political Economy, 62:124-142

http://www.fao.org/docrep/005/y3427e/y3427e03.htm

Narayanakumar, R. 2012. Maximum economic yield and its importance in fisheries management. Summer School on Advanced Methods for Fish Stock Assessment and Fisheries Management, 252-257.

Schaefer, M., 1957."Some Considerations of Population Dynamics and Economics in relation to the Management of the Commercial Marine Fisheries". *Journal of the Fisheries Research Board of Canada*, 14:69-681.

Sissenwine, M.P., Kirkley, J.E. 1980. Fishery Management Techniques, A Review. NOAA Technical Memorandum NMFS-F/NEC-4.